Radar and RF Front End System Designs for Wireless Systems

Shilpa Mehta
Auckland University of Technology, New Zealand

Rupesh Kumar
SRM University, India

A volume in the Advances in Wireless Technologies and Telecommunication (AWTT) Book Series

IGI Global
PUBLISHER of TIMELY KNOWLEDGE

Published in the United States of America by
IGI Global
Information Science Reference (an imprint of IGI Global)
701 E. Chocolate Avenue
Hershey PA, USA 17033
Tel: 717-533-8845
Fax: 717-533-8661
E-mail: cust@igi-global.com
Web site: http://www.igi-global.com

Library of Congress Cataloging-in-Publication Data

CIP Data Pending
ISBN: 979-8-3693-0916-2
eISBN: 979-8-3693-0917-9

This book is published in the IGI Global book series Advances in Wireless Technologies and Telecommunication (AWTT) (ISSN: 2327-3305; eISSN: 2327-3313)

British Cataloguing in Publication Data
A Cataloguing in Publication record for this book is available from the British Library.

Advances in Wireless Technologies and Telecommunication (AWTT) Book Series

ISSN:2327-3305
EISSN:2327-3313

Editor in Chief: Xiaoge Xu, University of Nottingham Ningbo China, China

MISSION

The wireless computing industry is constantly evolving, redesigning the ways in which individuals share information. Wireless technology and telecommunication remain one of the most important technologies in business organizations. The utilization of these technologies has enhanced business efficiency by enabling dynamic resources in all aspects of society.

The **Advances in Wireless Technologies and Telecommunication Book Series** aims to provide researchers and academic communities with quality research on the concepts and developments in the wireless technology fields. Developers, engineers, students, research strategists, and IT managers will find this series useful to gain insight into next generation wireless technologies and telecommunication.

COVERAGE

- Telecommunications
- Mobile Technology
- Network Management
- Grid Communications
- Mobile Web Services
- Cellular Networks
- Digital Communication
- Wireless Technologies
- Broadcasting
- Global Telecommunications

IGI Global is currently accepting manuscripts for publication within this series. To submit a proposal for a volume in this series, please contact our Acquisition Editors at Acquisitions@igi-global.com or visit: http://www.igi-global.com/publish/.

The Advances in Wireless Technologies and Telecommunication (AWTT) Book Series (ISSN 2327-3305) is published by IGI Global, 701 E. Chocolate Avenue, Hershey, PA 17033-1240, USA, www.igi-global.com. This series is composed of titles available for purchase individually; each title is edited to be contextually exclusive from any other title within the series. For pricing and ordering information please visit http://www.igi-global.com/book-series/advances-wireless-technologies-telecommunication/73684. Postmaster: Send all address changes to above address.

Titles in this Series

For a list of additional titles in this series, please visit:
http://www.igi-global.com/book-series/advances-wireless-technologies-telecommunication/73684

Metamaterial Technology and Intelligent Metasurfaces for Wireless Communicatin Systems
Shilpa Mehta (Auckland University of Technology, New Zealand) and Arij Naser Abougreen (University of Tripoli, Libya)
Information Science Reference • © 2023 • 364pp • H/C (ISBN: 9781668482872) • US $225.00

Opportunities and Challenges of Industrial IoT in 5G and 6G Networks
Poshan Yu (Soochow University, China & Australian Studies Centre, Shanghai University, China) Xiaohan Hu (Shanghai University, China) Ajai Prakash (University of Lucknow, India) Nyaribo Wycliffe Misuko (KCA University, Kenya) and Gu Haiyue (Shanghai University, China)
Information Science Reference • © 2023 • 304pp • H/C (ISBN: 9781799892663) • US $250.00

Applications of Artificial Intelligence in Wireless Communication Systems
Karan Kumar (Maharishi Markandeshwar University (Deemed), Mullana, India)
Information Science Reference • © 2023 • 284pp • H/C (ISBN: 9781668473481) • US $250.00

Designing and Developing Innovative Mobile Applications
Debabrata Samanta (Rochester Institute of Technology, Kosovo)
Information Science Reference • © 2023 • 426pp • H/C (ISBN: 9781668485828) • US $225.00

Multidisciplinary Applications of Computer-Mediated Communication
Hung Phu Bui (University of Economics, Ho Chi Minh City, Vietnam) and Raghvendra Kumar (GIET University, India)
Information Science Reference • © 2023 • 322pp • H/C (ISBN: 9781668470343) • US $250.00

For an entire list of titles in this series, please visit:
http://www.igi-global.com/book-series/advances-wireless-technologies-telecommunication/73684

IGI Global
PUBLISHER of TIMELY KNOWLEDGE

701 East Chocolate Avenue, Hershey, PA 17033, USA
Tel: 717-533-8845 x100 • Fax: 717-533-8661
E-Mail: cust@igi-global.com • www.igi-global.com

Table of Contents

Detailed Table of Contents

This chapter explores the synergy between radar and radio frequency (RF) front-end systems, ushering in a new era of wireless connectivity. It discusses the collaborative potential of radar and RF, emphasizing their role in enhancing security, reducing interference, and boosting adaptability. The chapter covers radar-based spectrum sensing, which enhances network efficiency, particularly in high-frequency scenarios like 5G. Radar and RF enable precise localization for IoT and autonomous vehicles, surpassing the capabilities of GPS. The chapter highlights radar's contributions to security, threat detection, and reducing signal interference. Radar-assisted RF improves vehicle communication, cooperative driving, and traffic management. In environmental monitoring and disaster management, radar augments RF for early warnings. This integration offers transformative potential, benefiting diverse applications and offering theoretical and practical insights for researchers and engineers. Radar and RF convergence offers a more connected, adaptable, and efficient wireless future.

The content of this chapter provides a thoughtful analysis of case studies that highlight the detection capabilities of FMCW radar systems operating in mmWave configurations. The case studies demonstrate how mmWave FMCW radar technology may be used to detect objects, motions, and changes in both line of sight and non-line of sight settings with accuracy, and efficiency. Each case study explores the unique difficulties presented by the application environment, which can include anything from identifying impediments in automotive safety systems to detecting minute movements for vital sign monitoring in healthcare. The steps for detecting different gesture recognition using IWR 1843 BOOST FMCW radar system and its processing are focussed upon. The document highlights the technology's excellent resolution, motion sensitivity, and adaptability to a variety of challenging environments in LoS and NLoS scenarios and with the technical details of operating mmWave radars, including signal processing methods, machine learning algorithms, and mitigating interference from surrounding objects.

This study conducts a comprehensive bibliometric analysis of research articles on reconfigurable antennas in radar applications retrieved from Google Scholar databases from 1989 to 2024. A total of 980 research articles are analysed, demonstrating significant interest and commitment to the topic, as evidenced by a total of 22.073 citations. Important metrics such as citations per year (649.21), citations per article (22.52), and citations per author (9,529). It highlight the importance and impact of research in this area. Collaboration is common, with an average of 410.95 articles per author and about three authors per article. The geographical analysis highlights the contributions of institutions in the United States and Switzerland. The study identifies emerging trends and future directions and highlights the need for dynamic antenna reconfiguration techniques, integration challenges, and energy-efficient solutions. The bibliometric analysis gives valuable insights into the current investigation landscape on reconfigurable antennas for radar applications.

The timely upgradation in 5G technology standards has already been released by third generation partnership project, which enabled the researchers to refine the research to improve the communication and support in the development. After development of fifth generation technologies, researchers, scientists, and engineers are looking for wide bandwidth which should improve wireless systems and devices to provide better services and experiences. Because of that, beyond 5G, a sixth generation is the next step to be widely used in early 2028 or intensively used around 2030 as per demand. Parameters like gain, bandwidth, efficiency, size of the antenna, technique of fabrication, and other useful quantities can bc modified and maximized, and also the cost of fabricating such antennas will be reduced and developed.

The evolution of wireless communication networks towards 6G demands innovative solutions to address the challenges of increasing data rates, network capacity, and energy efficiency. Reconfigurable intelligent surfaces (RIS) emerge as a promising technology that revolutionizes how wireless signals propagate and interact with the environment. In this chapter, the authors explore the integration of RIS with antennas for 6G wireless communications. RIS, comprised of programmable meta-surfaces, can dynamically manipulate electromagnetic waves to optimize signal quality, mitigate interference, and enhance network performance. By providing adaptive beam forming, beam steering, and signal enhancement capabilities, RIS-enabled antennas offer unprecedented flexibility and efficiency in communication systems. This chapter overviews RIS principles, discusses their integration with antennas, and highlights their potential to shape the future of 6G wireless communications.

In the rapidly evolving wireless communication landscape, antenna technology is indispensable. This chapter provides an in-depth review of recent advancements and trends in antenna technology transforming modern wireless communication systems. The authors delve into vital technological innovations, including massive MIMO, beamforming, metamaterial antennas, and reconfigurable intelligent surfaces, shedding light on their functions, potential benefits, and implications. The study underscores that the ongoing evolution in antenna technology holds immense potential to revolutionize wireless communication systems, enabling more efficient, high-speed, and sustainable networks. These insights will benefit communication engineers, researchers, and academicians, offering a broad understanding of antenna technology's current state and future trajectories in wireless communication systems.

Wireless technology now employs 5G (generation) communication, while 6G is the subject of increased research. Various G's have evolved due to the need to achieve high data transfer for reliable communication. A tremendous data transfer rate expressed in terabits per second (Tbps) is facilitated by the terahertz (THz) spectrum. The photonic crystal (PhC) patch antenna has been proposed in this chapter, and the consequences of a line defect were examined. The defect has been embedded in the perpendicular and parallel directions of the antenna substrates. The antenna's performance regarding return loss (RL), directivity, and voltage standing wave ratio (VSWR) is evaluated. The structure is simulated using CST, and the PhC structure results show excellent characteristics such as -48.02 dB RL, 1.007 VSWR, and 5.23 dB directivity at THz frequency.

This chapter shows and evaluates the design of a compact dual-band microstrip patch antenna working at 2.5 and 5.2 GHz for a number of important parameters. The antenna is designed to operate at these frequencies. By using not one but two distinct EBG ground planes, the proposed dual-band conventional ground plane antenna achieves a higher level of overall performance in comparison to its predecessor (in terms of directivity, gain, efficiency, and other features). It has been discovered that antennas modeled after EBGs perform much better than antennas constructed using conventional ground planes. In addition, the authors examined the performance of the recommended dual-band conventional ground-plane antenna under a variety of bending conditions.

Recent progress in microstrip antennas has improved their size and efficiency, particularly in mobile, satellite, and Wi-Max technologies. These enhancements are crucial for advancing wireless communication systems, benefiting applications like telemedicine and GPS technology. The chapter introduces a dual-band unit cell of frequency selective surface (FSS) resonating at 4 GHz and 5.5 GHz, using a combination of I-shaped and modified I-shaped metal strips for dual-band filtering. The research focuses on individual shape analyses and aims to contribute to the development of compact and efficient dual-band FSS designs. The research introduces a dual-band FSS unit cell resonating at 4 GHz and 5.5 GHz, combining I-shaped, H shaped, and modified I-shaped and H shaped metal strips for dual-band filtering. The proposed FSS design, with a 10x10 mm2 unit cell dimension and FR-4 dielectric, exhibits broad frequency band characteristics, aligning well with simulated and measured results.

There is a challenging era for wireless communication especially after commercialization of AI/ML tool. Specifically, the phenomenon of mobile communication has evolved completely from being used as a mere communicating device to computing device or small messaging medium to manifest unparalleled comfort like proposing and composing podcasts, presentations, and scripting social posts with the help of AI-enabled ChatGPT. This chapter explores AI-enabled 5G, its standardization, and leveraging AI/ML in 6G and beyond.

Rate splitting multiple access (RSMA) is a promising approach that has gained momentous attention in the field of wireless communications. With the increasing demand for higher data rates and reliable connections, traditional multiple access approaches like SDMA, NOMA have limitations in terms of interference management as well as channel state information uncertainty. The effectiveness of the RSMA has been recognized in addressing these limitations and providing improved performance in multi-antenna systems. RSMA ruptures user messages into common as well as private parts – the common part is decoded by all users to mitigate interference while the private part is decoded only by the intended user. This bridges orthogonal and non-orthogonal multiple access schemes. Benefits of RSMA include optimal performance in many scenarios, robustness against imperfect CSIT, lower feedback overhead and complexity compared to alternatives. RSMA has many potential applications in 6G use cases like mMTC, URLLC, V2X, and integrated sensing-communications. Ongoing research is still needed, especially on implementation and standardization. The current study focuses on the notion of rate splitting, the benefits of RSMA, working principle of RSMA, implementation challenges, RSMA's implementation in 5G and beyond, a comparison with other multiple access techniques, as well as its problems and future trends in wireless networks.

Foreword

Welcome to the Wireless Sensing System, where Radio Signals provide more information about our surroundings for a better experience of life! In this book, *Radar and RF Front-End System Designs for Wireless Systems*, we learn the important aspects of Radar and RF front-end design, especially the very foundation of sensing, detection, and identification as well as classification of the subject under test.

As we stand at the forefront of 6G technology, the significance of Radar and RF front-end technologies requires a special work plan to understand the importance of wireless-sensing, and reconfigurability of RF front-end antenna system designs because of its benefits and applications. These systems will play a crucial role in 6G implementation to facilitate the additional features from what we have up to 5G! This book provides a comprehensive guide, theoretical concepts, and realized work essential for understanding and developing radar system design needed toward the trends of wireless sensing by leveraging cutting-edge Radar and RF front-end systems.

The growth of wireless sensing systems has been a remarkable journey so far, with each advancement pushing the scope of what is possible beyond 5G and 6G. Radar technology has witnessed many breakthroughs to reach the present-day state-of-the-art technology, especially the RF front-end designs. This book provides insight into radar, front-end design, antenna design, signal processing, system integration, and machine learning approach which presents a holistic view of wireless sensing technologies.

Additionally, this book delves into the use of artificial intelligence (AI), and machine learning (ML) approaches for sensing, detecting, monitoring, and predicting the state of the subject under investigation related to various applications such as gesture recognition, image classification, pattern recognition and so on. Integration of AI/ ML in wireless technologies has further pushed the radar boundaries to a new era of cutting-edge autonomous applications related to defense as well as civil applications. This book gives a glimpse of what will shape the future of wireless sensing to a much-advanced level from the present.

It is with great pleasure that I introduce this book, a testament to the dedication, expertise, and passion of the authors who have meticulously crafted its contents.

Their commitment to excellence and their desire to share the knowledge shine through in these pages, leveraging readers as a valuable resource for learning and implementing the art of Radar and RF Front-end System designs.

As you embark on this exciting journey through the realm of Wireless Sensing, I encourage you to approach each chapter with the curiosity of exploring more on Radar and RF front-end System design. Let the rich information and contained knowledge guide you to an exciting world of imagination resonating with innovation and to the endless possibilities that these technologies would offer!

In closing, I extend my deep regard to all authors for their valuable contributions to the domain of Radar and RF frontend system design. This book presents the testimony of their expertise and dedication and undoubtedly inspires the present as well as future generations of engineers, researchers, and students to explore the unexplored territory of wireless sensing.

Join me on this remarkable journey through the realms of Radar and RF front-end system designs, where innovation meets the future of Wireless Sensing!

Mark Clemente
Universidad Nacional Tecnológica de Lima Sur, Peru

Mark Clemente *is a full professor at National University of South Lima UNTELS. He received his Ph.D. from Telecom Paristech in Telecommunications and Electronics. He has a M.Sc. degree on Communicating Objects from the Universite d'Aix Marseille. He also obtained a MEng degree in Microelectronics and Telecommunications from the Polytech Marseille. His B.S. degree is on Electronic Engineering at the Universidad Nacional de San Antonio Abad del Cusco. His research areas include RF/MW/mmW devices, antennas, metamaterials and complex materials for various applications such as 5G, radars, wireless sensors, satellites and EM modeling. IEEE senior member.*

Preface

As editors of this comprehensive reference book, *Radar and RF Front End System Designs for Wireless Systems*, it is our pleasure to present a compilation that delves into the intricate world of wireless technologies, particularly focusing on radar and RF front-end systems. The advent of wireless communication has ushered in a new era of connectivity, revolutionizing various sectors including healthcare, smart IoT systems, and sensing applications. In this context, the role of RF front-end systems, with their reconfigurable capabilities, has become increasingly vital.

The impetus behind this book stems from the remarkable surge in innovation witnessed in RF front-end systems. Researchers and practitioners alike have contributed a plethora of new configurations and design architectures, paving the way for unprecedented advancements in wireless systems. Our aim with this publication is to provide a platform for researchers to explore both theoretical insights and practical applications, thereby facilitating the dissemination of the latest trends and developments in the field.

The contents of this book cover a wide spectrum of topics, ranging from RF front-end antenna systems to the impact of artificial intelligence and machine learning in system design. With contributions from experts in academia and industry, readers can expect a comprehensive exploration of radar and antenna system design, modeling, and measurement techniques.

We envision this book serving as a valuable resource for students, researchers, scientists, and industry professionals seeking to deepen their understanding of RF front-end antenna and radar system designs. Whether it's exploring reconfigurable antenna systems for 5G/6G networks, delving into radar modeling and signal processing techniques, this book offers insights that are both timely and relevant.

We extend our gratitude to all the contributors whose expertise and dedication have made this publication possible. It is our sincere hope that *Radar and RF Front End System Designs for Wireless Systems* will inspire further innovation and serve as a catalyst for advancements in the ever-evolving landscape of wireless technologies.

Chapter 1: Engineering Next-Generation Wireless Experiences Through Radar and RF Front-End System Designs

Authored by Mangaiyarkkarasi J. and Shanthalakshmi Revathy J., this chapter elucidates the symbiotic relationship between radar and radio frequency (RF) front-end systems, shaping a transformative era in wireless connectivity. Emphasizing collaboration, the chapter underscores the pivotal roles of radar and RF in fortifying security, minimizing interference, and enhancing adaptability. It delves into radar-based spectrum sensing, elucidating its efficacy in optimizing network efficiency, particularly in high-frequency scenarios such as 5G deployments. Moreover, the chapter explores how radar and RF facilitate precise localization, benefiting IoT and autonomous vehicle applications. Highlighting radar's significance in security and threat detection, the authors showcase its contributions to improving vehicle communication, cooperative driving, and traffic management. In addition, radar-assisted RF integration augments environmental monitoring and disaster management efforts, providing early warnings for impending calamities. Through this exploration, the chapter unveils the transformative potential of radar and RF convergence, offering valuable insights and practical applications for researchers and engineers alike, while envisioning a more connected, adaptable, and efficient wireless future.

Chapter 2: Case Studies on Detection Using mmWave FMCW RADAR System

Authored by Gummadi Surya Prakash, W. Chandra, Shilpa Mehta, and Rupesh Kumar, this chapter presents a series of insightful case studies illuminating the detection capabilities of Frequency Modulated Continuous Wave (FMCW) radar systems operating in millimeter-wave (mmWave) configurations. The authors meticulously analyze scenarios encompassing object detection, motion tracking, and environmental monitoring, both in line of sight (LoS) and non-line of sight (NLoS) settings. Each case study offers a nuanced examination of the challenges posed by diverse application environments, ranging from automotive safety systems to healthcare monitoring. With a focus on the IWR 1843 BOOST FMCW Radar system, the chapter elucidates signal processing methodologies, machine learning algorithms, and strategies for mitigating interference. By showcasing the technology's remarkable resolution, motion sensitivity, and adaptability, the authors underscore its potential in addressing real-world challenges across various domains.

Chapter 3: Bibliometric Analysis of Reconfigurable Antennas in Radar Applications: Trends, Influential Authors, and Future Directions

In this chapter authored by Duygu Nazan Gençoglan, a comprehensive bibliometric analysis of research articles on reconfigurable antennas in radar applications is presented. Spanning from 1989 to 2024, the analysis encompasses 980 research articles, elucidating significant interest and impact in this domain, as evidenced by a substantial citation count. Key metrics such as citations per year, per article, and per author underscore the importance of research in this area. Moreover, collaboration trends and geographical distributions highlight the contributions of various institutions globally. The study identifies emerging trends and future directions, emphasizing the need for dynamic antenna reconfiguration techniques and energy-efficient solutions. Through this bibliometric analysis, readers gain valuable insights into the current landscape of research on reconfigurable antennas for radar applications, laying the groundwork for future advancements.

Chapter 4: Reconfigurable Antenna System for 5G/6G

Authored by Smrity Dwivedi, this chapter delves into the evolution of reconfigurable antenna systems within the context of 5G and beyond. With the advent of 5G technology standards and the ongoing quest for wide bandwidth, researchers are exploring novel approaches to enhance wireless communication systems. The chapter discusses the imperative of timely upgrades to meet escalating demands for high-speed communication and IoT applications. By elucidating the trajectory towards sixth-generation (6G) technologies, the author anticipates a future where reconfigurable antenna systems will play a pivotal role in enabling advanced wireless services and experiences.

Chapter 5: Integration of Reconfigurable Intelligent Surfaces With Antennas for 6G Wireless Communications

In this chapter authored by Pallavi Sapkale and Shilpa Mehta, the integration of Reconfigurable Intelligent Surfaces (RIS) with antennas for 6G wireless communications is explored. As wireless networks evolve towards 6G, the need for innovative solutions to enhance data rates, network capacity, and energy efficiency becomes paramount. RIS, comprising programmable meta-surfaces, emerges as a promising technology to address these challenges. The chapter elucidates how RIS-enabled antennas dynamically manipulate electromagnetic waves to optimize signal quality, mitigate interference, and enhance network performance. By providing adaptive beamforming, beam steering, and signal enhancement capabilities, RIS-enabled antennas offer unprecedented flexibility and efficiency in communication systems. Through a comprehensive overview of RIS principles and their integration with antennas, the chapter envisions a future where RIS technologies reshape the landscape of 6G wireless communications.

Chapter 6: Revolutionizing Wireless Communication: A Comprehensive Study on Modern Antenna Technologies

Authored by Pardhu Thottempudi and Vijay Kumar, this chapter provides a comprehensive review of modern antenna technologies and their transformative impact on wireless communication systems. As wireless communication continues to evolve rapidly, antennas play a crucial role in enabling efficient and high-speed connectivity. The chapter delves into key technological innovations such as Massive MIMO, Beamforming, Metamaterial Antennas, and Reconfigurable Intelligent Surfaces, elucidating their functions, benefits, and implications. By examining these advancements, the authors highlight the potential of antenna technology to revolutionize wireless communication systems, paving the way for more efficient and sustainable networks. This comprehensive study serves as a valuable resource for communication engineers, researchers, and academicians seeking to understand the current state and future trajectories of antenna technology in wireless communications.

Chapter 7: Design and Investigation of Line Defected Photonic Crystal Antenna for Outstanding Data Transmission

Authored by Sathish Kumar Danasegaran, Elizabeth Caroline Britto, S. Dhanasekaran, G. Rajalakshmi, and A. Sivasangari, this chapter explores the design and analysis of a novel Line Defected Photonic Crystal (LDPC) antenna for high-speed data transmission. In the current era of wireless communication, with the evolution from 5G to 6G technologies, the demand for reliable and high-speed data transmission has escalated. The proposed LDPC antenna, operating in the terahertz (THz) band, showcases exceptional performance characteristics, including low return loss, high directivity, and minimal voltage standing wave ratio (VSWR). Through meticulous simulation and analysis, the authors demonstrate the antenna's suitability for applications requiring outstanding data transmission capabilities. This research contributes to advancing wireless communication systems, facilitating seamless connectivity and high-speed data transfer.

Chapter 8: Design Considerations of EBG Structures for Wearable Applications

Authored by Mehaboob Mujawar and D Vijaya Saradhi, this chapter focuses on the design considerations of Electromagnetic Bandgap (EBG) structures for wearable applications. With the proliferation of wearable devices, there arises a need for compact and efficient antenna designs that can operate seamlessly in various environments. The chapter presents the design and evaluation of a dual-band microstrip patch antenna, leveraging EBG structures for enhanced performance. Through detailed analysis and experimentation, the authors demonstrate the superior performance of the proposed antenna in terms of directivity, gain, efficiency, and other key parameters. This research contributes to advancing wearable technology by providing insights into the design considerations and performance optimization of antennas for wearable applications.

Chapter 9: Design and Analysis of Dual Band Frequency Selective Surface

Authored by Devendra Somwanshi and Payal Bansal, this chapter focuses on the design and analysis of a dual-band Frequency Selective Surface (FSS) for wireless communication applications. As wireless communication systems continue to evolve, the demand for compact and efficient frequency-selective surfaces becomes increasingly crucial. The chapter introduces a novel dual-band FSS unit cell design resonating at 4 GHz and 5.5 GHz, utilizing a combination of I-shaped and modified I-shaped metal strips. Through detailed simulations and analyses, the authors demonstrate the efficacy of the proposed FSS design in achieving broad frequency band characteristics. This research contributes to advancing wireless communication systems by providing compact and efficient solutions for frequency-selective surface designs.

Chapter 10: Phenomenon of Excess of Artificial Intelligence, Quantifying the Native AI, Its Leverages in 5G/6G, and Beyond

Authored by SNEHASIS DEY, this chapter delves into the phenomenon of Artificial Intelligence (AI) and its significant impact on wireless communication systems, particularly in the context of 5G, 6G, and beyond. As wireless networks evolve, the integration of AI becomes increasingly essential for optimizing performance, enhancing user experiences, and enabling innovative applications. The chapter explores the concept of "AI native" in the context of 6G communication, highlighting the pervasive role of AI in various fields. Through a comprehensive analysis, the author elucidates the leveraging of AI and Machine Learning (ML) in 5G and anticipates their continued evolution in future wireless communication technologies. This chapter offers valuable insights into the intersection of AI and wireless communication, underscoring its transformative potential in shaping the future of connectivity.

Chapter 11: Past, Present, and Future of Rate Splitting Multiple Access for Wireless Networks: A Review

Authored by Krishanu Kundu, this chapter provides a comprehensive review of Rate Splitting Multiple Access (RSMA) and its applications in wireless communication networks. With the increasing demand for higher data rates and reliable connections, RSMA emerges as a promising technique for optimizing spectrum utilization and managing interference. The chapter explores the benefits of RSMA in various scenarios, highlighting its robustness against channel state information uncertainty and lower feedback overhead. Moreover, the author discusses the potential applications of RSMA in 5G and beyond, including its role in integrated sensing-communications and other emerging use cases. This review serves as a valuable resource for researchers and practitioners in the field of wireless communication, offering insights into the past, present, and future of RSMA and its implications for future network designs.

In the journey through the diverse chapters of *Radar and RF Front End System Designs for Wireless Systems*, we have explored the multifaceted realm of wireless technologies, witnessing the convergence of radar and RF front-end systems to redefine connectivity paradigms. From the foundational principles to cutting-edge applications, each chapter has offered a unique perspective, enriching our understanding of this dynamic field.

As editors, we are gratified to have curated a collection that not only reflects the current state of the art but also anticipates future directions in wireless systems. The collaboration between esteemed authors from academia has fostered a rich tapestry of insights, spanning topics such as spectrum sensing, radars, antenna design, and the integration of AI/ML.

Through this endeavor, we aim to provide a comprehensive resource for researchers, students, and practitioners alike, empowering them to navigate the complexities of radar and RF front-end systems with confidence. By bridging theory and practice, we aspire to catalyze innovation and drive advancements that will shape the wireless landscape for years to come.

We extend our heartfelt gratitude to all the contributors for their dedication and expertise in shaping this volume. It is our hope that *Radar and RF Front End System Designs for Wireless Systems* will serve as a beacon, illuminating pathways towards a more connected, efficient, and resilient wireless future.

Editors:

Shilpa Mehta
Auckland University of Technology, New Zealand

Rupesh Kumar
SRM University, India

Chapter 1
Engineering Next-Generation Wireless Experiences Through Radar and RF Front End System Designs

J. Mangaiyarkkarasi
https://orcid.org/0000-0003-1431-9584
N.M.S.S. Vellaichamy Nadar College, India

J. Shanthalakshmi Revathy
https://orcid.org/0000-0003-1724-7117
Velammal College of Engineering and Technology, India

ABSTRACT

This chapter explores the synergy between radar and radio frequency (RF) front-end systems, ushering in a new era of wireless connectivity. It discusses the collaborative potential of radar and RF, emphasizing their role in enhancing security, reducing interference, and boosting adaptability. The chapter covers radar-based spectrum sensing, which enhances network efficiency, particularly in high-frequency scenarios like 5G. Radar and RF enable precise localization for IoT and autonomous vehicles, surpassing the capabilities of GPS. The chapter highlights radar's contributions to security, threat detection, and reducing signal interference. Radar-assisted RF improves vehicle communication, cooperative driving, and traffic management. In environmental monitoring and disaster management, radar augments RF for early warnings. This integration offers transformative potential, benefiting diverse applications and offering theoretical and practical insights for researchers and engineers. Radar and RF convergence offers a more connected, adaptable, and efficient wireless future.

DOI: 10.4018/979-8-3693-0916-2.ch001

1. INTRODUCTION

1.1 Evolution of Wireless Technologies

The evolution of wireless technologies has been a remarkable journey that has fundamentally transformed the way we communicate, connect, and interact with the world around us. Starting from the early days of radio communication in the late 19th century, wireless technologies have progressed through numerous generations, each marked by significant advancements.

The first generation of wireless communication primarily focused on analog voice transmission, enabling long-distance communication via radio waves. With the advent of the second generation (2G), digital technologies emerged, allowing for more efficient voice transmission and the introduction of text messaging. However, it was the third generation (3G) that truly revolutionized the landscape by enabling mobile data connectivity, laying the groundwork for basic internet browsing and limited multimedia services on mobile devices (Gunasekaran & Harmantzis, 2007). The subsequent generations, including 4G and its advanced variant LTE (Long-Term Evolution), brought about substantial increases in data speeds and paved the way for video streaming, mobile applications, and the mobile-centric digital lifestyle we experience today. As demand for data-intensive applications soared, the industry responded with the ongoing deployment of the fifth generation (5G), promising ultra-high speeds, ultra-low latency, and the capacity to support the Internet of Things (IoT) on an unprecedented scale.

This evolutionary trajectory is not only defined by technological advancements but also by the ever-expanding scope of wireless applications. From personal communication to industrial automation, healthcare, transportation, and beyond, wireless technologies have woven themselves into the fabric of modern society. As we stand at the cusp of the sixth generation (6G), with visions of even faster speeds, seamless connectivity, and transformative use cases such as holographic communication and sentient environments, the journey of wireless evolution continues to unfold, promising a future where the boundaries of connectivity are pushed even further (Tightiz et al., 2020) .

1.2 Demand for Enhanced Wireless Experiences

The relentless surge in the demand for enhanced wireless experiences underscores the pivotal role that connectivity plays in our lives. As we navigate an increasingly digital world, the need for seamless, high-quality wireless interactions has become non-negotiable. Modern society relies on wireless technology not just for communication, but also for a spectrum of activities that range from remote work and virtual education to entertainment streaming and IoT-enabled smart living. This multifaceted reliance places tremendous pressure on wireless networks to deliver robust performance in terms of speed, reliability, and low latency.

Consumers expect to effortlessly connect with loved ones, colleagues, and information resources regardless of their physical location. This has led to an insatiable demand for faster data speeds that can support high-definition video streaming, online gaming, and real-time collaboration (Arbab-Zavar et al., 2019). Additionally, as evolving technologies viz augmented reality (AR), virtual reality (VR), and autonomous vehicles become more prominent, the necessity for wireless networks to handle massive data loads with minimal delay becomes paramount. The demand for enhanced wireless experiences isn't limited to individuals; industries are also capitalizing on wireless advancements to optimize processes, increase efficiency, and unlock new business models. Smart engineering, telemedicine, smart cities, and precision farming are just a few examples of sectors that rely heavily on seamless wireless connectivity.

In essence, the demand for enhanced wireless experiences is a reflection of our interconnected world's expectations and aspirations. As technology continues to evolve, wireless networks must rise to the challenge, continuously innovating to meet the soaring expectations of users and the expanding horizons of applications.

2. FUNDAMENTALS OF RADAR AND RF FRONT END SYSTEMS

2.1 Radar Principle

Radar (Radio Detection and Ranging) is a transformative technology based on the principle of using radio waves to detect and locate objects in its vicinity. Figure 1 explains the basic principle of Radar

Figure 1. Basic principle of radar

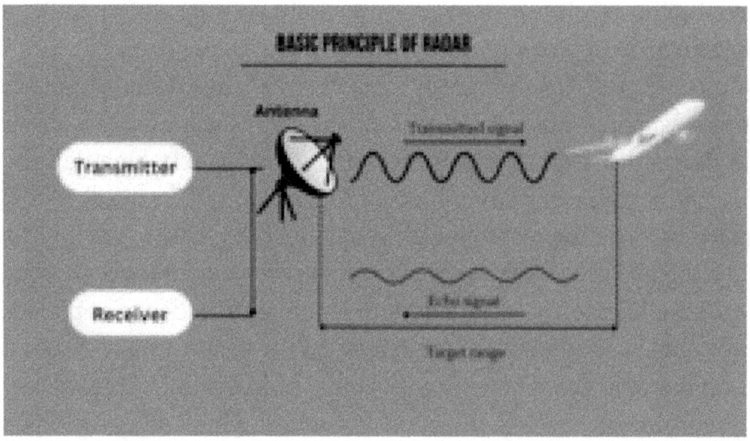

It operates by emitting radio frequency signals and then analyzing the echoes that bounce back from objects, allowing for the determination of their distance, speed, direction, and even size[(Ghorbanian et al., 2019). Range of radar is important in the designing of radar syatems which is expressed in equation (1)

$$R^4 = \sqrt{PsG^2\sigma\lambda^{2/}4\pi 3P_E} \qquad (1)$$

Where, Ps – transmitted power, σ- cross section of radar, G-Antenna gain, P_E – received power

Antenna gain is expressed in equation (2) given below,

$$\mathrm{G} = \frac{4\pi A k_a}{\lambda^2} \qquad (2)$$

Where, A – area of antenna, K_a – efficiency, λ -wavelength

2.2 Applications

Radar's applications are diverse and span multiple fields. In aviation, radar is instrumental for air traffic control, weather monitoring, and collision avoidance systems. In meteorology, it aids in tracking storms and assessing precipitation patterns. Military applications include surveillance, target tracking, and missile guidance. Additionally, radar plays a pivotal role in maritime navigation, helping ships and boats avoid collisions and providing information about sea conditions. Furthermore, radar has found its way into automotive safety, enabling features like adaptive cruise control and autonomous emergency braking. Its utility extends to geology, where ground-penetrating radar helps archaeologists and geologists discover subsurface features (Bruder, 2013). As technology advances, radar's versatility continues to expand, offering innovative solutions across domains and contributing significantly to our modern way of life.

2.3 Role of RF Front End Systems in Wireless Communication

The RF (Radio Frequency) front end is the gateway that bridges the physical world of electromagnetic signals with the digital realm of wireless communication systems. It plays a critical role in shaping, processing, and converting radio signals, forming a crucial interface between the transmitting and receiving elements of wireless devices.

In the realm of wireless communication, the RF front end system's significance cannot be overstated. It is accountable for responsibilities like signal amplification, filtering, modulation, and demodulation (Zhao et al., 2021). Figure 2 depicts the RF front end design.

Figure 2. RF front end design

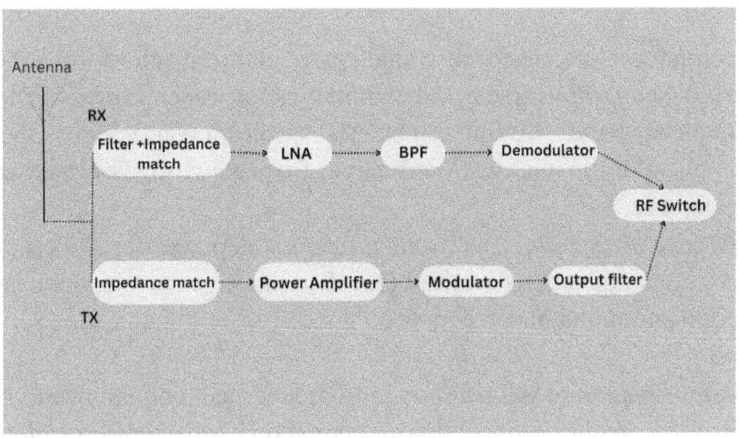

Signal amplification ensures that weak transmitted signals reach their intended destinations, while filtering helps eliminate unwanted noise and interference, enhancing signal quality. Modulation and demodulation processes allow data to be encoded onto carrier waves for transmission and decoded at the receiving end, enabling information exchange.

RF front end systems are tailored to the specific requirements of various wireless technologies, whether it's cellular networks, Wi-Fi, Bluetooth, or satellite communication. The design and capabilities of the RF front end directly influence the overall performance of wireless devices, determining factors like signal strength, data rate, range, and energy efficiency.

As wireless communication evolves to accommodate higher data rates, broader frequency ranges, and enhanced connectivity, the RF front end system's role becomes even more pivotal. The amalgamation of innovative technologies such as beamforming, massive MIMO, and millimeter-wave communications requires sophisticated RF front end designs to manage complex signal processing and ensure optimal performance (F. Liu et al., 2023) and hence the RF front end system serves as the cornerstone of wireless communication, enabling the translation of information between the analog and digital domains, and ultimately shaping the quality and efficiency of our modern wireless experiences.

3. INTEGRATION OF RADAR AND WIRELESS COMMUNICATION

3.1 Radar-Communication Coexistence Challenges

Radar – communication coexistence challenges are listed below.

***Spectrum Congestion*:** Radar and communication systems often operate within limited frequency bands, leading to potential spectrum congestion as the demand for both applications increases. Coordinating frequency usage becomes crucial to avoid interference.

***Interference Mitigation*:** The strong transmission power of radars can interfere with nearby communication devices. Innovative interference mitigation techniques are needed to ensure that radar signals do not disrupt communication services and vice versa.

***Dynamic Frequency Allocation*:** Both radar and communication systems require adaptable frequency allocation to accommodate changing operational needs. Developing dynamic spectrum sharing mechanisms is essential to prevent conflicts.

***Signal Isolation*:** Radar and communication systems emit strong electromagnetic signals that can unintentionally leak into neighboring frequency bands, causing cross-interference. Effective signal isolation techniques are necessary to maintain signal purity (Feng et al., 2020).

***Antenna Sharing*:** Efficient sharing of antennas for radar and communication can be challenging due to different radiation patterns and operational requirements.

Co-designing antennas to cater to both applications while minimizing interference is a complex task.

***Power Efficiency*:** Balancing power consumption is crucial as both radar and communication systems need to manage energy usage. Energy-efficient designs are required to ensure sustainable coexistence.

***Latency and Timing*:** Communication systems demand low latency, while radar applications require precise timing for target detection. Synchronizing these different timing requirements without compromising performance is a significant challenge.

***Protocol Compatibility*:** Radars and communication devices follow different protocols and communication standards. Ensuring that coexisting systems can understand each other's signals and protocols is essential for effective coordination.

***Regulatory Compliance*:** Regulatory frameworks and standards may differ for radar and communication technologies. Navigating these regulations while ensuring seamless coexistence can be complex, particularly in international deployments.

***Adaptive Algorithms*:** Developing adaptive algorithms that allow real-time adjustments in transmission parameters based on environmental conditions, interference levels, and operational needs is critical to address coexistence challenges effectively (R. Liu et al., 2021)

3.2 Spectrum Sharing Process and Managing Interference

Spectrum sharing process and interference are pivotal aspects of modern wireless ecosystems. As the demand for wireless services grows, effectively allocating and sharing the limited frequency spectrum becomes paramount. Sophisticated technologies and protocols are essential to ensure that different wireless systems, such as radar and communication, can coexist harmoniously without causing harmful interference. Figure 3 explains the process of spectrum sharing types.

Figure 3. Spectrum sharing types

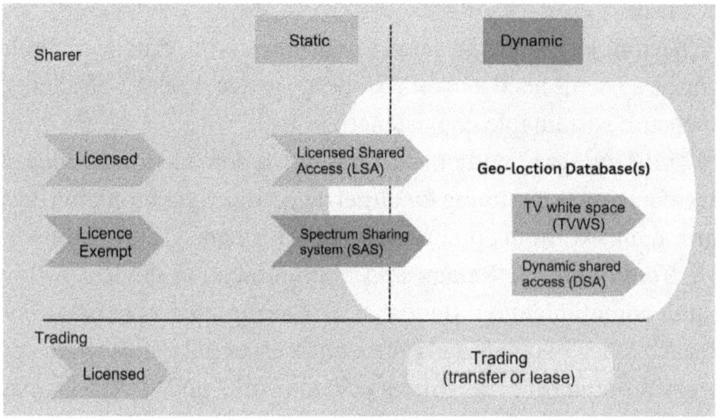

Dynamic spectrum admittance techniques, cognitive radio, and adaptive modulation are among the strategies employed to optimize spectrum utilization (Zheng et al., 2023). By intelligently managing interference and facilitating efficient spectrum sharing, these approaches enhance overall wireless performance, enabling diverse applications to thrive within the same frequency bands.

4. ADVANCED ANTENNA TECHNOLOGIES FOR NEXT-GENERATION WIRELESS

4.1 MIMO and Massive MIMO Systems

Multiple-Input Multiple-Output (MIMO) machinery has developed wireless communication by dramatically enhancing data rates, reliability, and spectral efficacy. MIMO systems employ several antennas at both the transmitting and receiving ends, exploiting spatial diversity to transmit and receive several data streams simultaneously over the similar frequency channels (Elijah et al., 2022). This technique mitigates signal fading, increases capacity, and improves overall system performance. Figure 4 explains Massive MIMO Techniques for 5G.

Figure 4. Massive MIMO systems for 5G

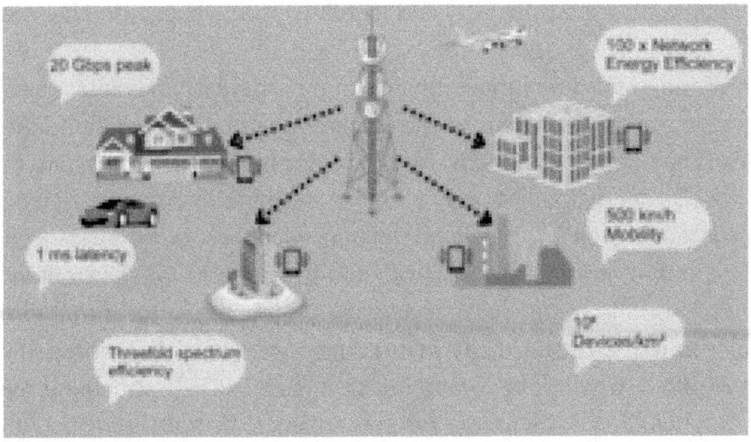

Massive MIMO takes MIMO to the next level by installing a large number of antennas at the base station. This architecture further boosts spectral efficiency and user experience. By utilizing spatial multiplexing techniques and beamforming, massive MIMO method increases signal-to-noise ratio, extends coverage, and reduces interference. The technology is particularly effective in crowded environments and for providing high-speed connectivity to a large number of users (Wei et al., 2023).

Massive MIMO's ability to serve multiple users concurrently, along with its resistance to fading and its potential for energy efficiency, makes it a cornerstone of future wireless networks. This technology aligns well with the demands of 5G and beyond, where the focus is on accommodating diverse applications such as augmented reality, Internet of Things (IoT), and ultra-high-definition streaming. As the wireless landscape continues to evolve, MIMO and massive MIMO systems will play a pivotal role in achieving the insatiable appetite for higher data rates, seamless connectivity, and enhanced user experiences.

4.2 Beamforming and Directional Antennas

Beamforming and directional antennas are instrumental technologies in wireless communication, allowing for targeted signal transmission and reception, improved link quality, and enhanced spectral efficiency. Beamforming involves shaping and directing radio frequency signals in specific directions, optimizing signal strength towards intended receivers while minimizing interference in other directions.

Directional antennas, a key component of beamforming, focus radio waves in a specific direction rather than uniformly radiating in all directions. This enables greater signal reach, reduced interference, and improved resistance to noise and fading. By concentrating signal energy towards the intended user or device, directional antennas enhance communication reliability and increase system capacity (Hassanien et al., 2016).

Beamforming and directional antennas are particularly advantageous in urban and dense environments where interference and signal degradation are common challenges. They enable better coverage, support higher data rates, and extend the reach of wireless networks. Moreover, they contribute to energy efficiency as less power is wasted on signals propagating in unneeded directions. These technologies find applications in various domains, from cellular networks to Wi-Fi routers, satellite communications, and even vehicular networks. They are integral to improving the quality of service, ensuring consistent connectivity, and supporting the diverse demands of modern wireless communication, such as IoT, streaming media, and real-time applications. As wireless systems advance, beamforming and directional antennas will continue to play a pivotal role in influencing the upcoming of wireless experiences

5. ADAPTIVE RF FRONT END DESIGNS

5.1 Dynamic Spectrum Access Techniques

Dynamic Spectrum Access (DSA) techniques are pivotal to understand the rising claim for the wireless communication by optimizing the utilization of available frequency spectrum. DSA allows for the efficient sharing of spectrum resources between various users, applications, and services in real-time, maximizing spectral efficiency while minimizing interference (Zhang et al., 2020). Figure 5 explains various dynamic Spectrum Access Techniques

Figure 5. Dynamic spectrum access techniques

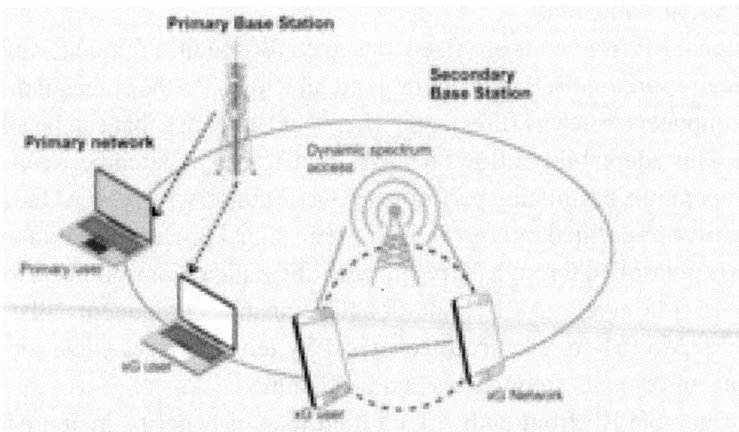

One key aspect of DSA is cognitive radio, where devices are equipped with the intelligence to sense and adapt to their surrounding radio environment. Cognitive radios can identify underutilized frequency bands or "white spaces" and opportunistically access them without causing interference to licensed users. This technology not only enhances spectrum utilization but also promotes the coexistence of diverse wireless services. Another approach is the use of spectrum databases that maintain information about available frequencies, regulatory rules, and usage patterns. When a device seeks to establish a connection, it queries the database for suitable frequency bands. This database-driven approach ensures compliance with regulatory requirements and minimizes conflicts.

DSA techniques hold immense potential for addressing spectrum scarcity and enabling innovative wireless applications. They are especially beneficial in scenarios where spectrum resources are unevenly distributed or underutilized. However, challenges remain, including interference management, accurate spectrum sensing, and ensuring fair access for various users (Y. Wu et al., 2023). As wireless technology evolves, DSA techniques are important in generating more flexible, efficient, and adaptable wireless communication schemes that cater to the dynamic demands of our interconnected world.

5.2 Reconfigurable RF Front End Architectures

Reconfigurable RF front end architectures are a transformative paradigm in wireless communication design, offering unprecedented flexibility and adaptability. These architectures enable wireless devices to dynamically adjust their radio frequency circuitry to accommodate varying operating conditions, frequency bands, and communication standards.

Traditional RF front ends are fixed, designed for specific frequency ranges and communication protocols. Reconfigurable architectures, on the other hand, integrate tunable components such as filters, amplifiers, and antennas that can be adjusted in real-time. This adaptability allows devices to efficiently switch between different modes of operation, optimizing performance while conserving power. One approach involves software-defined radios (SDRs), where signal processing and modulation schemes are controlled through software, enabling rapid reconfiguration for diverse applications (Ma et al., 2021). Additionally, tunable components like varactor diodes or MEMS (Micro-Electro-Mechanical Systems) devices allow for dynamic adjustments of resonant frequencies and bandwidths.

Reconfigurable RF front ends offer a multitude of benefits, including reduced manufacturing costs by catering to multiple frequency bands with the same hardware, enhanced coexistence in spectrum-sharing environments, and improved resilience against interference. They also play a crucial role in future-proofing devices as new wireless standards emerge.

6. MILLIMETER WAVE AND BEYOND FOR ENHANCED WIRELESS

6.1 Millimeter Wave Frequency Bands and Characteristics

Millimeter wave frequency bands, often abbreviated as mmWave bands, refer to a range of radio frequency (RF) wavelengths that fall within the millimeter range, typically between 30 GHz (gigahertz) and 300 GHz (Björnson et al., 2019). These frequencies occupy a higher portion of the electromagnetic spectrum than those commonly used for traditional wireless communication, such as cellular networks. Some of the well-known millimeter wave frequency bands include: 24 GHz to 100 GHz, 57 GHz to 86 GHz and 94 GHz to 300 GHz.

Millimeter wave frequencies offer the advantage of providing larger bandwidths, which can support extremely high data rates. This makes them suitable for next-generation wireless technologies, such as 5G and after that. However, millimeter waves have shorter propagation distances and can be more susceptible to atmospheric absorption and obstacles like buildings or foliage, which can limit their coverage area. To overcome these challenges, technologies like beamforming and advanced antenna designs are employed to direct and focus the signals (Kaushik et al., 2021). Millimeter wave frequencies are key players in unlocking the potential of ultra-fast wireless communication, enabling high-definition video systems, augmented reality systems, virtual reality systems, and Internet of Things in urban environments.

Millimeter wave frequency bands possess several distinctive characteristics that make them unique within the electromagnetic spectrum (Mishra et al., 2019). These characteristics shape their applications, advantages, and challenges. The important characteristics includes, high frequency, large Bandwidth, short wavelength, propagation challenges, spatial multiplexing, beamforming, line-of-sight dependency, high data rates and short range communications.

Millimeter wave frequency bands offer remarkable potential for high-speed, high-capacity wireless communication. While they come with unique challenges related to propagation and coverage, their advantages in terms of data rates, spatial multiplexing, and directional capabilities position them as a crucial component of the wireless technology landscape (Ahmed et al., 2021).

6.2 Benefits and Challenges of Millimeter Wave Communications

Millimeter wave technology presents a myriad of advantages that are shaping the landscape of wireless communication. First and foremost, its high data rates are a game-changer, addressing the insatiable appetite of data-intensive 4K video applications, virtual reality, and augmented reality systems. This is made possible by the massive bandwidth availability within the millimeter wave spectrum, which can put up the ever-growing number of connected devices and data-hungry services(Dokhanchi et al., 2019) . One of the remarkable features of millimeter waves is their efficient spatial reuse due to their short wavelengths. This enables multiple devices to operate in close proximity using the same frequency band without causing interference, greatly optimizing spectrum utilization. Moreover, their small antenna size, a direct consequence of the short wavelength, allows for the integration of multiple antennas in compact devices, fostering advanced beamforming techniques for improved signal quality and coverage. The ultra-low latency capabilities of millimeter wave communication are instrumental in online gaming, remote control surgery, and autonomous vehicles, where split-second decisions are critical. Additionally, their short-range and high directionality enhance security and privacy by reducing the risk of eavesdropping and unauthorized access, making them suitable for high-security applications.

Millimeter waves find utility in wireless backhaul solutions, offering high-capacity connections between cellular base stations and the core network, ensuring reliable and high-speed data transfer (Griffiths et al., 2015). Furthermore, this technology future-proofs networks, as it can easily accommodate the anticipated surge in data consumption as demands continue to escalate.

Some significant challenges of millimetre wave communications are propagation loss, atmospheric absorption, obstruction sensitivity, signal penetration, costly infrastructure, complex beamforming, device design challenges, regulatory hurdles, interference and crowding, trade-off between coverage and capacity (Cohen et al., 2018)

7. ENABLING TECHNOLOGIES FOR ENHANCED RADAR CAPABILITIES

7.1 Pulse-Doppler Techniques for Target Tracking

Pulse-Doppler techniques for target tracking involve advanced signal processing methods that utilize the Doppler frequency shift of radar echoes to accurately detect and track moving objects. Figure 6 shows Pulse Doppler processing of signal

Figure 6. Pulse doppler processing of signal

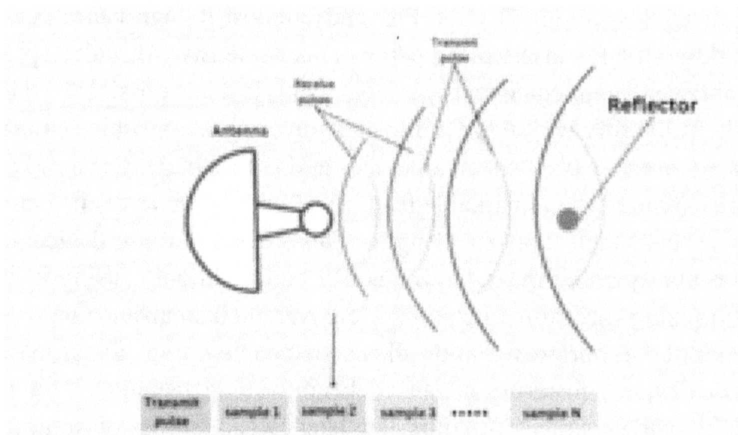

Here are some prominent Pulse-Doppler techniques used in target tracking:

Constant False Alarm Rate (CFAR): CFAR techniques adaptively adjust the detection threshold rely on the statistical properties of the received signal. This helps discriminate between true target echoes and clutter caused by stationary objects or environmental factors.

Moving Target Indication (MTI): MTI techniques employ multiple radar pulses to detect and track moving targets while suppressing the returns from stationary objects. Doppler filtering or cancellers are used to eliminate clutter returns, enhancing the visibility of moving targets.

***Pulse-Pair Techniques*:** Pulse-pair processing compares the phase difference between successive radar pulses to estimate the target's Doppler shift. This technique reduces the impact of Doppler ambiguities and improves velocity accuracy.

***Stretch Processing*:** Stretch processing involves using linear frequency modulation in transmitted pulses. This spreads the received echoes in the frequency domain, making Doppler shifts more discernible and improving target detection (Dokhanchi et al., 2018).

***Clutter Map Tracking*:** Clutter map tracking builds a spatial representation of the radar environment to distinguish between actual targets and clutter. This method enhances target tracking accuracy by effectively filtering out unwanted signals.

Doppler Ambiguity Resolution: Modern radar systems employ techniques like staggered pulse repetition frequency (PRF) to mitigate Doppler ambiguities, allowing accurate tracking of fast-moving targets.

***Space-Time Adaptive Processing (STAP)*:** STAP combines information from multiple radar snapshots to suppress clutter and interference, improving detection and tracking of low-observable targets in complex scenarios.

***Frequency Modulated Continuous Wave (FMCW) Radar*:** FMCW radars use continuous transmission and frequency modulation to measure both range and Doppler shift, enabling simultaneous target tracking and velocity estimation.

***Multi-Channel Processing*:** Multi-channel processing involves using multiple receive channels to enhance the signal-to-noise ratio and enable efficient Doppler processing for tracking.

***Kalman Filtering*:** Kalman filtering is a recursive estimation technique used for target tracking. It combines measurement data with a dynamic model to predict and update target state information, leading to accurate and smooth tracking results (Kumari et al., 2018).

These Pulse-Doppler techniques, among others, are crucial for extracting valuable information from radar echoes, allowing for the precise detection, localization, and tracking of moving objects even in challenging and cluttered environments.

7.2 Synthetic Aperture Radar (SAR) for Imaging

Synthetic Aperture Radar (SAR) is a powerful remote sensing technology that revolutionizes imaging in various fields, from earth observation and environmental monitoring to disaster assessment and military reconnaissance. Unlike optical imaging, SAR operates independently of daylight and weather conditions, making it an invaluable tool for obtaining detailed images regardless of environmental constraints. Figure 7 shows synthetic aperture radar.

Figure 7. Synthetic aperture radar

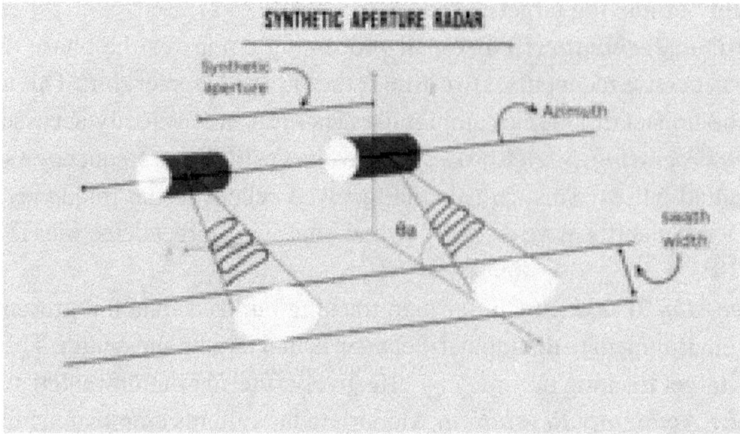

SAR imaging involves emitting microwave signals towards the Earth's surface from an airborne or space borne platform. The signals reflect off the target and return to the sensor, where they are recorded over multiple passes or pulses. By combining the information collected from these multiple perspectives, SAR effectively creates a "synthetic aperture," simulating a larger antenna that produces higher-resolution images (K. Wu et al., 2023). The key advantage of SAR lies in its ability to generate detailed, high-resolution images with consistent quality. It can penetrate cloud cover, haze, and even dense vegetation, providing critical information for disaster response, crop monitoring, deforestation assessment, and more. Additionally, SAR's capability to detect subtle changes over time makes it ideal for monitoring landscape alterations, tracking glacier movements, and studying ground deformation due to earthquakes or subsidence.

However, SAR imaging does present challenges such as complex signal processing, the need for accurate platform positioning, and issues related to speckle noise in resulting images. Nonetheless, its versatility, accuracy, and applicability across various domains have solidified Synthetic Aperture Radar as an indispensable tool for capturing detailed and reliable imagery of Earth's surfaces from space or the sky.

8. CO-DESIGN AND OPTIMIZATION OF RADAR-COMMUNICATION SYSTEMS

8.1 Cross-Layer Design Considerations

Cross-layer design considerations in radar communication involve optimizing interactions and trade-offs between different protocol layers to enhance overall system performance. Here are some key cross-layer design considerations in radar communication:

i. ***Physical Layer to Data Link Layer*:** Optimizing modulation schemes and coding techniques to suit the radar waveform characteristics while maintaining efficient data transmission and error correction mechanisms.
ii. ***Physical Layer to Network Layer*:** Coordinating radar transmission timing and power control to minimize interference with other communication systems and ensure efficient use of available spectrum resources.
iii. ***Physical Layer to Transport Layer*:** Balancing radar waveform characteristics, such as pulse repetition interval, with transport layer requirements to ensure timely and reliable data delivery (Zarakovitis et al., 2012).
iv. ***Physical Layer to MAC (Medium Access Control) Layer*:** Coordinating radar transmission schedules with MAC protocols to avoid collisions and ensure fair access for both radar and communication devices.
v. ***MAC Layer to Network Layer*:** Prioritizing radar signals over data communication during critical radar operations, and dynamically adjusting traffic prioritization based on network congestion and radar requirements.
vi. ***MAC Layer to Transport Layer*:** Adapting MAC layer parameters like access priority and contention resolution to optimize latency and reliability for both radar sensing and communication.
vii. ***Network Layer to Application Layer*:** Ensuring radar's quality of service requirements align with application-specific needs, such as tracking accuracy, object recognition, or target classification.
viii. ***Network Layer to Security Layer*:** Implementing security mechanisms to protect radar and communication data, considering the unique vulnerabilities and requirements of both systems.
ix. ***Network Layer to QoS (Quality of Service) Layer*:** Allocating network resources to ensure that both radar and communication systems meet their respective QoS targets while avoiding resource contention.
x. ***Network Layer to Energy Management Layer*:** Coordinating energy-efficient operation by optimizing the active and sleep periods of radar and communication devices to extend battery life.

By carefully addressing these cross-layer considerations, radar communication systems can achieve effective coexistence, minimize interference, and provide optimal performance for radar sensors and wireless systems.

8.2 Joint Optimization of Radar and Communication Performance

Joint optimization of radar and communication performance involves a holistic approach to designing systems that seamlessly integrate radar sensor and wireless communication functionalities. This approach recognizes the synergy between the two systems and leverages their interactions to achieve enhanced overall performance (Ashraf et al., 2023).

One key aspect of joint optimization is efficiently sharing spectrum resources between radar and communication systems. By dynamically allocating frequency bands and time slots based on real-time requirements, both radar and communication services can coexist without causing harmful interference. This shared spectrum usage optimizes spectral efficiency and maximizes the utility of the available bandwidth. Figure 8 clearly explains Joint Optimization of Radar and Communication

Figure 8. Joint optimization of radar and communication

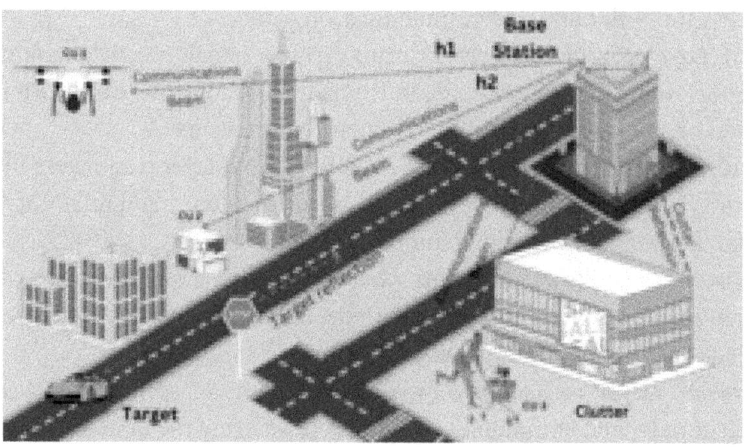

Furthermore, joint optimization considers the interchanges between radar sensing and communication tasks. For instance, radar can adapt its transmission parameters to minimize interference with communication devices during critical communication moments. Conversely, communication systems can prioritize data transmission while accommodating radar's sensing needs during less sensitive communication periods. In the context of infrastructure, co-locating antennas and sharing hardware components between radar and communication systems can lead to cost savings and efficient space utilization. This approach reduces the physical footprint and lowers deployment expenses.

Ultimately, joint optimization seeks to create a synergistic relationship where radar and communication systems not only coexist harmoniously but also enhance each other's performance. By sharing resources, adapting parameters, and aligning operational strategies, this approach leads to more capable, versatile, and responsive systems that address the demands of both radar sensing and wireless communication in a unified manner.

9 CASE STUDIES: INTEGRATING RADAR INTO WIRELESS SYSTEMS

9.1 Automotive Radar and V2X Communication

Automotive radar and Vehicle-to-Everything (V2X) communication are pivotal technologies reshaping the landscape of transportation safety and efficiency. These interconnected systems work in tandem to create smarter and safer road environments.

Automotive radar employs radar technology to enable vehicles to sense their surroundings. This includes detecting nearby vehicles, pedestrians, and obstacles, providing critical data for collision avoidance, adaptive cruise control, and autonomous driving functions. Radar sensors use the Doppler effect to measure relative speed and distance accurately, even in adverse weather conditions (Rauch et al., 2011).

V2X communication, on the other hand, leverages wireless communication to establish connections between vehicles and infrastructure elements such as traffic lights, road signs, and even pedestrians' smartphones. This bi-directional communication allows vehicles to share real-time information about their location, speed, and intentions, enhancing situational cognizance and enabling innovative safety structures like intersection collision evasion and cooperative adaptive cruise control. Figure 9 shows the v2x communication.

Figure 9. v2x communication

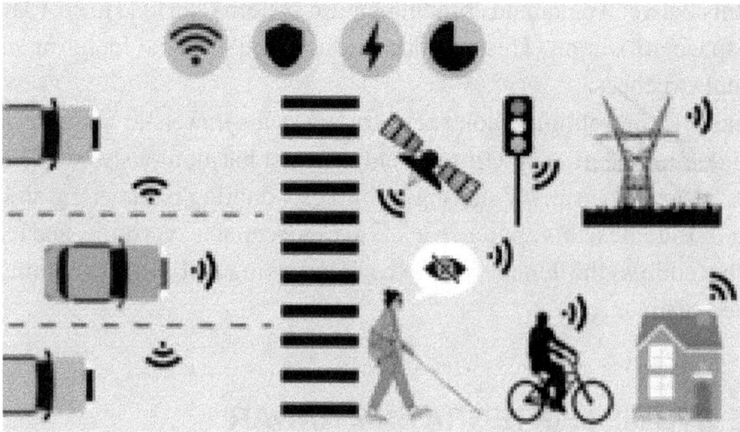

Together, automotive radar and V2X communication create a comprehensive safety net. Radar systems provide immediate object detection and short-range obstacle awareness, while V2X communication extends this awareness to encompass a broader environment, including non-line-of-sight scenarios. This collaborative approach enhances collision mitigation capabilities and supports the realization of fully autonomous vehicles.

Automotive manufacturers and researchers are continually advancing these technologies to ensure effective coexistence, optimal resource allocation, and reliable communication in complex driving scenarios. The combined power of automotive radar and V2X communication promises safer roads, reduced traffic congestion, and a future of smarter, more efficient transportation systems.

9.2 5G NR and Radar Integration for Industrial Applications

The integration of 5G New Radio (NR) and radar technologies holds significant promise for revolutionizing industrial applications, particularly in sectors like manufacturing, logistics, and automation. 5G NR's high data rates, low latency, and network sharing capabilities, coupled with radar's object detection and ranging prowess, create a potent synergy that can reshape industrial processes.

5G NR's ultra-reliable low-latency communication (URLLC) abilities enhance real-time control and communication between machines and systems, allowing for precise coordination and efficient resource management. This is essential for time-sensitive industrial tasks, such as robotics, where split-second decisions are crucial (Aijaz, 2020). Radar technology complements this by offering accurate object detection and tracking, enabling collision avoidance, automated guided vehicles (AGVs) routing, and asset tracking within industrial environments. Radar's resilience to adverse conditions like dust, smoke, and poor lighting further solidifies its reliability in demanding settings (Liang et al., 2023).

The integration of 5G NR and radar enables seamless communication between machines, devices, and central control units, resulting in improved production efficiency, reduced downtime, and enhanced worker safety. For instance, AGVs can benefit from radar-based navigation and collision avoidance systems while communicating critical status updates through 5G networks (Li et al., 2023).

As industries progressively embrace Industry 4.0 concepts, the amalgamation of 5G NR and radar is poised to create an ecosystem where machines collaborate intelligently, leading to heightened productivity, optimized processes, and safer work environments. The synergy between these technologies exemplifies the transformative potential of combining wireless communication and radar sensing in industrial applications (Qian et al., 2022).

9.3 Recent Advances in Radar Design

The enduring essence of the fundamental radar range equation continues to serve as a wellspring of inspiration for pioneering designs and innovations in both civilian and military radar systems (Brown & Li, 2023).

In the contemporary radar landscape, this enduring equation has given rise to a plethora of cutting-edge technologies. These encompass a range of technologies, such as Multiple Inputs, Multiple Output systems, Digital Beam Forming methods, Active Electronically Steered Array radar, millimeter-wave radar, Passive Coherent Location Radar Systems, semiconductor Power Amplifiers, and advanced techniques for signal coding and Digital Signal Processing in radar systems. These advanced technologies have played a pivotal role in shaping the landscape of modern radar systems (Tan et al., 2023). Researchers have introduced an innovative hybrid substrate integrated checkerboard metasurface (CMS) that combines resonance-based and dispersion techniques to enhance radar stealth, effectively reducing the radar cross section (RCS) of flat metallic targets (Wang et al., 2023). To attain a broader bandwidth, the researchers employed both single and dual resonant artificial magnetic conductor meta-atoms, featuring adjusted cell topologies. The investigation delved into the resonance mechanism, providing a comprehensive insight into the reflective behaviors of these meta-atoms. A prototype of the RCS reducer was fabricated and measured, aligning closely with theoretical simulations. The improved 8 dB RCS drop in experimental results was attributed to the dielectric substrate's dispersion effects. This innovative design sets a benchmark for economical CMS in radar evasion applications. Radar design development is shown in Figure 10.

Figure 10. Radar design development

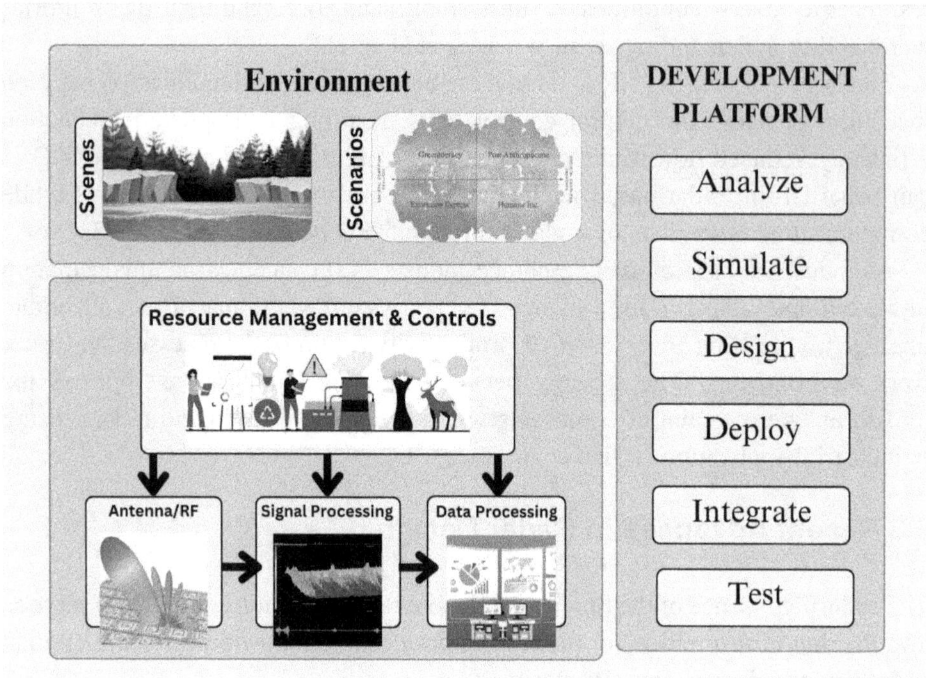

Nowadays, leveraging modeling and simulation tools within the radar system design process offers comprehensive benefits. By utilizing MATLAB and Simulink as the core platform for radar system development we can achieve better performance.

This comprehensive radar simulation and development framework offers a wide array of capabilities. It facilitates the interactive adjustment of radar parameters and waveform optimization, taking into account considerations based on the radar range equation and link budget. The framework expedites the development process with its extensive algorithm libraries with encompassing functions It supports modeling across a range of scenarios, including ground-based, airborne, ship-borne, space-based, and automotive radar systems, accommodating both moving targets and radar platforms. Additionally, it can create phased array systems, evaluate their presentation in diverse conditions using accurate synthetic data, and generate radar signals for machine and deep learning models. This framework seamlessly integrates RF component models and complex antenna designs, enhancing radar system design fidelity. It also streamlines the debugging and testing of complete radar models in the early project stages, minimizing the risk of costly redesigns. Furthermore, it automates the translation of MATLAB- Simulink methods into code for positioning on DSP and programmable gate array, making the implementation process more efficient.

10. OVERCOMING TECHNICAL AND REGULATORY HURDLES

10.1 Regulatory Framework for Radar and Wireless Integration

The integration of radar and wireless technologies necessitates a well-defined regulatory framework to ensure coexistence, minimize interference, and promote efficient spectrum utilization. Regulatory bodies worldwide play a vital role in establishing rules that govern the deployment and operation of such integrated systems.

One key challenge in radar and wireless integration lies in allocating spectrum resources in a way that prevents harmful interference between these systems. Moreover, regulatory frameworks need to address issues related to spectrum sharing and coordination. Dynamic spectrum access mechanisms, such as geolocation databases and spectrum sensing, can aid in avoiding interference between radar and wireless networks. Additionally, regulations must ensure that both radar and wireless communication systems comply with emission limits to prevent excessive interference (Huseien & Shah, 2022).

International harmonization is also essential. Radar and wireless technologies transcend geographical boundaries, making it crucial for regulatory frameworks to align internationally. This ensures consistency and interoperability, allowing integrated systems to function seamlessly across regions.

As technologies evolve, regulatory bodies must remain agile and adaptive to accommodate advancements in radar and wireless communication. They play a pivotal role in enabling innovation while safeguarding against potential disruptions and conflicts. By establishing clear guidelines and standards, regulatory frameworks foster a conducive environment for the successful integration of radar and wireless technologies to create safer, more efficient, and interconnected systems.

10.2 Addressing Privacy and Security Concerns

Addressing privacy and security concerns in the integration of radar with wireless systems is crucial to ensure the protection of sensitive information and maintain user trust. Some significant points are shown in figure 11.

Figure 11. Privacy and security concerns

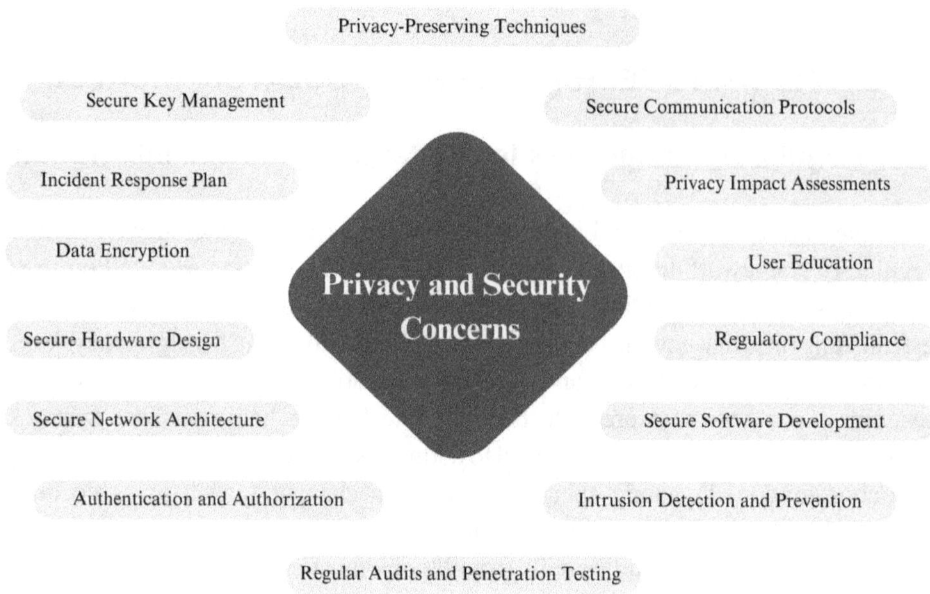

i. ***Data Encryption*:** Implement robust encryption mechanisms for both radar and wireless communication data to prevent unauthorized access and eavesdropping (You et al., 2021).
ii. ***Authentication and Authorization*:** Employ strong authentication methods to confirm the distinctiveness of devices and users before granting access to the network or radar system.
iii. ***Secure Key Management*:** Establish secure key management practices to safeguard encryption keys and ensure they are not compromised.
iv. ***Privacy-Preserving Techniques*:** Utilize privacy-preserving techniques like differential privacy to anonymize and aggregate data while still extracting meaningful insights.
v. ***Secure Software Development*:** Follow secure software development practices to minimize vulnerabilities and ensure that software and firmware updates are free from security flaws.
vi. ***Secure Communication Protocols*:** Use secure communication protocols that protect data integrity and confidentiality during transmission.

vii. ***Regular Audits and Penetration Testing*:** Conducting regular security audits and penetration testing to categorise vulnerabilities and flaws in the system's defenses.
viii. ***User Education*:** Educate users and stakeholders about the importance of security measures, best practices, and potential risks associated with the integrated system.
ix. ***Secure Hardware Design*:** Utilize hardware security features to prevent physical attacks and tampering of devices.
x. ***Secure Network Architecture*:** Design the network architecture with security in mind, realising firewalls, intrusion detection, and network division.
xi. ***Incident Response Plan*:** Develop a comprehensive incident reaction plan to handle security breaches and lessen the control on the integrated system.

By addressing these points, stakeholders can foster a secure and privacy-respecting environment for the incorporation of radar with wireless communication systems, mitigating risks and building user confidence in the technology's capabilities.

11.FUTURE DIRECTIONS IN RADAR-ENHANCED WIRELESS EXPERIENCES

11.1 Integration With 6G and Beyond

Integration with 6G and beyond represents a forward-looking approach to technology convergence, where radar and wireless communication systems synergistically coexist to shape the future of interconnected environments. As we approach the era beyond 5G, the integration of radar and wireless technologies holds immense potential to create innovative solutions that transcend traditional boundaries.

6G envisions a hyper-connected world, where data rates reach terabits per second, latency becomes nearly imperceptible, and diverse applications benefit from seamless connectivity. Integrating radar into this vision can lead to transformative advancements. For instance, combining radar's precise sensing capabilities with ultra-high-speed wireless communication can create unparalleled situational awareness for autonomous vehicles, enabling real-time collision avoidance even in complex urban scenarios(F. Liu et al., 2022). Moreover, radar's ability to penetrate obstacles and harsh weather conditions can enhance the reliability of communication networks. This is especially relevant for applications like disaster response, where integrated systems can ensure continuous connectivity in adverse situations.

The integration of radar and 6G wireless communication also has implications for security and privacy. Radar can be employed to detect and mitigate wireless signal jamming and cyberattacks, fortifying the resilience of communication networks against malicious actors.

As we look even further ahead to 7G and beyond, integration can amplify the potential for holistic sensing and communication ecosystems. Radar systems could play a pivotal role in creating intelligent environments that adapt in real-time based on both wireless communication and radar-sensed data. This could lead to urban infrastructure that self-adjusts traffic signals based on real-time vehicle movements, resulting in optimized traffic flow and reduced congestion.

However, this integration also introduces challenges such as harmonizing complex waveforms, addressing interference between radar and advanced wireless networks, and managing cross-layer optimizations effectively. Collaborative research development across disciplines will be essential to realize the potential of this integration (Shafie et al., 2022).

In conclusion, the integration of radar with 6G and beyond holds tremendous promise for reshaping our world's communication and sensing capabilities. As these technologies converge, they can revolutionize industries, enhance safety, and drive innovations that propel us toward a more interconnected and intelligent future.

11.2 AI-Driven Adaptations for Dynamic Environments

In dynamic and rapidly changing environments, the integration of artificial intelligence (AI) offers remarkable opportunities for adaptive systems that can respond intelligently to evolving conditions. AI-driven adaptations empower systems to make real-time decisions, optimize performance, and ensure efficiency in the face of uncertainties and fluctuations.

One of the basic strengths of AI lies in its skill to process large data from several sources. This data can be leveraged to generate comprehensive situational awareness, allowing systems to understand the environment's current state and predict its future dynamics. For instance, in autonomous vehicles, AI can process sensor data to assess road conditions, weather, traffic patterns, and even pedestrians' behavior, enabling the vehicle to adapt its driving strategy accordingly (Drinan et al., 2012).

Machine learning algorithms, a subset of AI, enable systems to learn patterns and trends from historical data. This enables AI-driven adaptations to recognize recurring scenarios and make informed decisions. In manufacturing, AI can optimize production processes by predicting equipment failures and adjusting operations in real-time to minimize downtime.

AI-driven adaptations are particularly relevant in sectors like healthcare and disaster management. In healthcare, AI can continuously monitor patients' vital signs and detect anomalies, triggering alerts to medical staff for timely intervention. In disaster management, AI can analyze incoming data from various sources to predict natural disasters and dynamically allocate resources to affected areas.

However, challenges exist, including the need for robust AI models that can generalize well in diverse scenarios and the ethical considerations around AI-driven decision-making.

Ultimately, AI-driven adaptations for dynamic environments promise to revolutionize how we interact with technology. As AI systems become more adept at processing data, learning from it, and making real-time adjustments, they will enable a new level of efficiency, safety, and responsiveness in sectors ranging from transportation to healthcare, from manufacturing to emergency response, and beyond. The synergy between AI and dynamic environments can reshape industries and improve quality of life on a global scale.

11.3 Achievements in Next-Gen Wireless Experiences

The realm of next-generation wireless experiences has seen remarkable achievements that are transforming the way we connect, communicate, and interact. These achievements are shaping a future where wireless technologies offer unprecedented capabilities and cater to the diverse demands of our interconnected world.

i. ***5G Revolution*:** One of the most notable achievements has been the rollout of 5G networks. With faster data rate, low latency, and massive device connections, 5G has unlocked new possibilities for real-time applications and the Internet of Things (IoT).
ii. ***Massive MIMO and Beamforming*:** Massive Multiple-Input Multiple-Output (MIMO) technology and advanced beamforming techniques have enabled networks to provide higher data rates and improved coverage, enhancing user experiences even in densely populated areas.
iii. ***Edge Computing*:** The integration of edge computing with wireless networks has brought computation nearer to data sources, decreasing latency and empowering time-sensitive applications like autonomous vehicles and industrial automation.
iv. ***IoT Connectivity*:** Next-gen wireless technologies have facilitated seamless IoT connectivity, allowing for a myriad of devices to communicate and interact with each other. This has led to advancements in smart homes, industrial IoT, and connected healthcare.
v. ***Enhanced Mobile Broadband*:** Faster and more reliable wireless connections have propelled enhanced mobile broadband experiences, enabling high-quality streaming, online gaming, and other bandwidth-intensive activities (Sahu et al., 2021).

vi. ***Wireless Power Transfer*:** Achievements in wireless power transfer have led to the development of technologies that wirelessly charge devices, from smartphones to electric vehicles, without the need for physical connectors.

vii. ***Smart Cities*:** Wireless technologies have been pivotal in the evolution of smart cities, enabling intelligent transportation systems, smart energy grids, and efficient urban management.

viii. ***Cognitive Radio and Spectrum Sharing*:** Cognitive radio techniques have facilitated dynamic spectrum sharing, optimizing spectrum utilization and addressing the challenge of spectrum scarcity.

ix. ***Wireless Healthcare*:** Wireless medical devices and telemedicine solutions are transforming healthcare delivery, allowing for remote monitoring, diagnosis, and treatment.

x. ***AI-Driven Wireless Networks*:** The incorporation of AI and machine learning with wireless networks has led to autonomous network management, predictive maintenance, and optimized resource allocation.

These achievements collectively mark a significant shift in the wireless landscape, fostering innovation, economic growth, and improved quality of life. As wireless technologies continue to advance, their transformative impact will likely extend to realms yet unexplored, paving the way for a future where connectivity is seamless, intelligent, and deeply integrated into our daily lives.

11.4 Potential Societal and Industrial Impact

The convergence of radar and wireless communication technologies holds immense potential for generating significant societal and industrial impact:

i. ***Safer Transportation*:** Integration can enhance collision avoidance systems, enabling safer roadways through improved radar-based object detection and real-time V2X communication for timely alerts to drivers.

ii. ***Smart Cities*:** Radar-wireless integration can contribute to efficient traffic management, waste collection, and energy distribution, transforming urban environments into smart and sustainable hubs.

iii. ***Industrial Automation*:** The combination can optimize manufacturing processes by using radar for precise object tracking and wireless communication for real-time data sharing, leading to increased productivity and reduced downtime.

iv. ***Agricultural Innovation*:** By providing accurate soil moisture data through radar and enabling remote control of machinery using wireless communication, agriculture can become more efficient and sustainable.

v. ***Healthcare Advancements*:** Integration can support wearable health devices that monitor patients' vital signs through radar and transmit data wirelessly for remote medical diagnosis and treatment (Lyu et al., 2021) .
vi. ***Emergency Response*:** Radar communication can assist in disaster management by enabling through-wall sensing for locating survivors, while wireless communication maintains connectivity for coordinated response efforts.
vii. ***Environmental Monitoring*:** Both technologies combined can facilitate real-time monitoring of pollution levels, weather patterns, and natural disasters, enhancing our understanding of and ability to respond to environmental changes.
viii. ***Supply Chain Efficiency*:** Radar can improve logistics by enabling precise asset tracking, while wireless communication supports real-time inventory management and coordination across the supply chain.
ix. ***Remote Sensing*:** The synergy can lead to more accurate weather forecasting, remote sensing for scientific research, and space exploration applications.
x. ***New Business Models*:** Radar-wireless integration can foster the creation of innovative services and business models, propelling economic growth and opening doors to new market opportunities.

The societal and industrial impact of integrating radar and wireless communication is profound. This synergy has the power to transmute various sectors, enhancing safety, efficiency, and class of life on a global scale while ushering in new possibilities for innovation and technological advancement.

CONCLUSION

In conclusion, the book chapter on "Engineering Next-Generation Wireless Experiences through Radar and RF Front End System Designs" sheds light on the synergistic potential of integrating radar and wireless communication technologies. This integration marks a pivotal step towards creating interconnected ecosystems that transcend traditional boundaries and enhance our ability to sense, communicate, and interact in dynamic environments.

Throughout the chapter, we have delved into the evolution of wireless technologies, the demand for enhanced wireless experiences, radar principles, RF front-end systems, and various cross-layer considerations. We've explored the challenges and benefits of coexistence between radar and communication, and examined advanced techniques like MIMO, beamforming, and dynamic spectrum access. Moreover, the chapter delved into reconfigurable RF front-end architectures and the regulatory aspects governing the integration of radar and wireless systems.

The societal and industrial impact of this integration is vast, with potential implications for transportation, healthcare, smart cities, agriculture, and beyond. This chapter underscores the crucial role that radar and RF front-end designs play in shaping the future of wireless technologies, opening avenues for innovation, economic growth, and improved quality of life.

As the world accelerates towards the realization of 5G, 6G, and beyond, the intersection of radar and wireless communication is poised to redefine how we perceive, connect, and interact with our environment. The book chapter underscores the importance of interdisciplinary collaboration and continuous research to harness the full potential of this integration, fostering a world where technology not only amplifies human capabilities but also ushers in a new era of connectivity and intelligence. Ultimately, by engineering next-generation wireless experiences through radar and RF front-end system designs, we pave the way for a future where communication and sensing are seamlessly intertwined, shaping a world that is safer, smarter, and more connected than ever before.

REFERENCES

Ahmed, A., Zhang, Y. D., & Hassanien, A. (2021). Joint Radar-Communications Exploiting Optimized OFDM Waveforms. *Remote Sensing (Basel)*, *13*(21), 4376. doi:10.3390/rs13214376

Aijaz, A. (2020). Private 5G: The Future of Industrial Wireless. *IEEE Industrial Electronics Magazine*, *14*(4), 136–145. doi:10.1109/MIE.2020.3004975

Arbab-Zavar, B., Palacios-Garcia, E., Vasquez, J., & Guerrero, J. (2019). Smart Inverters for Microgrid Applications: A Review. *Energies*, *12*(5), 840. doi:10.3390/en12050840

Ashraf, M., Tan, B., Moltchanov, D., Thompson, J. S., & Valkama, M. (2023). Joint Optimization of Radar and Communications Performance in 6G Cellular Systems. *IEEE Transactions on Green Communications and Networking*, *7*(1), 522–536. doi:10.1109/TGCN.2023.3234258

Björnson, E., Sanguinetti, L., Wymeersch, H., Hoydis, J., & Marzetta, T. L. (2019). Massive MIMO is a reality—What is next? *Digital Signal Processing*, *94*, 3–20. doi:10.1016/j.dsp.2019.06.007

Brown, M., & Li, C. (2023). Advancement of PMCW Radar and Its Board-Level Prototyping. *2023 IEEE 16th Dallas Circuits and Systems Conference (DCAS)*, 1–4. 10.1109/DCAS57389.2023.10130242

Bruder, J. A. (2013). IEEE Radar standards and the radar systems panel. *IEEE Aerospace and Electronic Systems Magazine*, *28*(7), 19–22. doi:10.1109/MAES.2013.6559377

Cohen, D., Mishra, K. V., & Eldar, Y. C. (2018). Spectrum Sharing Radar: Coexistence via Xampling. *IEEE Transactions on Aerospace and Electronic Systems*, *54*(3), 1279–1296. doi:10.1109/TAES.2017.2780599

Dokhanchi, S. H., Mysore, B. S., Mishra, K. V., & Ottersten, B. (2019). A mmWave Automotive Joint Radar-Communications System. *IEEE Transactions on Aerospace and Electronic Systems*, *55*(3), 1241–1260. doi:10.1109/TAES.2019.2899797

Dokhanchi, S. H., Shankar, M. R. B., Stifter, T., & Ottersten, B. (2018). OFDM-based automotive joint radar-communication system. *2018 IEEE Radar Conference (RadarConf18)*, 902–907. 10.1109/RADAR.2018.8378680

Drinan, D. P., Zale, A. V., Webb, M. A. H., Taper, M. L., Shepard, B. B., & Kalinowski, S. T. (2012). Evidence of Local Adaptation in Westslope Cutthroat Trout. *Transactions of the American Fisheries Society*, *141*(4), 872–880. doi:10.1080/00028487.2012.675907

Elijah, O., Abdul Rahim, S. K., New, W. K., Leow, C. Y., Cumanan, K., & Kim Geok, T. (2022). Intelligent Massive MIMO Systems for Beyond 5G Networks: An Overview and Future Trends. *IEEE Access : Practical Innovations, Open Solutions*, *10*, 102532–102563. doi:10.1109/ACCESS.2022.3208284

Feng, Z., Fang, Z., Wei, Z., Chen, X., Quan, Z., & Ji, D. (2020). Joint radar and communication: A survey. *China Communications*, *17*(1), 1–27. doi:10.23919/JCC.2020.01.001

Ghorbanian, M., Dolatabadi, S. H., Masjedi, M., & Siano, P. (2019). Communication in Smart Grids: A Comprehensive Review on the Existing and Future Communication and Information Infrastructures. *IEEE Systems Journal*, *13*(4), 4001–4014. doi:10.1109/JSYST.2019.2928090

Griffiths, H., Cohen, L., Watts, S., Mokole, E., Baker, C., Wicks, M., & Blunt, S. (2015). Radar Spectrum Engineering and Management: Technical and Regulatory Issues. *Proceedings of the IEEE*, *103*(1), 85–102. doi:10.1109/JPROC.2014.2365517

Gunasekaran, V., & Harmantzis, F. C. (2007). Emerging wireless technologies for developing countries. *Technology in Society*, *29*(1), 23–42. doi:10.1016/j.techsoc.2006.10.001

Hassanien, A., Amin, M. G., Zhang, Y. D., & Ahmad, F. (2016). Dual-Function Radar-Communications: Information Embedding Using Sidelobe Control and Waveform Diversity. *IEEE Transactions on Signal Processing*, *64*(8), 2168–2181. doi:10.1109/TSP.2015.2505667

Huseien, G. F., & Shah, K. W. (2022). A review on 5G technology for smart energy management and smart buildings in Singapore. *Energy and AI*, *7*, 100116. doi:10.1016/j.egyai.2021.100116

Kaushik, A., Masouros, C., & Liu, F. (2021). Hardware Efficient Joint Radar-Communications with Hybrid Precoding and RF Chain Optimization. *ICC 2021 - IEEE International Conference on Communications*, 1–6. 10.1109/ICC42927.2021.9500661

Kumari, P., Choi, J., Gonzalez-Prelcic, N., & Heath, R. W. (2018). IEEE 802.11ad-Based Radar: An Approach to Joint Vehicular Communication-Radar System. *IEEE Transactions on Vehicular Technology*, *67*(4), 3012–3027. doi:10.1109/TVT.2017.2774762

Li, H., Liu, Y., Liao, G., & Chen, Y. (2023). Joint Radar and Communications Waveform Design Based on Complementary Sequence Sets. *Remote Sensing (Basel)*, *15*(3), 645. doi:10.3390/rs15030645

Liang, C., Li, Y., Hu, X., Wang, Y., Zhang, L., Wang, M., & Guo, J. (2023). Coherent-on-Receive Synthesis Using Dominant Scatterer in Millimeter-Wave Distributed Coherent Aperture Radar. *Remote Sensing (Basel)*, *15*(6), 1505. doi:10.3390/rs15061505

Liu, F., Cui, Y., Masouros, C., Xu, J., Han, T. X., Eldar, Y. C., & Buzzi, S. (2022). Integrated Sensing and Communications: Toward Dual-Functional Wireless Networks for 6G and Beyond. *IEEE Journal on Selected Areas in Communications*, *40*(6), 1728–1767. doi:10.1109/JSAC.2022.3156632

Liu, F., Zheng, L., Cui, Y., Masouros, C., Petropulu, A. P., Griffiths, H., & Eldar, Y. C. (2023). Seventy Years of Radar and Communications: The road from separation to integration. *IEEE Signal Processing Magazine*, *40*(5), 106–121. doi:10.1109/MSP.2023.3272881

Liu, R., Li, M., Liu, Q., & Swindlehurst, A. L. (2021). Dual-Functional Radar-Communication Waveform Design: A Symbol-Level Precoding Approach. *IEEE Journal of Selected Topics in Signal Processing*, *15*(6), 1316–1331. doi:10.1109/JSTSP.2021.3111438

Lyu, Y., Zhang, Y., Liu, Y., Chen, W., Zhang, X., Xu, W., Wu, C., Wang, L., Zeng, H., Sheng, X., Yang, R., Wang, Z., Kuang, K., & Fei, W. (2021). Analysis of Potential Disruptive Technologies in the Electronics and Information Field Towards the Intelligent Society. *Engineering (Beijing), 7*(8), 1051–1056. doi:10.1016/j.eng.2021.05.004

Ma, D., Shlezinger, N., Huang, T., Liu, Y., & Eldar, Y. (2021). FRaC: FMCW-Based Joint Radar-Communications System Via Index Modulation. *IEEE Journal of Selected Topics in Signal Processing, 15*(6), 1348–1364. doi:10.1109/JSTSP.2021.3118219

Mishra, K. V., Bhavani Shankar, M. R., Koivunen, V., Ottersten, B., & Vorobyov, S. A. (2019). Toward Millimeter-Wave Joint Radar Communications: A Signal Processing Perspective. *IEEE Signal Processing Magazine, 36*(5), 100–114. doi:10.1109/MSP.2019.2913173

Qian, J., Liu, Z., Lu, Y., Zheng, L., Zhang, A., & Han, F. (2022). Radar and Communication Spectral Coexistence on Moving Platform with Interference Suppression. *Remote Sensing (Basel), 14*(19), 5018. doi:10.3390/rs14195018

Rauch, A., Klanner, F., & Dietmayer, K. (2011). Analysis of V2X communication parameters for the development of a fusion architecture for cooperative perception systems. *2011 IEEE Intelligent Vehicles Symposium (IV)*, 685–690. 10.1109/IVS.2011.5940479

Sahu, C. K., Young, C., & Rai, R. (2021). Artificial intelligence (AI) in augmented reality (AR)-assisted manufacturing applications: A review. *International Journal of Production Research, 59*(16), 4903–4959. doi:10.1080/00207543.2020.1859636

Shafie, A., Yang, N., Han, C., Jornet, J. M., Juntti, M., & Kurner, T. (2022). Terahertz Communications for 6G and Beyond Wireless Networks: Challenges, Key Advancements, and Opportunities. *IEEE Network*, 1–8. doi:10.1109/MNET.118.2200057

Tan, B., Ma, Z., Zhu, X., Li, S., Zheng, L., Huang, L., & Bai, J. (2023). Tracking of Multiple Static and Dynamic Targets for 4D Automotive Millimeter-Wave Radar Point Cloud in Urban Environments. *Remote Sensing (Basel), 15*(11), 2923. doi:10.3390/rs15112923

Tightiz, L., Yang, H., & Piran, M. J. (2020). A Survey on Enhanced Smart Micro-Grid Management System with Modern Wireless Technology Contribution. *Energies, 13*(9), 2258. doi:10.3390/en13092258

Wang, C., Wang, R.-Z., Zhang, S.-J., Wang, H., & Wang, W.-S. (2023). Reducing radar cross section of flat metallic targets using checkerboard metasurface: Design, analysis, and realization. *Journal of Applied Physics*, *134*(4), 044902. Advance online publication. doi:10.1063/5.0154916

Wei, Z., Wang, Y., Ma, L., Yang, S., Feng, Z., Pan, C., Zhang, Q., Wang, Y., Wu, H., & Zhang, P. (2023). 5G PRS-Based Sensing: A Sensing Reference Signal Approach for Joint Sensing and Communication System. *IEEE Transactions on Vehicular Technology*, *72*(3), 3250–3263. doi:10.1109/TVT.2022.3215159

Wu, K., Zhang, J. A., Huang, X., & Guo, Y. J. (2023). OTFS-Based Joint Communication and Sensing for Future Industrial IoT. *IEEE Internet of Things Journal*, *10*(3), 1973–1989. doi:10.1109/JIOT.2021.3139683

Wu, Y., Lemic, F., Han, C., & Chen, Z. (2023). Sensing Integrated DFT-Spread OFDM Waveform and Deep Learning-Powered Receiver Design for Terahertz Integrated Sensing and Communication Systems. *IEEE Transactions on Communications*, *71*(1), 595–610. doi:10.1109/TCOMM.2022.3225920

You, X., Wang, C.-X., Huang, J., Gao, X., Zhang, Z., Wang, M., Huang, Y., Zhang, C., Jiang, Y., Wang, J., Zhu, M., Sheng, B., Wang, D., Pan, Z., Zhu, P., Yang, Y., Liu, Z., Zhang, P., Tao, X., ... Liang, Y.-C. (2021). Towards 6G wireless communication networks: Vision, enabling technologies, and new paradigm shifts. *Science China. Information Sciences*, *64*(1), 110301. doi:10.1007/s11432-020-2955-6

Zarakovitis, C. C., Ni, Q., Skordoulis, D. E., & Hadjinicolaou, M. G. (2012). Power-Efficient Cross-Layer Design for OFDMA Systems With Heterogeneous QoS, Imperfect CSI, and Outage Considerations. *IEEE Transactions on Vehicular Technology*, *61*(2), 781–798. doi:10.1109/TVT.2011.2179817

Zhang, Z., Du, Z., & Yu, W. (2020). Mutual-Information-Based OFDM Waveform Design for Integrated Radar-Communication System in Gaussian Mixture Clutter. *IEEE Sensors Letters*, *4*(1), 1–4. doi:10.1109/LSENS.2019.2946735

Zhao, J., Li, J., & Yi, Y. (2021). Research on Radar Communication Integrated Signal Based on 64QAM-LFM. *2021 International Wireless Communications and Mobile Computing (IWCMC)*, 1151–1155. doi:10.1109/IWCMC51323.2021.9498961

Zheng, J., Chu, P., Wang, X., & Yang, Z. (2023). Inner-Frame Time Division Multiplexing Waveform Design of Integrated Sensing and Communication in 5G NR System. *Sensors (Basel)*, *23*(15), 6855. doi:10.3390/s23156855 PMID:37571638

Chapter 2
Case Studies on Detection Using mmWave FMCW RADAR System

Gummadi Surya Prakash
SRM University, India

Rupesh Kumar
SRM University, India

W. Chandra
SRM University, India

Shilpa Mehta
https://orcid.org/0000-0002-4691-3185
Auckland University of Technology, New Zealand

ABSTRACT

The content of this chapter provides a thoughtful analysis of case studies that highlight the detection capabilities of FMCW radar systems operating in mmWave configurations. The case studies demonstrate how mmWave FMCW radar technology may be used to detect objects, motions, and changes in both line of sight and non-line of sight settings with accuracy, and efficiency. Each case study explores the unique difficulties presented by the application environment, which can include anything from identifying impediments in automotive safety systems to detecting minute movements for vital sign monitoring in healthcare. The steps for detecting different gesture recognition using IWR 1843 BOOST FMCW radar system and its processing are focussed upon. The document highlights the technology's excellent resolution, motion sensitivity, and adaptability to a variety of challenging environments in LoS and NLoS scenarios and with the technical details of operating mmWave radars, including signal processing methods, machine learning algorithms, and mitigating interference from surrounding objects.

DOI: 10.4018/979-8-3693-0916-2.ch002

1. INTRODUCTION

Millimeter wave (mmWave) radar is a type of radar technology that employs electromagnetic waves. Radar systems send out electromagnetic wave, which are subsequently reflected off nearby objects. A radar system can facilitate the measurement of range, velocity, and angle by processing the received signal (Jankiraman, 2018).

Radars operating at millimeter wavelengths [30 gigahertz (GHz) to 300 GHz] are known as mmWave radars (Jankiraman, 2018). The benefits of using these short wavelengths in the electromagnetic spectrum include high resolution, penetration through materials, high data rates etc. High precision of short wavelengths is also another benefit that can be achieved through several key factors like short wavelengths, high bandwidth, advanced signal processing, MIMO [*Multi-Input-Multi-Output*] antenna configurations, adaptive beamforming, high-speed data processing making it a valuable technology for a wide range of applications. For example, a wavelength of around 4 mm mmWave system operating at 76–81 GHz will be able to detect movements as small as a fraction of a millimeter as detailed in (Instruments, T., 2020).

A mmWave signal processing requires a compact system architecture design. mmWave radar system consists of radio frequency (RF) components for transmit and receive (TX and RX), analog components for clocking, and digital components like digital signal processors (DSPs), microcontrollers (MCUs), analog-to-digital converters (ADCs) and antenna design. The design of each component is complex and distinct in design. The use of discrete components in the implementation of these systems in the past raised power consumption and system costs (Jankiraman, 2018). Additionally, system design is challenging due to its complexity and high frequency. Texas Instruments (TI)- *XWR 1843 BOOST (Amar et al., 2021, pp.1-11), XWR 1642 BOOST (Gao et al., 2019, pp.1-6),* Infineon Technologies- *BGT23MTR12 (Ahmed et al., 2021)*, Novelda- *NVA6100, X4*(Borzov et al., 2017) and few other companies have addressed these issues by developing mmWave radar devices based on complementary metal-oxide semiconductor (CMOS) technology. TI Instruments (TI) devices incorporate TX-RF (transmit chain) and RX-RF (receiver chain) analog components such as clocks, as well as digital components such as the ADC, MCU, and hardware accelerator. To enhance their signal-processing capabilities, certain mmWave sensor models from TI incorporate a DSP (Instruments, T., 2020).

While mmWave technology has a wide range of applications, including telecommunications and imaging, one such application that utilizes its characteristics is radar that is frequency-modulated continuous-wave (FMCW), which uses mmWave frequencies for precise and efficient sensing as discussed by (Tang et al., 2023).

TI devices have demonstrated FMCW (Frequency-Modulated Continuous Wave), a type of millimeter-wave technology in the form of evaluation kits such as IWR 1843 BOOST, IWR 1642 BOOST, AWR 1443 (Phaisiri, K. et al., 2023) etc., FMCW radar uses a frequency-modulated signal that is continually transmitted to measure range in addition to angle and velocity. On the other hand, traditional pulsed-radar systems send out brief pulses on a regular basis, and the distance to the target is determined by the reception echo. Based on differences in time, the transmitted signal and the echo signal are separated (Jankiraman, 2018).

Rapid advancements in FMCW radar technology have enabled a broader range of applications, including smart home devices (Rodríguez Lasso, N. A, 2023), industrial monitoring (Ahmad, W.A., et al., 2020) and even healthcare (Ahmed, S., et al., 2023)· in addition to its traditional use in navigation and automotive radar systems. This chapter explores the principles of FMCW radar operation and defining these concepts using detection case studies. We explore different experimental configurations to test detection under different scenarios, utilizing two popular radar evaluation modules from Texas Instruments: the IWR1642 Boost and IWR1843 Boost both working at frequency ranging from 76-81GHz coverage with continuous bandwidth of 4GHz and the operating frequency may slightly vary depending on the configuration and application requirements detailed in (Instruments, T., 2020).

2. FMCW RADAR BASICS

The fundamental concept in radar systems is the transmission of an electromagnetic signal that objects reflect in its path. In the signal used in FMCW radars, the frequency increases linearly with time. This type of signal is also called a chirp (Instruments, T., 2020) as shown in Figure 1.

Figure 1. A chirp signal with amplitude and time as a function (Instruments, T., 2020)

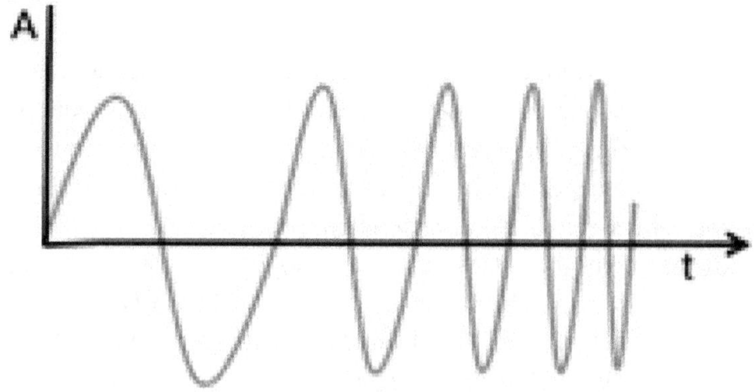

FMCW Radar system transmits a chirp signal and captures the signals reflected by the objects present in the path. An FMCW sensor uses a combination of RF, analog, and digital electrical components to estimate the range, velocity, and angle of nearby objects as many researchers have reported that in (Instruments, T., 2020) (Hao, Z., et al., 2023) (Soumya, et al., 2023).

Figure 2. Components of FMCW radar

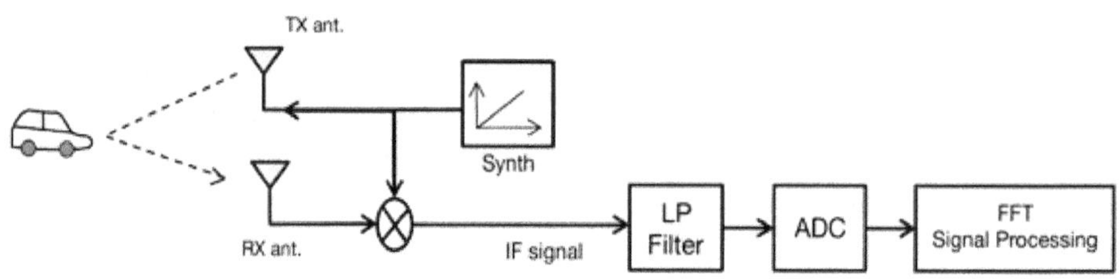

An FMCW Radar system operation and parameters that effect the operation at every level are as follows:

- A synthesizer (synth) generates a chirp: Generating a chirp signal presents challenges particularly in achieving precise frequency modulation, frequency range, bandwidth, and managing power consumption.
- The chirp is transmitted by the TX antenna, that is reflected off an object and the reflected chirp is received at the RX antenna: signal amplification, SNR, and dynamic range pose challenges in the exchange of chirp between transmitter and receiver.

- The RX signal and TX signal are 'mixed', and the resulting signal is called an 'IF signal': The challenges associated with the mixing of signals are signal leakage, interference, frequency offset, etc.,
- Complex Signal Processing: The Signal processing involves RF, Analog, and digital processing.

Operating the FMCW while addressing these challenges requires careful system design, shielding, filtering, and isolation techniques.

FMCW radar distinguishes itself by transmitting a continuous wave, frequency-modulated over time, allowing for the simultaneous measurement of the range and velocity of targets. The key advantage of FMCW radar is its high resolution and accuracy in measuring short distances, making it ideal for sensing applications (Wolff, n.d.).

3. SENSING

77GHz radar systems are excellent at picking up on minute details and accurately differentiating objects. These radar sensors are appropriate for applications where accuracy and proximity are essential due to their high frequency (77 GHz), which allows them to give precise spatial resolution (Tang et al., 2023).

Working at 77GHz, the system provides imaging capabilities with high resolution that are appropriate for recording minute details in human movements (Hao, Z., et al., 2023). Radar systems operating at 77GHz for close objects evaluate data by examining reflected signals to derive useful details about the targets they have spotted. Sophisticated algorithms like LSTM, CNN etc., are used to interpret radar data (Kim et al., 2017) (Leem et al., 2019) (Zhang, Z., et al., 2018) remove noise and create precise reconstructions of nearby objects. Researchers can acquire precise measurements of the near objects' distance, velocity, and spatial information by efficiently processing the radar data as mentioned in the earlier works of (Toker, O., et al., 2020) (Molchanov, P., et al., 2015).

After the data is gathered, it is processed using a variety of imaging algorithms to produce images, which are then examined for gesture recognition covered by several researchers in (Tang et al., 2023) (Lv, Q., et al., 2014) (Zhang, J., et al., 2023).

Now with the basic understanding of FMCW and sensing using 77 GHz radar sensor let's move into the practical case study that demonstrates the line-of- sight scenarios.

Human-computer interaction and surveillance systems have been completely transformed by radar technology's ability to detect human movements through walls and recognize human gestures. Researchers have created novel techniques to detect and analyze human gestures and motions by utilizing radar sensors accurately and efficiently in the works of (Ahmed et al., 2021) (Peng, Z., et al., 2017). To facilitate effortless interaction between people and technology, human gesture recognition entails the interpretation of hand and body movements (Hazra & Santra, 2018) (Li et al., 2021). Researchers have been able to record minute details of gestures, such as hand locations, finger movements, and body gestures, using sophisticated radar sensors like frequency-modulated continuous wave (FMCW) radar performed by (Wang et al., 2020) (Khan, F., & Cho, et al., 2017) (Ryu et al., 2018). These systems are capable of classifying and interpreting motions for use in virtual reality, gaming, robotics, and other fields by employing advanced signal processing algorithms and radar data (Jhaung et al., 2022). The sensitivity of the sensor and signal degradation brought on by objects interacting with the radar waves are challenges in radar-based gesture detection. The system's capacity to reliably recognize gestures can be impacted by the material qualities of objects, which can have an impact on signal quality (Attygalle et al., 2021). Furthermore, in order to maximize efficiency and guarantee accurate gesture recognition, sensor calibration is essential (Khan et al., 2017). Radar technology's capacity to track human movement through walls is a major development for security and surveillance systems (Li, G., et al., 2020). Ultra-wideband (UWB) radar is one type of radar sensor that can pass through walls and other obstructions to identify minute motions linked to human activity (Li et al., 2014) (Yang et al., 2021). These systems can distinguish between moving and stationary people behind walls by evaluating radar cross-section data and Doppler changes clearly explained as in (Kim et al., 2023) (Saeed, T. R., et al, 2019) (Singh et al., 2011). This provides important information for applications in law enforcement, search, and rescue, and building security (Deng, H., et al., 2022).

4. LINE OF SIGHT (LOS) APPLICATIONS

4.1 Case Study: Human Gesture Recognition

4.1.1 Objective

To demonstrate the IWR1843 Boost's ability to recognize specific human gestures in real time as performed earlier by (Soumya, et al., 2023).

An organized method incorporating hardware design, software setup, data collecting, processing algorithms, and machine learning methodologies is needed to show off the IWR1843 Boost's capacity to identify certain human movements in real-time. This section provides a high-level overview of these components and a design for creating a gesture identification system using the Texas Instruments IWR1843 Boost mmWave radar sensor.

Methodology: The location of the radar module is such that it faces the region where the motions will be executed. Custom signal processing methods are designed to interpret radar reflections and distinguish between distinct gestures as implemented by (Oh et al., 2019).

4.1.2 Hardware and Software Requirements

IWR1843 Boost Radar Sensor: A high precision mmWave sensor capable of capturing high-resolution radar signals.

DCA1000EVM Data Capture Card (optional): For real-time data capture and streaming (Oh et al., 2019).

Figure 3. DCA 1000 EVM interfaced with IWR 1843 BOOST

Radar Studio or mmWave SDK: For configuring the radar sensor and initial data capture.

MATLAB or Python: For data processing and algorithm development, including machine learning model training and testing.

Once the hardware requirements are established, the subsequent steps are followed up from human gesture recognition, signal processing, and classification algorithms to translate raw data into gestures.

4.1.3 The Steps Involved in Performing Human Gesture Recognition Using IWR 1843 BOOST (Attygalle et al., 2021)

Performing human gesture recognition involves precise measures in setting experimental setup of the radar, distance of the hand from the sensor, data acquisition, implementation of algorithm for detecting the gesture.

A thorough investigation was conducted into the commonly used techniques for data collection, processing, and classification in gesture recognition in (Tang et al., 2023) (AWR1843 Data Sheet, Product Information and Support I TI.com, n.d.) (Kim & Toomajian, 2016).

Various radar sensors that can be used for human gesture recognition are discussed in Table2.

Let us observe the steps and parameters required for performing the Human Gesture Recognition (HGR)

Step 1: Radar Configuration:

Frequency Setup:

Set the IWR1843's frequency band to function in a way that offers the greatest combination of penetration and range resolution for gesture detection.

Resolution and Range:

Adjust the chirp parameters to improve the radar's resolution and ensure it can distinguish between various human gestures.

Data Capture Rate:

Select an acceptable frame rate to record movement fluently while avoiding excessive data that may overload processing capabilities.

Step 2: Data Acquisition

Gesture Dataset Creation: Create a gesture dataset by performing a sequence of specified gestures in front of the radar to collect raw signal data. To accommodate for heterogeneity in gesture execution, each gesture should be included numerous times in this dataset by various people (Khan et al., 2017).

The Corresponding heatmaps can be converted to the format required for implementing machine learning algorithms depending on the requirement of the platform being utilized for signal processing (MATLAB/Python). For example, If MATLAB is being used as the processing tool, then the heatmaps can be converted to *'.mat'* or *'.dat'* files as MATLAB trains the algorithm using numeric from the converted files (Islam, M. Z., et al., 2019).

Real-Time Streaming: If available, stream data straight to a PC for processing by using the DCA1000EVM for high-speed data capture.

*Heatmaps of different gestures and the difference in the gestures are highlighted by circle.

(The above experimental setup and gesture recognition have been performed in Wireless Sensing and Imaging (WSI) Lab, SRM University AP)

Figure 4. Gestures dataset with their corresponding heatmaps

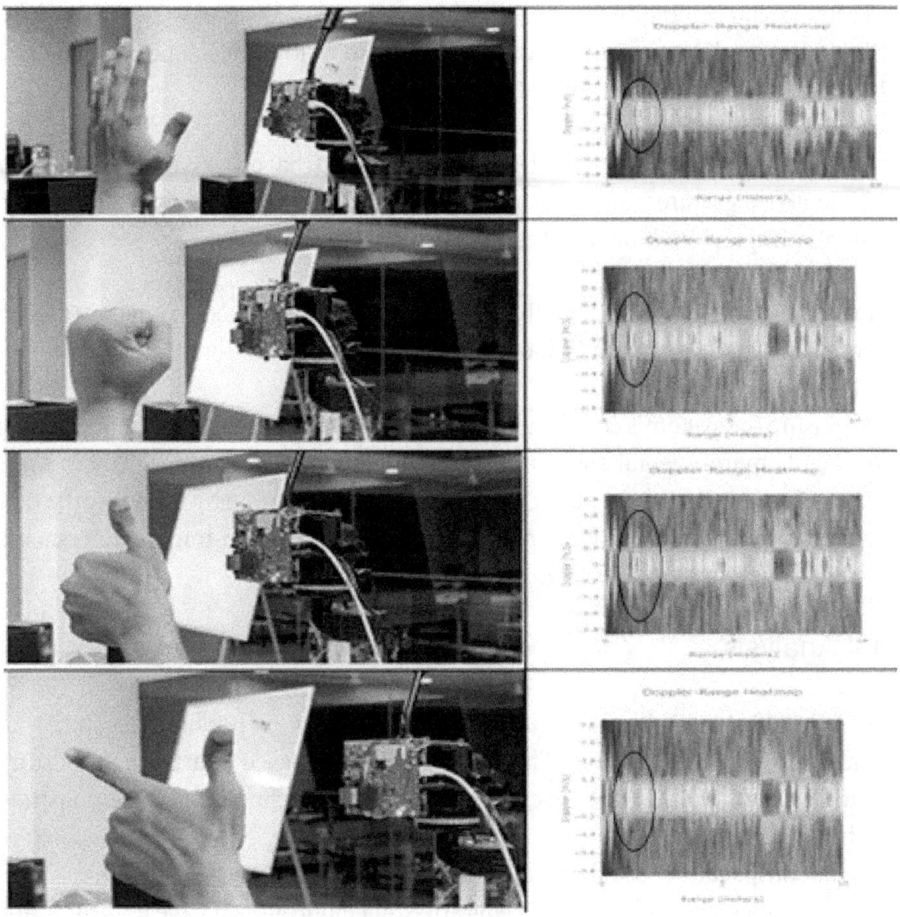

Step 3: Signal Processing

Pre-processing: Apply filters and noise reduction techniques to clean the raw radar data.

Feature Extraction: Convert the processed radar information into meaningful features that accurately reflect the gestures. This could include extracting data from time-frequency domains, range-Doppler maps, or micro-Doppler signatures (Deng, H., et al., 2022).

Step 4: Machine Learning Model

Model Selection:

Select a suitable machine learning method for the classification of gestures. Convolutional Neural Networks (CNNs) are useful for spatial data analysis (Oh et al., 2019), but Recurrent Neural Networks (RNNs) or Long Short-Term Memory (LSTM) networks are best suited for time-series data (Zhang, Z., et al., 2018).

Training and Validation:

Train the model using the processed dataset, employing techniques like cross-validation to evaluate its performance and avoid overfitting.

Real-Time Inference:

Integrate the learned model into a real-time system that processes incoming radar data, identifies gestures, and returns classification results immediately.

Step 5: Implementation and Testing

Integration:

Develop a software interface that integrates the radar sensor, signal processing pipeline, and machine learning model to ensure continuous real-time operation.

Testing:

To ensure the system's correctness and resilience, thoroughly test it under a range of user and environmental circumstances.

Optimization: Based on testing feedback, refine the system by re-tuning the radar parameters, modifying the signal processing algorithms, or re-training the machine learning model for better performance.

4.1.4 Analysis

The IWR1843 Boost's high-resolution capabilities make it possible to identify and categorize a variety of motions. The potential of radar-based interfaces in consumer electronics, smart home appliances, and accessibility technologies is demonstrated by this application.

Radar propagation is relatively simple in line-of-sight circumstances, but it becomes much more difficult in non-line-of-sight environments due to obstructions and reflections, which complicate signal transmission and reception. Researchers aim to create a strong through-wall human recognition system that can precisely track and identify people in real-time, even when they are hidden from direct view, by utilizing the unique features of the IWR 1843 Boost.

5. NON-LINE OF SIGHT (NLOS) APPLICATIONS

5.1 Case Study: Through-Wall Human Activity Monitoring (Wang et al., 2021)

Operating at millimeter-wave frequencies, the IWR 1843 Boost radar system provides high-resolution sensing capabilities that allow it to see through thin layers, like walls, to identify movement and human presence. Several research papers can be studied which are mentioned in the (Phaisiri, K. et al., 2023) (Oh et al., 2019) (Islam, M. Z., et al., 2019) can be studied for understanding the working conditions and challenges for performing this case study.

5.1.1 Objective

To assess the capability of the IWR1843 Boost to detect and monitor human activity through obstacles such as walls.

Detect Human Presence: Determine the presence of humans behind walls or obstacles.

Activity Classification: Classify basic human activities, such as walking, standing, or falling.

Tracking: Monitor the movement paths of individuals within a room.

Methodology: The radar is placed on one side of a standard building wall, and movements are performed on the opposite side. Advanced signal processing techniques are applied to filter out the wall's reflections and isolate the movement signatures (Kim & Toomajian, 2016).

5.1.2 Hardware and Software Requirements

are the same as discussed in 4.1.2 in this chapter. In addition, cardboard can also be used in place of a wall as an obstacle.

Detecting human presence behind walls or obstacles using technologies like the IWR1843 BOOST mmWave radar involves a sophisticated blend of hardware capabilities and signal processing techniques.

5.1.3 The Steps Involved in Performing Through-Wall Human Activity Monitoring Using IWR 1843 BOOST

Step 1. Radar System Setup

Hardware Configuration: Use the IWR1843 BOOST mmWave radar module, which has a fine resolution suitable for detecting minute motions through walls, such as human breathing. It operates in the 76-81 GHz band.

Radar Placement and Orientation: Place the radar strategically to cover the region of interest. The wall or obstruction that you want to detect should be in the radar's field of vision.

Step 2. Data Acquisition

Radar Operation Mode: Select a mode of operation for the radar that is appropriate for through-wall detection. For high-resolution detection, this may entail programming the radar to continuously produce chirps over a large bandwidth.

*Heatmaps of human detection through a) Wall and b) Cardboard

The presence of human movements is highlighted by circle.

(The above experimental setup and human recognition have been performed in Wireless Sensing and Imaging (WSI) Lab, SRM University AP)

Figure 5. Through-wall human detection using IWR 1843 BOOST

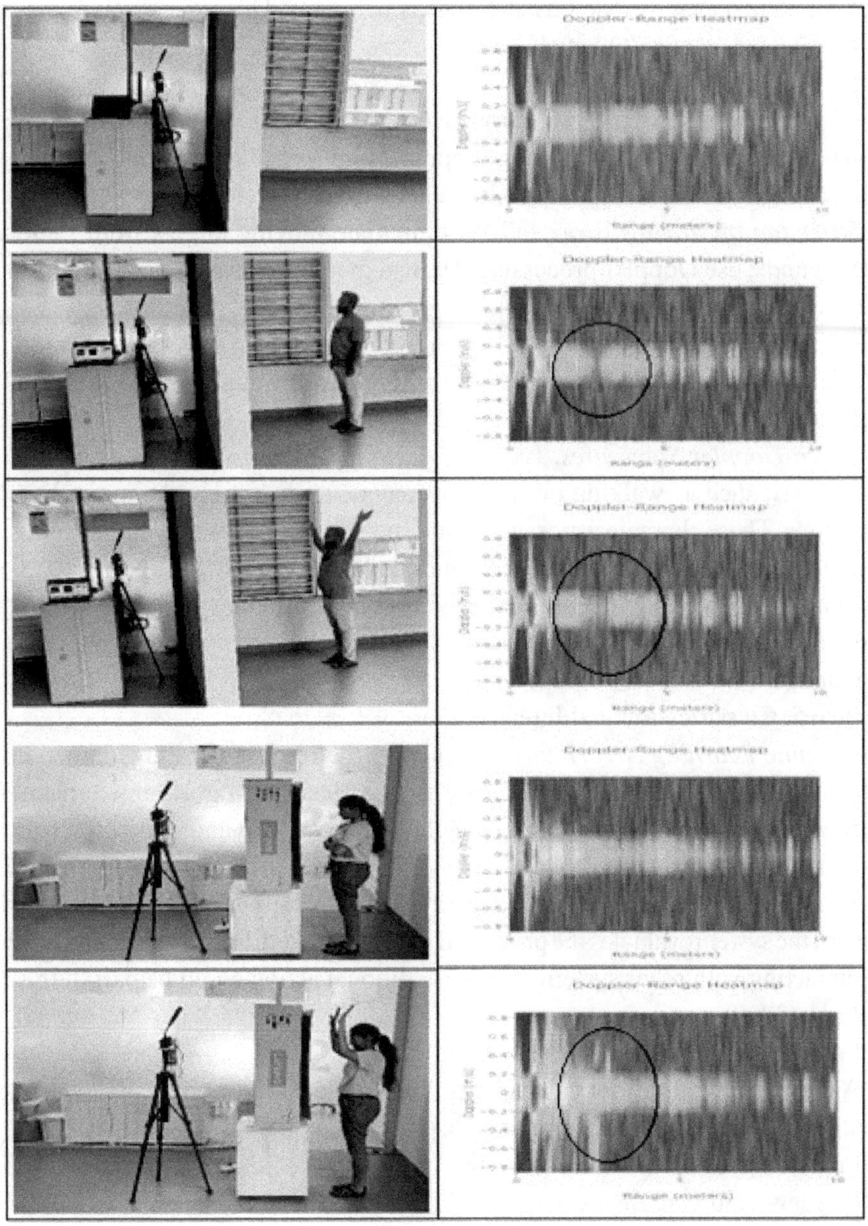

a) The first three heatmaps are for the human detection through wall with human separation of 187cm away from radar. The thickness of the wall is 25cm.

b) The last two heatmaps are for the human detection through cardboard with human separation of 125cm away from radar.

Collect Raw Signals: The electromagnetic waves that the radar emits pass through the wall and bounce off persons and objects behind it. The radar receiver picks up these reflected signals (Oh et al., 2019).

Step 3. Signal Processing

Clutter Removal: Static reflections from the wall and other stationary objects must be eliminated as the initial stage of processing. Techniques such as background removal or the use of a high-pass filter can be used.

Detection of Moving Targets: To differentiate moving targets from stationary backgrounds, use Doppler processing. Human motions cause a Doppler shift in the reflected signal, even small ones like breathing.

Range Estimation: Determine the movement's distance from the radar by calculating the interval between the sent and received signals (Kim et al., 2022).

Step 4. Feature Extraction

Micro-Doppler Signatures: Examine the Doppler shift fluctuations brought on by motions, such as walking or arm movements, to determine the micro-Doppler signatures. These characteristics, which are specific to various movement kinds, can be utilized to deduce human presence mentioned by (Scheer, J., & Holm, et al., 2010).

Vital Sign Detection: In more advanced systems, the radar system can identify the minute movements in the chest brought on by breathing to determine the presence of a human.

Step 5. Classification and Interpretation

Machine Learning Algorithms: To classify the retrieved features into signs of human presence, apply machine learning or deep learning algorithms. A dataset of labelled instances of various settings, including those with and without humans, is needed to train these algorithms.

Real-Time Decision Making: To decide in real-time whether there are humans present, the system evaluates the processed and categorized data. Sending notifications or interacting with other systems to take additional action could be part of this (Li, G., et al., 2020).

Step 6. Privacy Considerations

Anonymity: Unlike optical cameras, mmWave radar devices do not collect identifying features, therefore privacy is preserved.

Processing of Data: Make sure that processing of data conforms with local privacy laws and guidelines.

5.1.4 Analysis

The IWR1843 Boost can detect and tracking human activity on the other side, despite the difficulties caused by signal attenuation and reflection from the wall.

The potential of radar in security, surveillance, and search-and-rescue missions is demonstrated in this case study.

Table 1. Comparison between case studies of LOS and NLOS with IWR 1843 BOOST

S.No	Parameter	Gesture Recognition (Line-of-Sight)	Through-Wall Detection (Non-Line-of-Sight)
1	Operating Principle	Analyze Doppler and range profiles	Detect human movements through wall reflections
2	Frequency Range	77 GHz	77 GHz
3	Bandwidth	4 GHz	4 GHz
4	Signal Processing	CNN, K-means Clustering, LSTM	CNN,K-means clustering, LSTM
5	Environmental Interference	Low	High (Due to obstacles like Wall, Cardboard, Glass etc.,)
6	Maximum Detection Range	Upto 100m	Upto 300 cm* (based on the used wall thickness)

Table 1 gives the comparison between the case studies of Lineof Sight (LOS) and Non-Line of Sight (NLOS). IWR 1843 BOOST is used for both the case studies. The frequency of operation starts from 77GHz to 81 GHz. The detection range of HGR is from 100 to 200m as it is performed in LOS without any obstacle whereas for through wall detection the obstacle dimensions will play a major role in detection range.

As we navigate from case studies of line-of sight and non-line-of-sight scenarios, it becomes essential to understand the specifications and capabilities of different radar sensors. Below is a thorough table that lists all the characteristics that radar sensors must have to function properly in a variety of environmental settings.

Table 2. Summary of radar sensors that can be used for detection based on characteristics of operation

Reference Paper No.	RADAR used	Radar Type	Company Name	Frequency Range	Bandwidth	Algorithm used	Target Accuracy
15	NVA6201 (Novelda)	IR-UWB	Novelda, Oslo, Gjerdrums vei 8 0484 Oslo, Norway	Not Specified	Not Specified	CNN	90%
16	X4 (Novelda, Norway)	IR-UWB	Novelda, Oslo, Gjerdrums vei 8 0484 Oslo, Norway	7.29GHz	1.4GHz	CNN	---
17	BGT23MTR12	FMCW	Infineon Technologies, Neubiberg, Germany	25GHz	4GHz	---	---
19	77GHz FMCW TI XWR 1443, XWR 1642	FMCW	Texas-Instrument (TI) Dallas, Texas 75243, USA	77GHz	4GHz	---	---
40,46	Bumblebee Doppler radar	FMCW	Samraksh Co. Ltd., Dublin OH 43017, Ireland	5.8GHz	Not Specified	DCNN	85.6%
41	Ancortek SDR-kit 2500B	CW K-band radar	Ancortek Radars, Fairfax, VA 22030, United States	25GHz	---	HGMM	---
47	Xethru X2	IR-UWB	Novelda, Oslo, Gjerdrums vei 8 0484 Oslo, Norway	6.8GHz	2.3GHz	K-means algorithm	---
48	NVA6100 (Novelda)	IR-UWB	Novelda, Oslo, Gjerdrums vei 8 0484 Oslo, Norway	Not Specified	Not Specified	---	96%
49,50	BGT24MRT12 2	FMCW	Infineon Technologies, Neubiberg, Germany	24GHz	4GHz	3D-CNN, LSTM, CTC	93.3%

In the above Table:2 the summary of different radar sensors used for detection and their frequency range, bandwidth, and algorithms used for signal processing, etc., in different applications and scenarios has been produced. CNN algorithm has been reported by many researchers (Zhang, Z., et al., 2018) (Shi et al., 2022) that provides a good option to 77GHz FMCW radar sensors for better accuracy in target classification. From the earlier research on detection of human presence and gesture recognition, the radar sensors mentioned in Table2 have produced reliable results and accuracy.

The Bumble bee Doppler radar which is operated in 5.8GHz with high resolution in detection uses DCNN algorithm for interpreting the data (Kim & Toomajian, 2016) (Kim, Y., & Toomajian, B., 2017). The k-means algorithm is used in IR-UWB radar sensor Xethru X2 with its frequency range of 6.8GHz and bandwidth of 2.3GHz (Ahmed & Cho, 2020). In (Ren et al., 2016) The human hand's direction and distance changes, as well as its frontal surface area changed in the direction of the radar sensor, are the basis of the gesture identification algorithm. The program can identify around six different types of hand gestures by integrating these changes throughout gesture execution using NVA6100. A hand gesture recognition sensor using reflected impulses by NVA6201 can be studied in (Kim et al., 2017). For understanding the implementation of the CNN algorithm, the researcher can view (Kim et al., 2017) (Leem et al., 2019) and (Zhang, Z., et al., 2018). Mid-air digit gesture recognition using an IR-UWB X4 radar sensor is studied in (Leem et al., 2019). A CW-K Band radar Ancortek SDR-kit 2500B gives dynamic Hand Gesture Recognition Based on Micro-Doppler Radar Signatures using Hidden Gauss-Markov Models in (Wang et al., 2021). 25 GHz radar sensors BGT23MTR12 and BGT24MRT122 of Infineon Technologies are used for the applications of short-range detection with machine learning algorithms like CNN for hand gesture sensing has been discussed in (Zhang, Z., et al., 2018) (Molchanov, P., et al., 2015) (Dekker, B., et al., 2017). Human Gesture Recognition using 77 GHz mmWave technology, steps and measures involved for performing the application, data acquisition, and interpretation has been widely discussed in (Tang et al., 2023) (Li, G., et al., 2020) (Shastri et al., 2022).

CONCLUSION

In conclusion, this book chapter has given a general introduction of FMCW (frequency-modulated continuous wave) radar technology, emphasizing its foundations, principles of operation, essential elements, benefits, and uses. Target distance and velocity can be measured with great accuracy and resolution using FMCW radar, which is a flexible and strong tool. Range and velocity can be measured simultaneously thanks to coherent signal processing made possible by its continuous wave transmission. We have covered the value of FMCW radar throughout the chapter in relation to several industries, including automotive, industrial, healthcare, and environmental monitoring. FMCW radar has proven useful in a variety of applications, including non-contact vital sign monitoring in healthcare facilities and collision avoidance systems in automobiles. Although FMCW radar offers remarkable capabilities for near-field detection, several challenges remain, particularly in NLoS applications. These include signal attenuation, environmental noise, and the need for complex signal processing algorithms. Ongoing research and development efforts are focused on overcoming these obstacles, with promising approaches involving machine learning and adaptive signal processing.

REFERENCES

Ahmad, W. A., Wessel, J., Ng, H. J., & Kissinger, D. (2020, December). IoT-ready millimeter-wave radar sensors. In *2020 IEEE Global Conference on Artificial Intelligence and Internet of Things (GCAIoT)* (pp. 1-5). IEEE.

Ahmed, S., & Cho, S. H. (2020). Hand gesture recognition using an IR-UWB radar with an inception module-based classifier. *Sensors (Basel)*, *20*(2), 564. doi:10.3390/s20020564 PMID:31968587

Ahmed, S., & Cho, S. H. (2023). Machine learning for healthcare radars: Recent progresses in human vital sign measurement and activity recognition. *IEEE Communications Surveys and Tutorials*.

Ahmed, S., Kallu, K. D., Ahmed, S., & Cho, S. H. (2021). Hand gestures recognition using radar sensors for human-computer-interaction: A review. *Remote Sensing (Basel)*, *13*(3), 527. doi:10.3390/rs13030527

Amar, R., Alaee-Kerahroodi, M., & Shankar, M. B. (2021, June). FMCW-FMCW interference analysis in mm-wave radars; an indoor case study and validation by measurements. In *2021 21st International Radar Symposium (IRS)* (pp. 1-11). IEEE.

Attygalle, N. T., Leiva, L. A., Kljun, M., Sandor, C., Plopski, A., Kato, H., & Čopič Pucihar, K. (2021). No interface, no problem: Gesture recognition on physical objects using radar sensing. *Sensors (Basel)*, *21*(17), 5771. doi:10.3390/s21175771 PMID:34502662

AWR1843 data sheet, product information and support | TI.com. (n.d.). https://www.ti.com/product/AWR1843

Borzov, A. B., Likhoedenko, K. P., Seregin, G. M., & Suchkov, V. B. (2017, August). Applications of short-range radars on the base of ultrashort pulse single chip. In *2017 3rd IEEE International Conference on Control Science and Systems Engineering (ICCSSE)* (pp. 461-464). IEEE. 10.1109/CCSSE.2017.8088041

Dekker, B., Jacobs, S., Kossen, A. S., Kruithof, M. C., Huizing, A. G., & Geurts, M. (2017, October). Gesture recognition with a low power FMCW radar and a deep convolutional neural network. In *2017 European Radar Conference (EURAD)* (pp. 163-166). IEEE. 10.23919/EURAD.2017.8249172

Deng, H., Guo, S., Jin, L., Cao, X., & Xu, X. (2022, April). A human gesture recognition method for through-the-wall MIMO radar. In *2022 7th International Conference on Computer and Communication Systems (ICCCS)* (pp. 239-244). IEEE. 10.1109/ICCCS55155.2022.9846748

Gao, X., Xing, G., Roy, S., & Liu, H. (2019, November). Experiments with mmWave automotive radar test-bed. In *2019 53rd Asilomar conference on signals, systems, and computers* (pp. 1-6). IEEE. 10.1109/IEEECONF44664.2019.9048939

Hao, Z., Wang, R., Peng, J., & Dang, X. (2023). Static Hand Gesture Recognition Based on Millimeter-Wave Near-Field FMCW-SAR Imaging. *Electronics (Basel)*, *12*(19), 4013. doi:10.3390/electronics12194013

Hazra, S., & Santra, A. (2018). Robust gesture recognition using millimetric-wave radar system. *IEEE Sensors Letters*, *2*(4), 1–4. doi:10.1109/LSENS.2018.2882642

Instruments, T. (2020). *The fundamentals of millimeter wave radar sensors*. Academic Press.

Islam, M. Z., & Hossain, M. S., ul Islam, R., & Andersson, K. (2019, May). Static hand gesture recognition using convolutional neural network with data augmentation. In *2019 Joint 8th International Conference on Informatics, Electronics & Vision (ICIEV) and 2019 3rd International Conference on Imaging, Vision & Pattern Recognition (icIVPR)* (pp. 324-329). IEEE.

Jankiraman, M. (2018). *FMCW radar design*. Artech House.

Jhaung, Y. C., Lin, Y. M., Zha, C., Leu, J. S., & Köppen, M. (2022). Implementing a hand gesture recognition system based on range-doppler map. *Sensors (Basel)*, *22*(11), 4260. doi:10.3390/s22114260 PMID:35684880

Khan, F., & Cho, S. H. (2017). Hand based Gesture Recognition inside a car through IR-UWB Radar. 대한전자공학회 학술대회, 154-157.

Khan, F., Leem, S. K., & Cho, S. H. (2017). Hand-based gesture recognition for vehicular applications using IR-UWB radar. *Sensors (Basel)*, *17*(4), 833. doi:10.3390/s17040833 PMID:28398267

. Kim, B. S., Jin, Y., Lee, J., & Kim, S. (2022). FMCW radar estimation algorithm with high resolution and low complexity based on reduced search area. *Sensors*, *22*(3), 1202.

Kim, G. W., Lee, S. W., Son, H. Y., & Choi, K. W. (2023). A study on 3D human pose estimation using through-wall IR-UWB radar and transformer. *IEEE Access : Practical Innovations, Open Solutions*, *11*, 15082–15095. doi:10.1109/ACCESS.2023.3244017

Kim, S. Y., Han, H. G., Kim, J. W., Lee, S., & Kim, T. W. (2017). A hand gesture recognition sensor using reflected impulses. *IEEE Sensors Journal*, *17*(10), 2975–2976. doi:10.1109/JSEN.2017.2679220

Kim, Y., & Toomajian, B. (2016). Hand gesture recognition using micro-Doppler signatures with convolutional neural network. *IEEE Access : Practical Innovations, Open Solutions*, *4*, 7125–7130. doi:10.1109/ACCESS.2016.2617282

Kim, Y., & Toomajian, B. (2017, March). Application of Doppler radar for the recognition of hand gestures using optimized deep convolutional neural networks. In *2017 11th European conference on antennas and propagation (EUCAP)* (pp. 1258-1260). IEEE. 10.23919/EuCAP.2017.7928465

Leem, S. K., Khan, F., & Cho, S. H. (2019). Detecting mid-air gestures for digit writing with radio sensors and a CNN. *IEEE Transactions on Instrumentation and Measurement*, *69*(4), 1066–1081. doi:10.1109/TIM.2019.2909249

Li, G., Zhang, Z., Yang, H., Pan, J., Chen, D., & Zhang, J. (2020, March). Capturing human pose using mmWave radar. In *2020 IEEE International Conference on Pervasive Computing and Communications Workshops (PerCom Workshops)* (pp. 1-6). IEEE. 10.1109/PerComWorkshops48775.2020.9156151

Li, X., Liang, Q., & Lau, F. C. (2014). Sense-through-wall human detection using the UWB radar with sparse SVD. *Physical Communication*, *13*, 260–266. doi:10.1016/j.phycom.2013.12.002

Li, Z., Jin, T., Dai, Y., & Song, Y. (2021). Through-wall multi-subject localization and vital signs monitoring using UWB MIMO imaging radar. *Remote Sensing (Basel)*, *13*(15), 2905. doi:10.3390/rs13152905

Lv, Q., Ye, D., Qiao, S., Salamin, Y., Huangfu, J., Li, C., & Ran, L. (2014). High dynamic-range motion imaging based on linearized Doppler radar sensor. *IEEE Transactions on Microwave Theory and Techniques*, *62*(9), 1837–1846. doi:10.1109/TMTT.2014.2342663

Molchanov, P., Gupta, S., Kim, K., & Pulli, K. (2015, May). Short-range FMCW monopulse radar for hand-gesture sensing. In *2015 IEEE Radar Conference (RadarCon)* (pp. 1491-1496). IEEE. 10.1109/RADAR.2015.7131232

Oh, B. S., Guo, X., & Lin, Z. (2019). A UAV classification system based on FMCW radar micro-Doppler signature analysis. *Expert Systems with Applications*, *132*, 239–255. doi:10.1016/j.eswa.2019.05.007

Peng, Z., Li, C., Muñoz-Ferreras, J. M., & Gómez-García, R. (2017, May). An FMCW radar sensor for human gesture recognition in the presence of multiple targets. In *2017 First IEEE MTT-S International Microwave Bio Conference (IMBIOC)* (pp. 1-3). IEEE. 10.1109/IMBIOC.2017.7965798

. Phaisiri, K., Boonpoonja, A., Kittiyanpunya, C., Kaemarungsi, K., & Chudpooti, N. (2023, August). Through-Wall Human Sensing Using mmWave FMCW Radar. In *2023 Research, Invention, and Innovation Congress: Innovative Electricals and Electronics (RI2C)* (pp. 1-4). IEEE.

Ren, N., Quan, X., & Cho, S. H. (2016). Algorithm for gesture recognition using an IR-UWB radar sensor. *Journal of Computer and Communications*, *4*(3), 95–100. doi:10.4236/jcc.2016.43015

Ryu, S. J., Suh, J. S., Baek, S. H., Hong, S., & Kim, J. H. (2018). Feature-based hand gesture recognition using an FMCW radar and its temporal feature analysis. *IEEE Sensors Journal*, *18*(18), 7593–7602. doi:10.1109/JSEN.2018.2859815

Saeed, T. R., Mutashar, S., Abed, A. N., Al-Muifraje, M. H., Shuraiji, A. L., & Abd-Elghany, S. (2019). Human motion detection through wall based on micro-doppler using kalman filter combined with convolutional neural network. *International Journal of Intelligent Engineering & Systems*, *12*(4), 317–327. doi:10.22266/ijies2019.0831.29

Scheer, J., & Holm, W. A. (2010). *Principles of modern radar*. Academic Press.

Shastri, A., Valecha, N., Bashirov, E., Tataria, H., Lentmaier, M., Tufvesson, F., Rossi, M., & Casari, P. (2022). A review of millimeter wave device-based localization and device-free sensing technologies and applications. *IEEE Communications Surveys and Tutorials*, *24*(3), 1708–1749. doi:10.1109/COMST.2022.3177305

Shi, C., Zheng, Z., Pan, J., Ni, Z. K., Ye, S., & Fang, G. (2022). Multiple stationary human targets detection in through-wall UWB radar based on convolutional neural network. *Applied Sciences (Basel, Switzerland)*, *12*(9), 4720. doi:10.3390/app12094720

Singh, S., Liang, Q., Chen, D., & Sheng, L. (2011). Sense through wall human detection using UWB radar. *EURASIP Journal on Wireless Communications and Networking*, *2011*(1), 1–11. doi:10.1186/1687-1499-2011-20

Soumya, A., Krishna Mohan, C., & Cenkeramaddi, L. R. (2023). Recent Advances in mmWave-Radar-Based Sensing, Its Applications, and Machine Learning Techniques: A Review. *Sensors (Basel)*, *23*(21), 8901. doi:10.3390/s23218901 PMID:37960603

Tang, G., Wu, T., & Li, C. (2023). Dynamic Gesture Recognition Based on FMCW Millimeter Wave Radar: Review of Methodologies and Results. *Sensors (Basel)*, *23*(17), 7478. doi:10.3390/s23177478 PMID:37687932

Toker, O., Alsweiss, S., & Abid, M. (2020, March). A computer vision-based testbed for 77 GHz mmWave radar sensors. In *2020 SoutheastCon* (pp. 1-7). IEEE.

Wang, Y., Wang, W., Zhou, M., Ren, A., & Tian, Z. (2020). Remote monitoring of human vital signs based on 77-GHz mm-wave FMCW radar. *Sensors (Basel), 20*(10), 2999. doi:10.3390/s20102999 PMID:32466309

Wang, Z., Li, G., & Yang, L. (2020). Dynamic hand gesture recognition based on micro-Doppler radar signatures using hidden Gauss–Markov models. *IEEE Geoscience and Remote Sensing Letters, 18*(2), 291–295. doi:10.1109/LGRS.2020.2974821

Wolff, D. I. C. (n.d.). *Radartutorial.* https://www.radartutorial.eu/index.en.html

Yang, D., Zhu, Z., Zhang, J., & Liang, B. (2021). The overview of human localization and vital sign signal measurement using handheld IR-UWB through-wall radar. *Sensors (Basel), 21*(2), 402. doi:10.3390/s21020402 PMID:33430061

Zhang, J., Xi, R., He, Y., Sun, Y., Guo, X., Wang, W., Na, X., Liu, Y., Shi, Z., & Gu, T. (2023). A survey of mmWave-based human sensing: Technology, platforms and applications. *IEEE Communications Surveys and Tutorials, 25*(4), 2052–2087. doi:10.1109/COMST.2023.3298300

Zhang, Z., Tian, Z., & Zhou, M. (2018). Latern: Dynamic continuous hand gesture recognition using FMCW radar sensor. *IEEE Sensors Journal, 18*(8), 3278–3289. doi:10.1109/JSEN.2018.2808688

Chapter 3
Bibliometric Analysis of Reconfigurable Antennas in Radar Applications:
Trends, Influential Authors, and Future Directions

Duygu Nazan Gençoğlan
https://orcid.org/0000-0001-5014-9514
Department of Electrical-Electronics Engineering, Adana Alparslan Türkeş Science and Technology University, Adana, Turkey

ABSTRACT

This study conducts a comprehensive bibliometric analysis of research articles on reconfigurable antennas in radar applications retrieved from Google Scholar databases from 1989 to 2024. A total of 980 research articles are analysed, demonstrating significant interest and commitment to the topic, as evidenced by a total of 22.073 citations. Important metrics such as citations per year (649.21), citations per article (22.52), and citations per author (9,529). It highlight the importance and impact of research in this area. Collaboration is common, with an average of 410.95 articles per author and about three authors per article. The geographical analysis highlights the contributions of institutions in the United States and Switzerland. The study identifies emerging trends and future directions and highlights the need for dynamic antenna reconfiguration techniques, integration challenges, and energy-efficient solutions. The bibliometric analysis gives valuable insights into the current investigation landscape on reconfigurable antennas for radar applications.

DOI: 10.4018/979-8-3693-0916-2.ch003

1. INTRODUCTION

Bibliometric analysis is one of the techniques applied for evaluation and analyse the academic literature along with the desired field or any scientific discipline (Donthu et al., 2021). These methods underpin the quantitative analysis of publications, such as journal articles, conference papers, and patents so as to determine the trends, key contributors, research gaps, and impacts on research in a particular area (Xiao & Hao, 2021; Yu & Hayes, 2018). In the context of reconfigurable antennas in radar applications, bibliometric analysis plays crucial role due to the myriad of reasons (Haider et al., 2013; Motovilova & Huang, 2020; Iqbal et al., 2019). Bibliometric analysis paves the way for identifying the current trends and emerging technologies within the field of reconfigurable antennas for radar applications. This analysis contributes to the exploration of new design methodologies, materials, and signal processing techniques. Addition to this, by analysing the literature, researchers and scientists are able to visualize and outline the research landscape in terms of geographical distribution, collaboration networks, and institutional contributions (Dixit et al., 2020; Shevada et al., 2020). This information is paramount to understand the global impact of research in the field (Raut et al., 2021). Bibliometric analysis allows researchers to identify key contributors to the field, including authors, institutions, and research groups. This information is also useful for collaboration opportunities and for recognizing expertise in specific subfields. Focusing on the citation impact of research papers provides insights into the influence of specific studies and methodologies. Highly cited papers often indicate important contributions to the field and can be used to gauge the overall impact of a particular research topic. Bibliometric analysis allows for benchmarking and comparative analysis of different approaches and technologies. Researchers can assess the performance of various reconfigurable antenna designs in radar applications based on their publication and citation metrics. By analysing the existing literature, researchers can identify gaps in knowledge and potential areas for future research. This helps guide researchers in focusing on unexplored aspects and advancing the state-of-the-art in reconfigurable antennas for radar applications. Policymakers and funding agencies can use bibliometric analysis to make informed decisions about resource allocation and policy development. Understanding the research landscape helps in directing funding towards areas with significant potential and impact (Kounta et al., 2022). In summary, bibliometric analysis plays a crucial role in providing a systematic and data-driven understanding of the state of research in reconfigurable antennas for radar applications. It aids researchers, institutions, and policymakers in making informed decisions, fostering collaboration, and advancing the field.

The use of the "Publish or Perish" software developed by Professor Anne-Wil Harzing in a study to analyse existing literature offers a reliable method for carrying out bibliometric analyses (Harzing, 2007). Publish or Perish is widely recognized and used by researchers, scholars, and institutions for its ability to effectively analyse academic citations and publication rates. The software leverages data from leading sources such as Google Scholar, Microsoft Academic, and others to produce citation metrics and bibliometric analyses for various entities, including individual researchers, research groups, journals, and institutions. One of the main advantages of Publish or Perish is the ability to provide a range of citation metrics, such as total citations, h-index, g-index and citations per article. These metrics provide valuable information about the impact and impact of a researcher's publications in their field, enabling a comprehensive assessment of research results. Additionally, the software enables detailed analysis of authors and journals, allowing users to assess the reputation and impact of specific researchers or identify top journals in specific fields. Additionally, Publish or Perish simplifies the search for keywords and citations in scientific publications and allows researchers to analyse citation patterns for specific topics or research areas. This feature is particularly useful for identifying emerging trends, evaluating the impact of a concept or methodology, and gaining a deeper understanding of the scientific landscape in a particular area. The ability to export citation data and analysis results also increases the utility for researchers. This feature allows users to seamlessly integrate citation metrics into a variety of academic documents, such as tenure and promotion documents and grant applications, increasing the credibility and impact of their research efforts. Overall, the reliability and versatility of Publish or Perish make it a valuable tool for conducting bibliometric analyses and evaluating existing research in the literature. Its wide acceptance in the academic community underscores its effectiveness in helping researchers navigate the complex research and publishing landscape. Therefore, the use of the Publish or Perish tool in this study increases the credibility and accuracy of the bibliometric analysis performed and provides valuable insights into the existing literature.

2. METHODOLOGY

There is a myriad of database choices to read and research about the reconfigurable antenna structures for radar applications. These crucial databases are considered as IEEE Explore, PubMed, Scopus, ScienceDirect, IET Digital Library, Google Scholar, and Web of Science (WoS) etc. In the context of the aforementioned database structures, the term called "keywords" as "radar applications" and "reconfigurable antenna" is searched via database called Google Scholar. Google Scholar database is utilized to analysis the bibliometric characteristics of the proposed study. By combining these keywords in the proposed research, a comprehensive understanding of the role and significance of reconfigurable antenna structures has been gained in radar applications, which covers both technical aspects and practical implementations. By analysing the database, the important and crucial points to include or exclude the searching terms are taken into consideration with respect to the keywords.

This chapter provides a comprehensive overview of the research landscape related to reconfigurable antennas for radar applications using various databases and bibliometric analysis. The integration of several databases such as IEEE Xplore, PubMed, Scopus, ScienceDirect, IET Digital Library, Google Scholar and Web of Science (WoS) ensures a well-founded analysis of existing literature. Additionally, using Google Scholar for bibliometric analysis provides valuable insight into citation patterns, research impact, and author characteristics in a particular field. Data collection and inclusion/exclusion criteria were clearly defined and specified the use of the keywords "radar applications" and "reconfigurable antenna" to search for relevant publications. This chapter effectively presents metrics such as total citations, citations per year, citations per article, and citations per author, thereby providing quantitative measurements of research impact and impact. Additionally, the graphical representation of the total number of studies and the distribution of the citations over time provides a visual representation of research trends and growing interest in reconfigurable antennas for radar applications. The analysis of the ten most frequently cited publications provides an overview of innovative work and influential contributions in this field. By highlighting key articles and their citation frequency, this chapter identifies key research topics and topics that have generated significant interest in the academic community. Furthermore, the geographical and institutional analysis enriches the understanding of research collaboration networks and affiliations and highlights the interdisciplinary nature of research in this area. The inclusion of authorship analysis, citation analysis, and citation pattern visualization further increases the depth of chapter exploration across the research landscape. Overall, the chapter effectively illustrates the issues, issues, and trends associated with reconfigurable antennas for radar applications through a systematic analysis of existing literature and bibliometric data.

2.1. Data Collection and Inclusion/Exclusion Criteria

Research study data is retrieved from Google Scholar Databases among years from 1989 to 2024 along with the keywords as "radar applications" & "reconfigurable antenna". During the literature survey, citations and patents are also taken into consideration in this study. Based on Google Scholar analysis, 980 research papers including conferences, articles, patents, and proceedings are determined based on the proposed keywords. The metrics provide a comprehensive overview of the research impact, citation patterns, and authorship characteristics of the individual or group being evaluated. Hence, metrics parameters are also investigated to outline the importance of the reconfigurable antenna in radar applications in bibliometrics to assess the significance and influence of academic work. It is deduced from Google Scholar database that the total number of citations is 22073, the citations per year are 649.21, the citations per paper is 22.52, and citations per author is 9529.3, respectively. It is clear from citation results that the reconfigurable antenna structures have been drawn attention by the researchers, scientists, and authors since 1989. Addition to citation metrics; the papers per author is determined as 410.95. Furthermore, authors per paper is analysed in terms of mean, median, and mode, respectively. The corresponding values are determined as 3.05, 3.0, 3.0, respectively. The detailed metric analysis is also tabulated in Table 1.

Table 1. Detailed metrics analysis

Publication Years: 1989-2024
Citation years: 34 (1989-2023)
Papers: 980
Citations: 22073
Citations/year: 649.21 (acc1=521, acc2=367, acc5=201, acc10=98, acc20=30)
Citations/paper: 22.52
Citations/author: 9529.30
Papers/author: 410.95
Authors/paper: 3.05/3.0/3 (mean/median/mode
Age-weighted citation rate: 3870.00 (sqrt=62.21), 1617.87/author
Hirsch h-index: 64 (a=5.39, m=1.88, 12640 cites=57.3% coverage
Egghe g-index: 125 (g/h=1.95, 15632 cites=70.8% coverage
PoP hI,norm: 41
PoP hI,annual: 1.21
Fassin hA-index: 24

The total number of research papers or publications related to reconfigurable antennas in radar applications from 1989 to 2024 is shown in Figure 1. From 1989 to 2024, the number of studies conducted in this area has increased significantly, indicating an increase in interest and research activity in the area of reconfigurable antennas for radar applications. The number of studies shows fluctuations over the years, with some years showing a more significant increase or decrease than others. An upward trend in the number of studies has been observed since the beginning of the 21st century, with a particularly significant increase from around 2009. The number of studies more than doubled between 2009 (26 studies) and 2021 (93 studies), showing the rapid development of research in this area. In recent years, from 2020 to 2024, the number of studies appears to have decreased from the peak in 2021. However, note that the 2024 data may be incomplete or only cover part of the year. Overall, the data shows strong and ongoing interest in reconfigurable antennas for radar applications, with more and more studies being conducted each year, particularly over the last decade.

Figure 1. The total number of studies based on the years from 1989 to 2024

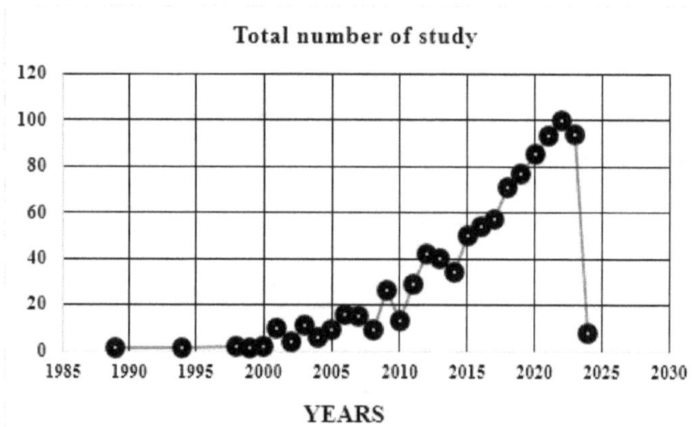

2.2. Analytical Tools and Techniques

As of analytical tools for reconfigurable antennas in radar applications, MATLAB is a powerful numerical computing environment widely used in the design and simulation of reconfigurable antennas for radar applications. Its rich set of functions and toolboxes facilitate the implementation of complex algorithms for electromagnetic analysis, allowing engineers to model and optimize antenna parameters efficiently. MATLAB's versatility makes it a preferred choice for exploring various design configurations and conducting parametric studies to enhance the reconfigurability of antennas in radar systems. Additionally, HFSS (High-Frequency Structure Simulator) for Electromagnetic Simulation a specialized electromagnetic simulation tool that excels in analysing high-frequency structures, making it invaluable for the study of reconfigurable antennas in radar applications. With HFSS, engineers can perform accurate electromagnetic simulations to assess the impact of frequency changes on antenna performance. This tool aids in understanding the intricate electromagnetic interactions within the antenna structure, ensuring that the reconfigurable design meets the stringent requirements of radar systems. On the other hand, CST Microwave Studio is a comprehensive simulation platform tailored for electromagnetic and multi-physics simulations, making it an essential tool for modelling reconfigurable antennas in radar applications. Its intuitive interface allows engineers to construct detailed three-dimensional models of antennas, enabling precise analysis of their electromagnetic behaviour. CST Microwave Studio facilitates the exploration of various design configurations, ensuring that reconfigurable antennas are optimized for radar applications in terms of radiation patterns, gain, and impedance matching. In radar applications, the structural and thermal integrity of reconfigurable antennas are paramount for reliable performance. ANSYS, a leading simulation tool, excels in structural and thermal analysis, ensuring that antennas can withstand environmental conditions and operate within specified temperature ranges. By employing ANSYS, engineers can assess the mechanical robustness and thermal stability of reconfigurable antennas, guaranteeing their effectiveness and longevity in radar systems. As a conclusion, the combination of MATLAB, HFSS, CST Microwave Studio, and ANSYS provides a comprehensive analytical toolkit for the design, simulation, and optimization of reconfigurable antennas in radar applications. These tools collectively enable engineers to explore the electromagnetic behaviour, structural integrity, and thermal performance of antennas, ensuring their seamless integration into radar systems with enhanced adaptability and reliability.

Table 2. The commonly used analytical tools

MATLAB for Antenna Simulation and Design
HFSS (High-Frequency Structure Simulator) for electromagnetic simulation
CST Microwave Studio for antenna modelling
ANSYS for structural and thermal analysis

3. PUBLICATION TRENDS

The retrieve relevant publications from selected databases using the specified keywords and inclusion criteria are visualized in Figure 2, respectively. By simultaneously analysing keywords as reconfigurable antenna and radar applications, 10 most cited publications are visualized in Figure 3. The most cited paper was titled as “RF-MEMS switches for reconfigurable integrated circuits” with the 944 cites by Brown in 1998(Brown, 1998). The cites per year is determined as 37.76 as well.

Secondly, the paper titled as "Reconfigurable reflect arrays and array lenses for dynamic antenna beam control: A review" by Hum and Carrier are cited with the number of 825 as depicted in Figure 3(Hum & Carrier, 2013). Figure 4 outlines the percentage of the most cited papers with respect to years as well.

Figure 2. Count of title by type

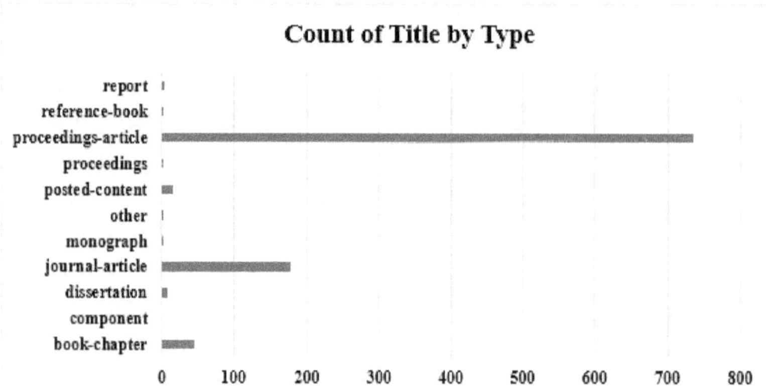

Figure 3. Count of title by type

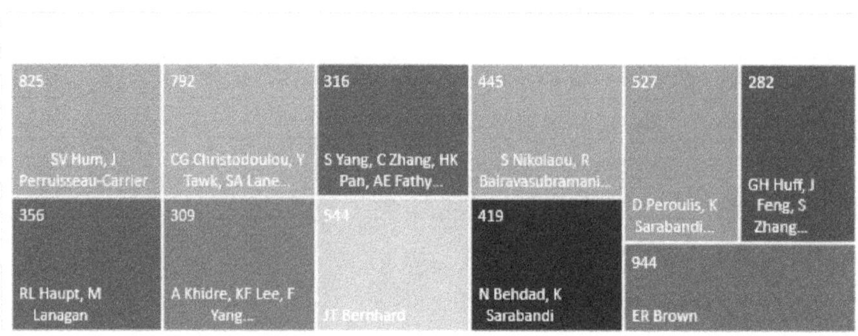

Table 3. 10 of most cited articles in the literature

Article Title	Year	Cited by
RF-MEMS switches for reconfigurable integrated circuits (Brown, 1998)	1998	944
Reconfigurable reflect arrays and array lenses for dynamic antenna beam control: A review (Hum & Carrier, 2013)	2013	825
Reconfigurable antennas for wireless and space applications (Christodoulou, 2012)	2012	792
Reconfigurable multifunction antennas: Next steps for the future (Bernhard, 2007)	2007	544
Design of reconfigurable slot antennas (Peroulis et al., 2005)	2005	527
Pattern and frequency reconfigurable annular slot antenna using PIN diodes (Nikolaou et al., 2006)	2006	445
Dual-band reconfigurable antenna with a very wide tunability range (Behdad & Sarabandi, 2006)	2006	419
Reconfigurable antennas (Haupt & Lanagan, 2013)	2013	356
Frequency-reconfigurable antennas for multi radio wireless platforms (Yang et al., 2009)	2009	316
Circular polarization reconfigurable wideband E-shaped patch antenna for wireless applications (Khidre et al., 2012)	2012	309

Figure 4. The Percentage of 10 most cited articles

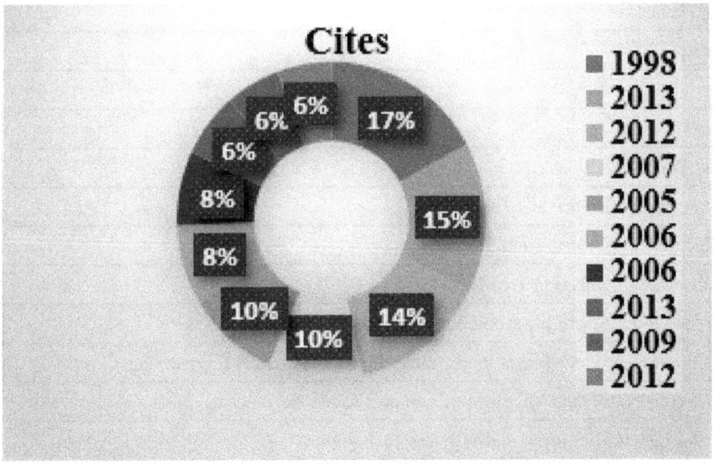

The most suitable publication outlet depends on your specific research, target audience, and publication goals. Consider the impact factor, review process, and publication speed of each journal or conference before submitting your work.

Networking at conferences and workshops can be valuable for connecting with other researchers and identifying potential collaboration opportunities. While patents offer intellectual property protection, they may not be the best way to disseminate your research findings to the wider scientific community.

Table 3. The most commonly journals which articles published

Telecommunications and Radio Engineering
Journal of Integrated Science and Technology
IETE Journal of Research
IEEE Transactions on Antennas and Propagation
IEEE Access
Electronics
Advances in Microwave Engineering
Wireless Personal Communications
US Patent
TELKOMNIKA
Telecommunications and Radio Engineering
Research
Radio Science
Progress in Electromagnetics Research C
Óptica Pura y Aplicada
Open Engineering
Nano Select
Micromachines
Measurement: Sensors
Journal of Engineering Science and Technology
Journal of Electrical Engineering and Technology
Journal of Advanced Engineering Trends
Iraqi Journal for Electrical and Electronic Engineering
Iranian Journal of Science and Technology, Transactions of Electrical Engineering
International Journal of RF and Microwave Computer-Aided Engineering
International Journal of Microwave and Wireless Technologies
Indonesian Journal of Electrical Engineering and Computer Science
El-Cezeri Journal of Science and Engineering
Bulletin of Electrical Engineering and Informatics
Applied Sciences
AEU-International Journal of Electronics and Communications

4. GEOGRAPHICAL AND INSTITUTIONAL ANALYSIS

Table 4 depicts the most cited articles along with the reconfigurable antenna and radar applications. The investigation is analysed by using author information, company or university, city, and country. From the information provided, it appears that the most cited articles on reconfigurable antennas in radar applications come from authors from various institutions and companies in the United States and Switzerland. The collaboration includes scientists from renowned academic institutions such as the University of California, EPFL and the University of New Mexico, as well as contributions from the Space Vehicle Directorate. This interdisciplinary collaboration is expected to bring together expertise from different areas to accelerate research into reconfigurable antennas for radar applications. The author list shows that the most cited article on reconfigurable antennas for radar applications is probably the result of a collaborative effort by researchers from several universities, companies and institutions in the USA and Switzerland. This research collaboration brings together a diverse group of experts from renowned academic institutions, such as: University of California, Los Angeles (UCLA), Federal Polytechnic School of Lausanne (EPFL), University of New Mexico, University of Illinois Urbana-Champaign, University of Michigan, Georgia Institute of Technology, NASA Glenn Research Centre, Pennsylvania State University, University of Tennessee, University of Mississippi etc.

This also includes contributions from companies such as Intel company, many employees and public institutions such as Spacecraft directorate This interdisciplinary team is likely to draw on knowledge from diverse fields such as electrical engineering, materials science, electromagnetism and aerospace engineering. This different perspective could lead to significant advances in reconfigurable antenna technology for radar applications, enabling features such as: dynamic beam steering for better target tracking, multi-frequency operation for a larger detection range, reconfigurable polarization for better signal reception.

Table 4. The most cited paper along with author, company or university, city, and country

Authors	Company	City	Country
E.R. Brown	University of California	Los Angeles	USA
Sean Victor Hum	École Polytechnique Fédérale de Lausanne (EPFL)	Lausanne	Switzerland
Julien Perruisseau-Carrier	École Polytechnique Fédérale de Lausanne (EPFL)	Lausanne	Switzerland
Christos G. Christodoulou	University of New Mexico	Albuquerque	USA
Youssef Tawk	University of New Mexico	Albuquerque	USA
Steven A. Lane	Space Vehicles Directorate	Albuquerque	USA
Scott R. Erwin	Space Vehicles Directorate	Albuquerque	USA
J. T. Bernhard	University of Illinois	Urbana-Champaign	USA
D. Peroulis	University of Michigan	Ann Arbor	USA
K. Sarabandi	University of Michigan	Ann Arbor	USA
L.P.B. Katehi	University of Michigan	Ann Arbor	USA
Symeon Nikolaou	Georgia Institute of Technology	Atlanta	USA
R. Bairavasubramanian	Georgia Institute of Technology	Atlanta	USA
C. Lugo	Georgia Institute of Technology	Atlanta	USA
D.C. Thompson	Georgia Institute of Technology	Atlanta	USA
J. Papapolymerou	Georgia Institute of Technology	Atlanta	USA
M.M. Tentzeris	Georgia Institute of Technology	Atlanta	USA
G.E. Ponchak	NASA Glenn Research Center	Cleveland	USA
N. Behdad	University of Michigan	Ann Arbor	USA
K. Sarabandi	University of Michigan	Ann Arbor	USA
Randy L. Haupt	Haupt Associates	Boulder	USA
Michael Lanagan	The Pennsylvania State University	University Park	USA
Songnan Yang	Intel Corporation	Santa Clara	USA
Helen K. Pan	Intel Corporation	Santa Clara	USA
Vijay K. Nair	Intel Corporation	Santa Clara	USA
Chunna Zhang	University of Tennessee	Knoxville	USA
Aly E. Fathy	University of Tennessee	Knoxville	USA
Ahmed Khidre	The University of Mississippi	University	USA
Kai-Fong Lee	The University of Mississippi	University	USA
Fan Yang	The University of Mississippi	University	USA
Atef Z. Elsherbeni	The University of Mississippi	University	USA

To sum up, there are several notable collaborations. Research collaborations play a key role in promoting knowledge and innovation by pooling different skills and resources. There are some notable collaborations been observed:

Interdisciplinary Collaboration: Many projects involved researchers from different disciplines, such as electrical engineering, materials science, electromagnetism, and aerospace engineering. This interdisciplinary approach is evident, among other things, in the collaboration between institutions such as the University of California, EPFL, the University of New Mexico and the NASA Glenn Research Center. Such collaborations are expected to leverage expertise from diverse fields to address the complex challenges of reconfigurable antenna technology for radar applications.

University-Industry Collaboration: Collaboration between academic institutions and industry players such as Intel Corporation highlights the importance of bridging the gap between theoretical research and practical applications. This collaboration often leads to research results with real-world relevance and commercialization potential. For example, collaborations between universities and companies such as Intel Corporation could focus on developing innovative antenna technologies for wireless communications and radar systems.

International Collaboration: Collaboration between researchers and institutions from different countries contributes to the global exchange of knowledge and experience. Notable collaborations include partnerships between institutions in the United States, Switzerland, and other countries. This international collaboration facilitates the exchange of best practices, methodologies, and resources, thereby improving the quality and impact of research results.

Academic Collaborations: Collaborations between academic institutions such as the University of California, EPFL, the University of New Mexico and the Georgia Institute of Technology demonstrate the collaborative nature of academic research. These collaborations often include joint research projects, co-authored publications and knowledge sharing activities. By pooling resources and expertise, academic collaborations accelerate progress in reconfigurable antenna research and foster a collaborative culture within the academic community. In conclusion, this extraordinary collaboration demonstrates the importance of collaboration to promote innovation and advance research in reconfigurable antennas for radar applications. By leveraging diverse expertise, resources and perspectives, collaborative efforts have the potential to address complex challenges and accelerate the development of transformative technologies.

5. AUTHORSHIP ANALYSIS

Figure 5 shows the number of authors associated with publications on reconfigurable antennas for radar applications each year. The data set covers the years 1989 to 2024. The number of authors can be viewed as an indication of the level of research activity or interest in the field in a given year. The data shows that the number of authors fluctuated over the years, suggesting different research activities or interest in reconfigurable antennas for radar applications. To gain deeper insight into trends from this data, consider time trends, benchmarking, event correlation, deeper regional analysis, and collaboration models. Temporal trends can analyse how the number of authors changes over time, what there are significant increases or decreases in some years, what these fluctuations could have caused by? The researcher should also compare the number of authors from year to year to identify periods of significant growth or decline in research activity. Researchers should determine whether particular events or advances in radar technology coincided with changes in research activity in the field and whether the distribution of the authors is concentrated in particular regions or countries, which could provide insight into areas where research in this area is needed is more active. Furthermore, examining collaboration patterns between authors is a crucial task for understanding the network of researchers working on reconfigurable antennas for radar applications.

Figure 5. The total number of authors for the studies with respect to years

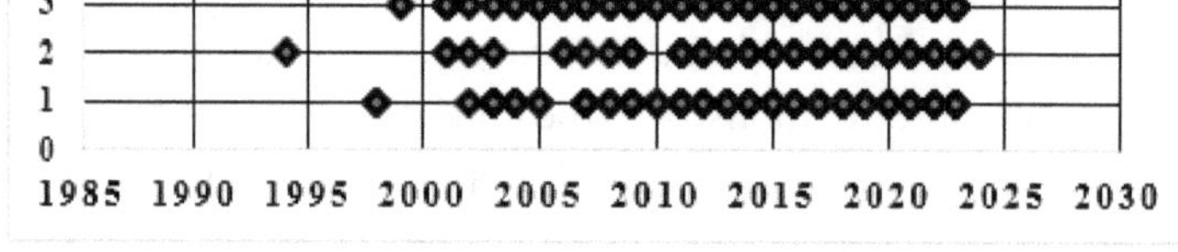

6. CITATION ANALYSIS

Figure 6 appears the dataset containing information about the number of citations per author for various years ranging from 1989 to 2024. For each year, there's a corresponding number indicating the number of citations per author. Some years have multiple entries, indicating that there were multiple authors with different numbers of citations. The number of citations per author varies widely across different years, ranging from 0 to as high as 946. This information obtained by literature analyses can give the researchers insights into the distribution and trends of citations over time as well. The researchers and scientists can observe how the number of citations per author has changed over time by analysing the trends across different years. This can help identify periods of growth, stagnation, or decline in citation activity within the field. Looking at the range of values within each year, it can be determined the variability in citation counts among authors. Years with a wide range of values suggest a diverse range of citation counts among authors, while years with a narrow range indicate more uniformity. On the other hand, identification of outlier years or extreme values can provide insights into exceptional circumstances or events that may have influenced citation patterns. For example, a sudden spike in citations in a particular year could be attributed to a ground-breaking discovery or a highly influential publication. Furthermore, utilizing historical citation data, researchers can develop predictive models to forecast future citation trends, identify emerging research topics, or predict the impact of specific research initiatives or policies.

Figure 6. Cites per author

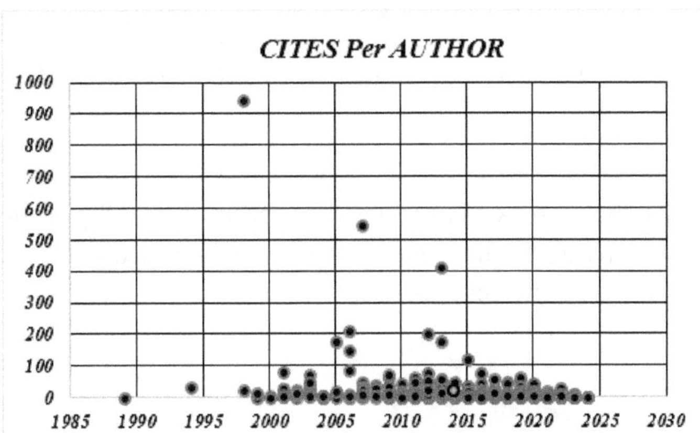

The data set about cites per year ranges from 1989 to 2024 as given in Figure 7 and covers a wide annual range. The number of citations varies greatly from year to year, with some years receiving significantly more citations than others. Overall, there appears to be an increasing trend in the number of citations over time, with occasional fluctuations. Some years are characterized by particularly high citation numbers, for example 2013, in which there were several high citation numbers. On the other hand, there are also years in which the number of citations is very low, in some cases even zero, indicating periods of lower influence or perhaps fewer publications in the field at that time. Dataset ends in 2024, meaning citations are down significantly compared to previous years. Because, it is important to note that this may be due to incomplete data for this year. Overall, the data shows varying but generally increasing levels of interest and influence in the area of reconfigurable antennas for radar applications over the years.

Figure 7. Cites per year

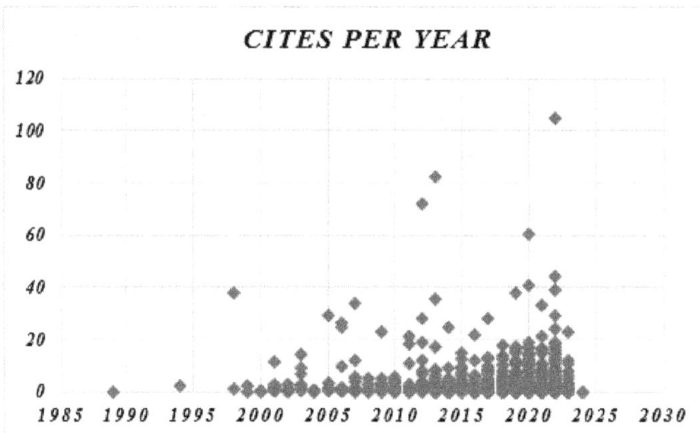

The dataset about years vs citations depicted in Figure 8 provides insight into the citation patterns across different years, which can be useful for analysing trends in research output and impact over time.

Figure 8. Years vs. citations

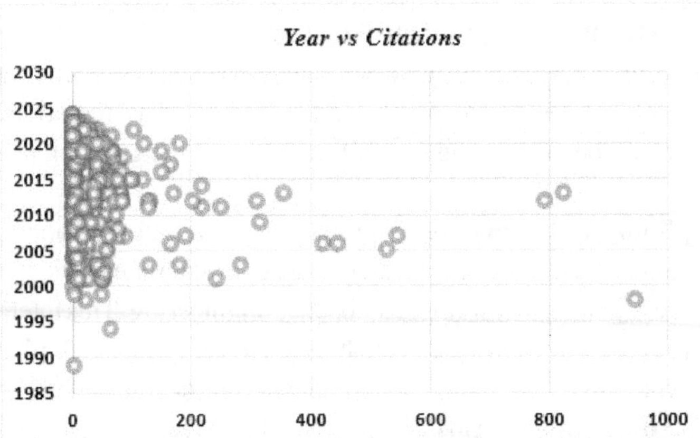

A higher number of citations per year suggests that the research has significant impact and has generated significant interest in the academic community. It reflects the continued importance and impact of research over time. A higher number of citations per article indicates that the specific research findings are widely recognized and cited by other researchers in related research papers. It serves as an indicator of the quality, significance, and impact of research results. Articles with a high citation count of per article are often considered seminal contributions to the field and can shape future research directions. Authors with a high number of citations per author are often considered experts or leading authorities in a particular field. This measure is useful for assessing each researcher's overall scientific influence and reputation.

7. EMERGING TRENDS AND FUTURE DIRECTIONS

A bibliometric analysis of research articles on reconfigurable antennas for radar applications revealed several key trends and patterns that provided valuable insights into the state and evolution of the industry. There are the key trends observed:

Increase in search activity: One noticeable trend is an overall increase in search activity over time. The analysis showed a steady increase in the number of publications related to reconfigurable antennas for radar applications, indicating growing interest and investment in this research area.

Emergence of new technologies: The analysis highlights the emergence of new technologies and methods in the design of reconfigurable antennas for radar applications. Scientists are exploring innovative approaches to increase the performance, versatility, and adaptability of reconfigurable antennas, including new materials, manufacturing techniques and signal processing algorithms.

Focus on Performance Optimization: Another notable trend is the focus on performance optimization of reconfigurable antennas for radar applications. Scientists are trying to improve key performance indicators such as radiation efficiency, bandwidth, gain and beam steering capabilities to meet the growing needs of radar systems.

Interdisciplinary Collaboration: The analysis highlights collaboration between disciplines, with contributions from researchers in areas such as electrical engineering, materials science, signal processing and aerospace engineering. This interdisciplinary collaboration promotes innovation and enables the development of holistic solutions to the complex challenges associated with radar antenna design and implementation.

Global distribution of research: The analysis reveals the global distribution of research activities involving researchers and institutions from different countries and regions. Although some countries such as the United States and Switzerland boast higher research achievements, collaborations and contributions from different geographical regions have highlighted the international nature of research in this area.

Influential publications: Some publications have proven to be highly influential in their field and have received significant citations and recognition in the research community. This innovative work often addresses fundamental concepts, proposes innovative solutions, or provides comprehensive analysis that determines the direction of research and influences future development of reconfigurable antennas for radar applications.

Technological Trends and Innovations: The analysis identified specific technological trends and innovations underlying research on reconfigurable antennas for radar applications. These include improvements in RF-MEMS switches, reflector systems, multi-function antennas, frequency-reconfigurable antennas, and dynamic beam steering techniques. These innovations reflect ongoing efforts to meet the changing needs and challenges of radar systems. In summary, the bibliometric analysis provided valuable insights into the current status, trends, and research directions of reconfigurable antennas for radar applications. These findings not only contribute to the collective understanding of the field, but also shape future research priorities, collaborations, and technological developments in the design and implementation of radar antennas.

Based on analysis, several important research gaps in reconfigurable antennas for radar applications are identified. These gaps represent areas where further research and innovation is required to address existing challenges and provide innovative solutions. The detailed explanation of each suggestion is given as follows:

- ❖ Selection reconfigurable antennas that preserve performance and efficiency while potentially radiating over a wider frequency range.
- ❖ Development of new and essential techniques for dynamic antenna reconfiguration in response to quickly shifting radar environments.
- ❖ Investigation opportunities and challenges for integration when implementing reconfigurable antennas in advanced radar systems, like MIMO or phased array radar.
- ❖ Studies should aim at improving reconfigurable antennas' resilience and dependability, particularly in demanding applications and harsh environments.
- ❖ Thorough trade-off analysis between different reconfiguration parameters (radiation pattern, polarization, frequency, etc.) to guide the design process based on specific application requirements.
- ❖ Find ways to reduce the size of a reconfigurable antenna without sacrificing performance, which is important in space-limited applications.
- ❖ Promotion greater interdisciplinary research to solve complex problems and develop global solutions, involving antenna engineers, signal processing experts and radar system designers.
- ❖ Research on energy-efficient reconfigurable antennas to optimize power consumption, especially in portable and power-sensitive applications.
- ❖ Exploration of real-world obstacles and limitations encountered in implementing reconfigurable antennas in radar systems, considering issues such as cost, ease of integration, and manufacturability.
- ❖ The aim of the research was to improve the security of reconfigurable antennas by developing methods to reduce interference and prevent illegal reconfigurations.

By conclusion these research gaps, scientists and researchers can promote innovation, expand knowledge, and contribute to the development of robust and efficient reconfigurable antennas for radar applications. These suggestions provide a roadmap for future research efforts aimed at overcoming existing challenges and expanding the boundaries of antenna technology.

The development of dynamic antenna reconfiguration techniques and energy saving solutions poses several challenges. A key obstacle is maintaining antenna performance during reconfiguration, which requires advanced design methods and optimization techniques. Furthermore, ensuring dynamic adaptation to the evolving radar environment requires the implementation of adaptive real-time algorithms and efficient control systems. Integrating reconfigurable antennas into advanced radar systems like the, such as MIMO or phased array radar, requires collaborative interdisciplinary research to address compatibility and optimization challenges. Additionally, increasing resilience and reliability in harsh environments requires robust design techniques, including redundancy and fault tolerance mechanisms.

Finally, conducting a thorough analysis of the trade-offs between reconfiguration parameters is critical to optimal design and guides researchers to solutions that meet specific application requirements.

8. CONCLUSION

The analysis of research articles on "radar applications" and "reconfigurable antennas" provides valuable insights into the dynamics of this area in science. The total number of citations of 22,073 indicates significant interest and engagement in the topic of reconfigurable antennas for radar applications within the academic community. With an average number of citations per year of 649.21, it is clear that the importance and impact of research in this area has been consistently recognized and cited over time, demonstrating its continued relevance to the field. The average number of citations per article, calculated at 22.52, underlines the quality and impact of individual research contributions. This indicator shows that, on average, each article has generated significant interest in the scientific community. The high average citations per author (9529.3) highlight the significant influence of researchers and authors contributing to this field. This suggests that individual authors have made significant contributions that have resonated widely within the academic community. With an average of 410.95 articles per author, it is clear that many authors have been actively involved in research related to reconfigurable antennas for radar applications. This indicator shows the breadth and depth of expertise of the academic community working in this area. The analysis of authors per article provides insight into the collaboration patterns within the research community. Mean, median and mode values (3.05, 3.0, 3.0) indicate that each work is created on average by around three people. This suggests a common trend towards collaborative efforts and interdisciplinary research in this area.

REFERENCES

Abbas, A., Sufian, M. A., Awan, W. A., Hussain, N., Lee, J., & Kim, N. (2023, March). A Reconfigurable Dual Sharp Notch-band UWB Antenna. In *2023 17th European Conference on Antennas and Propagation (EuCAP)* (pp. 1-5). IEEE. 10.23919/EuCAP57121.2023.10133304

Abdulhussein, A. J., Farhan, M. J., & Ali, G. M. (2023). Design and implementation of a frequency reconfigurable antenna using PIN switch for sub-6 GHz applications. *Open Engineering*, *13*(1), 20220453. doi:10.1515/eng-2022-0453

Abdulzahra, D. H., Alnahwi, F. M., & Abdullah, A. S. (2023). Design of Compact Wideband/Bi-Band Frequency Reconfigurable Antenna for IoT Applications. *Iraqi Journal for Electrical and Electronic Engineering*, *19*(2), 100–109. doi:10.37917/ijeee.19.2.12

Ahmad, A., Lee, G. O., & Choi, D. Y. (2023). Design and Performance Evaluation of a Compact Frequency-Reconfigurable Coplanar-Waveguide-Fed Slotted Patch Antenna for Multi-Band Wireless Communication. *Electronics (Basel)*, *12*(18), 3889. doi:10.3390/electronics12183889

Alieldin, A., Eid, A. M., Salama, A. A., & El-Akhdar, A. M. (2023, February). A Reconfigurable Cosecant-Squared/Pencil Beam Antenna Array for Radar Systems. In *2023 International Microwave and Antenna Symposium (IMAS)* (pp. 50-52). IEEE. 10.1109/IMAS55807.2023.10066936

Allanic, R., Le Berre, D., Quendo, C., Ball, E. A., Ng, J. S., Huang, G., ... Merlet, T. (2023, November). Monolithic Frequency-Reconfigurable Antenna on Silicon Carbide for Constrained Environments. In *2023 IEEE Conference on Antenna Measurements and Applications (CAMA)* (pp. 649-652). IEEE. 10.1109/CAMA57522.2023.10352802

Alsharif, F., & Kurnaz, C. (2023, July). Design and analysis frequency reconfigurable coplanar waveguide antenna for S/C bands satellite applications. In *2023 46th International Conference on Telecommunications and Signal Processing (TSP)* (pp. 70-73). IEEE.

Amri, M., & Gharsallah, A. (2023, July). Design of Frequency-Pattern Reconfigurable Patch Array Antenna for Radar Application. In *2023 3rd International Conference on Electrical, Computer, Communications and Mechatronics Engineering (ICECCME)* (pp. 1-7). IEEE. 10.1109/ICECCME57830.2023.10253299

Anwar, M. S. (2023). *Microfluidics Based 3D Printed Reconfigurable Antenna for K-Band Applications* (Doctoral dissertation).

Apparao, M., & Karunakar, G. (2023). Design and analysis of frequency reconfigurable antenna in S-band. *Bulletin of Electrical Engineering and Informatics*, *12*(5), 2878–2886. doi:10.11591/eei.v12i5.4052

Bahrami, S., Lee, D., Hong, W., & Song, H. J. (2023). Analysis of On-Antenna Power Combining for Realization of Pattern-Reconfigurable Multifeed Microstrip Antennas. *IEEE Transactions on Antennas and Propagation*, *71*(4), 3091–3103. doi:10.1109/TAP.2023.3240073

Bai, Y. Y., Xiao, S., Tang, M. C., Ding, Z. F., & Wang, B. Z. (2011). Wide-angle scanning phased array with pattern reconfigurable elements. *IEEE Transactions on Antennas and Propagation*, *59*(11), 4071–4076. doi:10.1109/TAP.2011.2164176

Bammidi, D., & Kota, C. B. R. (2023). Design and validation of frequency reconfigurable multiband antenna using varied current distribution method. *Measurement. Sensors*, *29*, 100844. doi:10.1016/j.measen.2023.100844

Battaglia, G. M., Isernia, T., Palmeri, R., & Morabito, A. F. (2023). Four-Beams-Reconfigurable Circular-Ring Array Antennas for Monopulse Radar Applications. *Radio Science*, *58*(9).

Behdad, N., & Sarabandi, K. (2006). Dual-band reconfigurable antenna with a very wide tunability range. *IEEE Transactions on Antennas and Propagation*, *54*(2), 409–416. doi:10.1109/TAP.2005.863412

Bekali, Y. K., Zugari, A., El Bhiri, B., Ennasar, M. A., Elmajid, H., Khalladi, M., & Elghandour, A. (2023). A New Design for a Novel Compact Frequency Reconfigurable Antenna. *International Journal of Microwave & Optical Technology*, *18*(4).

Bellapukonda, V. K., Muduli, A., & Nayak, D. K. (2023, April). Reconfigurable T-shaped discrete tunable Antenna for Cognitive Radio and Natural Sciences Applications. In *2023 International Conference on Recent Advances in Electrical, Electronics, Ubiquitous Communication, and Computational Intelligence (RAEEUCCI)* (pp. 1-4). IEEE. 10.1109/RAEEUCCI57140.2023.10134509

Benny, S., & Sahoo, S. (2023, June). Polarization switching frequency reconfigurable dual polarization antenna element for phased array weather applications. In *2023 IEEE Wireless Antenna and Microwave Symposium (WAMS)* (pp. 1-5). IEEE. 10.1109/WAMS57261.2023.10242976

Berhab, S., Annou, A., & Chebbara, F. (2023). Reconfigurable Low-Profile Antenna-Based Metamaterial for on/off Body Communications. In *Handbook of Research on Emerging Designs and Applications for Microwave and Millimeter Wave Circuits* (pp. 166–200). IGI Global. doi:10.4018/978-1-6684-5955-3.ch007

Bernhard, J. T. (2007, August). Reconfigurable multifunction antennas: Next steps for the future. In *2007 International Symposium on Microwave, Antenna, Propagation and EMC Technologies for Wireless Communications* (pp. K2-1). IEEE. 10.1109/MAPE.2007.4393804

Bikrat, M., & Bri, S. (2023). A bandwidth reconfigurable antenna for devices in low UWB-applications. *TELKOMNIKA*, *21*(2), 272–279. doi:10.12928/telkomnika.v21i2.24256

Bildik, S., Dieter, S., Fritzsch, C., Menzel, W., & Jakoby, R. (2014). Reconfigurable folded reflectarray antenna based upon liquid crystal technology. *IEEE Transactions on Antennas and Propagation*, *63*(1), 122–132. doi:10.1109/TAP.2014.2367491

Brown, E. R. (1998). RF-MEMS switches for reconfigurable integrated circuits. *IEEE Transactions on Microwave Theory and Techniques*, *46*(11), 1868–1880. doi:10.1109/22.734501

Bshara, O. (2023). *Millimeter wave coarse beamforming using outband sub-6GHz reconfigurable antennas* (U.S. Patent No. 11,843,434). United States Patent and Trademark Office.

Buzzi, S., Grossi, E., Lops, M., & Venturino, L. (2022). Foundations of MIMO radar detection aided by reconfigurable intelligent surfaces. *IEEE Transactions on Signal Processing*, *70*, 1749–1763. doi:10.1109/TSP.2022.3157975

Cai, Y., Guo, Y. J., & Bird, T. S. (2012). A frequency reconfigurable printed Yagi-Uda dipole antenna for cognitive radio applications. *IEEE Transactions on Antennas and Propagation*, *60*(6), 2905–2912. doi:10.1109/TAP.2012.2194654

Chakrabarti, S., & Chakraborty, A. (2024). A Pattern Reconfigurable Antenna for Ka-band Application. *Journal of the Institution of Electronics and Telecommunication Engineers*, 1–13. doi:10.1080/03772063.2024.2303994

Chen, M., Zhang, Z. Q., Li, X. D., Liu, H. L., Chen, R. S., Shao, Q., & Huang, G. L. (2023, May). A Beam Reconfigurable Antenna Array Based on Highly Conductive Graphene-Assembled Films. In *2023 International Conference on Microwave and Millimeter Wave Technology (ICMMT)* (pp. 1-3). IEEE. 10.1109/ICMMT58241.2023.10276949

Chen, Y., Guo, H., Liu, Y., Li, J., Zhan, Y., Wu, Q., & Li, M. (2023). An L-Slot Frequency Reconfigurable Antenna Based on MEMS Technology. *Micromachines*, *14*(10), 1945. doi:10.3390/mi14101945 PMID:37893382

Chidurala, S., & Rao, A. P. (2023, July). Frequency Reconfigurable Antenna with Partial Ground for WLAN, WiMAX, C-band, and Sub-6GHz Wireless Communications. In *2023 14th International Conference on Computing Communication and Networking Technologies (ICCCNT)* (pp. 1-4). IEEE.

Choudhury, S. (2023). *Reconfigurable Antennas: Trends and applications*. IoP Series in Electromagnetics and Metamaterials. doi:10.1088/978-0-7503-5457-8

Christodoulou, C. G., Tawk, Y., Lane, S. A., & Erwin, S. R. (2012). Reconfigurable antennas for wireless and space applications. *Proceedings of the IEEE, 100*(7), 2250–2261. doi:10.1109/JPROC.2012.2188249

Clemente, A., Dussopt, L., Sauleau, R., Potier, P., & Pouliguen, P. (2012). 1-Bit reconfigurable unit cell based on PIN diodes for transmit-array applications in X -Band. *IEEE Transactions on Antennas and Propagation*, *60*(5), 2260–2269. doi:10.1109/TAP.2012.2189716

Correas-Serrano, D., Gomez-Diaz, J. S., Alù, A., & Melcón, A. Á. (2015). Electrically and magnetically biased graphene-based cylindrical waveguides: Analysis and applications as reconfigurable antennas. *IEEE Transactions on Terahertz Science and Technology*, *5*(6), 951–960. doi:10.1109/TTHZ.2015.2472985

Dixit, A. S., Shevada, L. K., Raut, H. D., Malekar, R. R., & Kumar, S. (2020). Fifth generation antennas: A bibliometric survey and future research directions. *Library Philosophy and Practice*, *2020*, 1–24.

Donthu, N., Kumar, S., Mukherjee, D., Pandey, N., & Lim, W. M. (2021). How to conduct a bibliometric analysis: An overview and guidelines. *Journal of Business Research, 133*, 285–296. doi:10.1016/j.jbusres.2021.04.070

Elhemaily, H. (2023). Reconfigurable Monopole Antenna Using Graphene Material for Wireless Applications. *Journal of Advanced Engineering Trends*, *42*(2), 173–179. doi:10.21608/jaet.2022.111686.1134

Fang, Z., & Lu, C. (2023, November). Design of a 1-Bit Reconfigurable Low-Profile UWB Antenna Element. In *2023 IEEE MTT-S International Microwave Workshop Series on Advanced Materials and Processes for RF and THz Applications (IMWS-AMP)* (pp. 1-2). IEEE. 10.1109/IMWS-AMP57814.2023.10381461

Fang, Z., Yang, H., Gao, Y., Zhai, F., Wu, J. W., Cheng, Q., & Cui, T. J. (2023). Design of a 2-Bit Reconfigurable UWB Planar Antenna Array for Beam Scanning Application. *IEEE Open Journal of Antennas and Propagation*, *4*, 91–96. doi:10.1109/OJAP.2023.3234541

Fathy, A. E., Rosen, A., Owen, H. S., McGinty, F., McGee, D. J., Taylor, G. C., Amenta, R., Swain, P. K., Perlow, S. M., & ElSherbiny, M. (2003). Silicon-based reconfigurable antennas-concepts, analysis, implementation, and feasibility. *IEEE Transactions on Microwave Theory and Techniques*, *51*(6), 1650–1661. doi:10.1109/TMTT.2003.812559

Frist, Y., Elhabchi, M., & Srifi, M. N. (2023, October). A Novel Frequency Reconfigurable Antenna Based on Metamaterial for Millimeter-Wave 5G Applications. In *2023 10th International Conference on Wireless Networks and Mobile Communications (WINCOM)* (pp. 1-5). IEEE.

Galvan, L. A. H. (2023). *Simple Mechanically Reconfigurable Patch Antennas* [Doctoral dissertation]. The University of Texas Rio Grande Valley.

Gençoğlan, D. N., Çolak, Ş., & Palandöken, M. (2023). Spiral-resonator-based frequency reconfigurable antenna design for sub-6 GHz applications. *Applied Sciences (Basel, Switzerland)*, *13*(15), 8719. doi:10.3390/app13158719

Gençoğlan, D. N., Palandöken, M., & Çolak, Ş. (2023). Novel Frequency-Reconfigurable Antennas with Ring Resonators and RF Switches: Enhancing Versatility and Adaptability in Wireless Communication Systems. *Applied Sciences (Basel, Switzerland)*, *13*(18), 10237. doi:10.3390/app131810237

Ghimire, J., Khan, D., & Choi, D. Y. (2023). Microstrip to Slot-Line-Fed Microstrip Patch Antenna with Radiation Pattern Diversity for X-Band Application. *Electronics (Basel)*, *12*(17), 3672. doi:10.3390/electronics12173672

Gök, Ç., & Alatan, L. (2023, March). A Compact Pattern Reconfigurable Antenna with 4 Different Radiation Modes. In *2023 17th European Conference on Antennas and Propagation (EuCAP)* (pp. 1-5). IEEE.

Gross, F. B. (2015). *Smart antennas with MATLAB*. McGraw-Hill Education.

Haider, N., Caratelli, D., & Yarovoy, A. (2013). Recent developments in reconfigurable and multiband antenna technology. *International Journal of Antennas and Propagation*, *2013*, 1–14. doi:10.1155/2013/869170

Haider, N., Caratelli, D., & Yarovoy, A. G. (2013). Recent developments in reconfigurable and multiband antenna technology. *International Journal of Antennas and Propagation*, *2013*, 1–14. doi:10.1155/2013/869170

HarzingA. W. (2007). *Publish or Perish*. Available from http://www.harzing.com/pop.htm

Hasan, L., Nassr, M., Anbar, M., & Alasadi, H. A. A. (2023). Inverted U-shaped Frequency Reconfigurable Microstrip patch antenna for 5G communication systems. *Óptica Pura y Aplicada*, *56*(3).

Haupt, R. L., & Lanagan, M. (2013). Reconfigurable antennas. *IEEE Antennas & Propagation Magazine*, *55*(1), 49–61. doi:10.1109/MAP.2013.6474484

Hernandez Galvan, L. A. (2023). Simple Mechanically Reconfigurable Patch Antennas. *IEEE Antennas and Wireless Propagation Letters*, *17*(7), 1186–1189.

Hossain, M. A., Rahman, M. A., Murshed, A. H., Nishiyama, E., & Toyoda, I. (2023). A simple design and fabrication of polarization reconfigurable antenna for industrial scientific and medical-band applications. International. *Journal of Electrical and Computer Engineering*, *13*(2), 1580.

Huang, R., Luo, Y., Zhao, S., Yan, N., An, W., & Ma, K. (2023, August). A Compact Anti-Interference Pattern Reconfigurable Quasi-Yagi Antenna. In *2023 International Applied Computational Electromagnetics Society Symposium (ACES-China)* (pp. 1-3). IEEE. 10.23919/ACES-China60289.2023.10250083

Huff, G. H., Feng, J., Zhang, S., & Bernhard, J. T. (2003). A novel radiation pattern and frequency reconfigurable single turn square spiral microstrip antenna. *IEEE Microwave and Wireless Components Letters*, *13*(2), 57–59. doi:10.1109/LMWC.2003.808714

Hum, S. V., & Perruisseau-Carrier, J. (2013). Reconfigurable reflect arrays and array lenses for dynamic antenna beam control: A review. *IEEE Transactions on Antennas and Propagation*, *62*(1), 183–198. doi:10.1109/TAP.2013.2287296

Hussain, M., Awan, W. A., Alzaidi, M. S., Hussain, N., Ali, E. M., & Falcone, F. (2023). Metamaterials and their application in the performance enhancement of reconfigurable antennas: A review. *Micromachines*, *14*(2), 349. doi:10.3390/mi14020349 PMID:36838049

Iqbal, A., Smida, A., Abdulrazak, L. F., Saraereh, O. A., Mallat, N. K., Elfergani, I., & Kim, S. (2019). Low-profile frequency reconfigurable antenna for heterogeneous wireless systems. *Electronics (Basel)*, *8*(9), 976. doi:10.3390/electronics8090976

Ismail, M. M., Elwi, T. A., & Salim, A. J. (2023). Reconfigurable composite right/left-handed transmission line antenna-based Hilbert/Minkowski stepped impedance resonator for wireless applications. *Radioelectronic and Computer Systems*, (1), 77–91. doi:10.32620/reks.2023.1.06

Jabbar, A., Jamshed, M. A., Abbasi, Q., Imran, M. A., & Ur-Rehman, M. (2023). Leveraging the Role of Dynamic Reconfigurable Antennas in Viewpoint of Industry 4.0 and Beyond. *Research, 6*, 0110.

Janagouda, M., Badagavi, P., & Iyer, S. (2023, October). Comparative Study of Origami Enabled Helical and Curlicue Reconfigurable Antennas. In *2023 IEEE International Conference on Distributed Computing, VLSI, Electrical Circuits and Robotics (DISCOVER)* (pp. 269-274). IEEE. 10.1109/DISCOVER58830.2023.10316702

Kamran Shereen, M., Khattak, M. I., Basit, A., & Ahmad, G. (2023). Intelligent Frequency, Radiation Pattern and Polarization Reconfigurable Antenna for 5G Applications. *Journal of Electrical Engineering & Technology*, 1–15. doi:10.1007/s42835-023-01718-2

Kandasamy, K., Majumder, B., Mukherjee, J., & Ray, K. P. (2015). Low-RCS and polarization-reconfigurable antenna using cross-slot-based metasurface. *IEEE Antennas and Wireless Propagation Letters*, *14*, 1638–1641. doi:10.1109/LAWP.2015.2415585

Kang, M. J., Kim, T. H., Choi, Y. S., & Lee, W. S. (2023, July). Low-Profile Beam-Reconfigurable Antenna Using Multiple Parasitic Elements for UAV Applications. In *2023 IEEE International Symposium on Antennas and Propagation and USNC-URSI Radio Science Meeting (USNC-URSI)* (pp. 621-622). IEEE. 10.1109/USNC-URSI52151.2023.10238150

Karra, J. M., & Eroglu, A. (2023, July). Performance Improvement of Wideband Reconfigurable Planar Antennas. In *2023 IEEE International Symposium on Antennas and Propagation and USNC-URSI Radio Science Meeting (USNC-URSI)* (pp. 1075-1076). IEEE. 10.1109/USNC-URSI52151.2023.10237838

Karthika, K., Anusha, K., Kavitha, K., & Geetha, D. M. (2024). Optimization Algorithms for Reconfigurable Antenna Design: A Review. *Advances in Microwave Engineering*, 85-103.

Karthika, K., & Kavitha, K. (2023). Design and Development of Parasitic Elements Loaded Quadband Frequency and Pattern Reconfigurable Antenna. *International Journal of RF and Microwave Computer-Aided Engineering*, *2023*, 2023. doi:10.1155/2023/4034241

Karthikeyan, M., Pradeep, J., Harikrishnan, M., Sitharthan, R., Mishra, N., Rajesh, M., & Sivanesan, T. (2023). Recent developments in flexible and printed reconfigurable antennas for medical and Internet of things applications. *Advances in Flexible and Printed Electronics: Materials, fabrication, and applications*, 9-1.

Karuppanan, S., Vinothini, V. R., & Chenguttuvan, E. (2023). Reconfigured antenna with switchable polarization for S Band Wireless applications. *El-Cezeri Journal of Science and Engineering*, *10*(2), 231–239. doi:10.31202/ecjse.1137882

Khidre, A., Lee, K. F., Yang, F., & Elsherbeni, A. Z. (2012). Circular polarization reconfigurable wideband E-shaped patch antenna for wireless applications. *IEEE Transactions on Antennas and Propagation*, *61*(2), 960–964. doi:10.1109/TAP.2012.2223436

Kounta, C. A. K. A., Arnaud, L., Kamsu-Foguem, B., & Tangara, F. (2022). Review of ai-based methods for chatter detection in machining based on bibliometric analysis. *International Journal of Advanced Manufacturing Technology*, *122*(5-6), 2161–2186. doi:10.1007/s00170-022-10059-9

Kumar, P. R., Sunitha, P., & Prasad, M. V. (2023). Compact Reconfigurable Patch Antenna for Wireless Applications. *Progress in Electromagnetics Research, C*, 138, 161–174.

Kumar, S., Rayavarapu, N., & Naidu, P. V. (2023). A novel frequency reconfigurable antenna for smart grid applications in TV white space band. *Iranian Journal of Electrical and Computer Engineering*, *13*(1), 611–618. doi:10.11591/ijece.v13i1.pp611-618

Lin, T. C., & Hung, C. M. (2023). Reconfigurable RF front end and antenna arrays for radar mode switching (U.S. Patent No. 11,835,645). *United States Patent and Trademark Office*.

Lin, W., Wong, H., & Ziolkowski, R. W. (2017). Wideband pattern-reconfigurable antenna with switchable broadside and conical beams. *IEEE Antennas and Wireless Propagation Letters*, *16*, 2638–2641. doi:10.1109/LAWP.2017.2738101

Liu, D. (2023, June). Design of reconfigurable antenna based on Si/Ge/Si Heterogeneous SPiN diodes. In *Sixth International Conference on Intelligent Computing, Communication, and Devices (ICCD 2023)* (Vol. 12703, pp. 486-493). SPIE. 10.1117/12.2682890

Ma, X., Han, J., Li, G., Chang, M., Tian, Y., Chen, S., Liu, H., Li, L., & Cui, T. J. (2023). A Wideband 1-Bit Reconfigurable Electromagnetic Surface for Monopulse Radar Applications. *IEEE Transactions on Antennas and Propagation*, *71*(6), 5475–5480. doi:10.1109/TAP.2023.3263204

Mackertich-Sengerdy, G., Campbell, S. D., Werner, P. L., & Werner, D. H. (2023, March). Ruggedized Reconfigurable Antennas through the Implementation of Innovative Mechanical Techniques. In *2023 International Applied Computational Electromagnetics Society Symposium (ACES)* (pp. 1-2). IEEE. 10.23919/ACES57841.2023.10114782

Mandal, P., Kashyap, C. S., Roy, L. P., & Das, S. K. (2023, January). Reconfigurable Radar Antenna Design for UAV Application. In *2023 International Conference for Advancement in Technology (ICONAT)* (pp. 1-6). IEEE. 10.1109/ICONAT57137.2023.10080058

Manikandan, T., Gopinath, P., & Aravindh, K. (2023, April). Design and Parametric Analysis of Reconfigurable Antenna. In *2023 International Conference on Inventive Computation Technologies (ICICT)* (pp. 1668-1674). IEEE. 10.1109/ICICT57646.2023.10134170

Milias, C., Andersen, R. B., Jørgensen, T. H., Lazaridis, P. I., Zaharis, Z. D., Muhammad, B., & Hermansen, D. D. (2023, May). Metasurface-based, Pattern-reconfigurable, Wide-angle Scanning Antenna Array for UAV-borne Radar. In *2023 IEEE Radar Conference (RadarConf23)* (pp. 1-6). IEEE. 10.1109/RadarConf2351548.2023.10149666

Motovilova, E., & Huang, S. Y. (2020). A review on reconfigurable liquid dielectric antennas. *Materials (Basel)*, *13*(8), 1863. doi:10.3390/ma13081863 PMID:32316173

Mu, Y., Han, J., Xia, D., Liu, H., & Li, L. (2024). A Self-Decoupling Wide-Angle Scanning Pattern Reconfigurable Antenna Array Using Parasitic Comb Structures. *IEEE Transactions on Antennas and Propagation*, *72*(2), 1212–1223. doi:10.1109/TAP.2023.3347869

Nataraj, B., Prakash, A. G., Prabha, K. R., Jayavarshan, M., & Kanna, K. G. (2023, September). Compact Reconfigurable Antenna for 5G Bands. In *2023 4th International Conference on Smart Electronics and Communication (ICOSEC)* (pp. 327-333). IEEE.

Navarajan, J., Jebarani, M. E., & Krishnan, V. G. (2023). Design of frequency reconfigurable antenna based on μC-μEMS switch. *AEÜ. International Journal of Electronics and Communications*, *171*, 154911. doi:10.1016/j.aeue.2023.154911

Ndimubandi, S. D. J., Ding, X., & Zeng, Y. (2023). A Wide-Angle Scanning Planar Array Based on Four-Mode Pattern-Reconfigurable Elements. *International Journal of RF and Microwave Computer-Aided Engineering*, *2023*, 2023. doi:10.1155/2023/7871837

Nikolaou, S., Bairavasubramanian, R., Lugo, C., Carrasquillo, I., Thompson, D. C., Ponchak, G. E., Papapolymerou, J., & Tentzeris, M. M. (2006). Pattern and frequency reconfigurable annular slot antenna using PIN diodes. *IEEE Transactions on Antennas and Propagation*, *54*(2), 439–448. doi:10.1109/TAP.2005.863398

Ojaroudi Parchin, N., Jahanbakhsh Basherlou, H., Al-Yasir, Y. I., Abd-Alhameed, R. A., Abdulkhaleq, A. M., & Noras, J. M. (2019). Recent developments of reconfigurable antennas for current and future wireless communication systems. *Electronics (Basel)*, *8*(2), 128. doi:10.3390/electronics8020128

Ojaroudi Parchin, N., Jahanbakhsh Basherlou, H., Al-Yasir, Y. I., & Abdulkhaleq, M., A., & A. Abd-Alhameed, R. (. (2020). Reconfigurable antennas: Switching techniques—A survey. *Electronics (Basel)*, *9*(2), 336. doi:10.3390/electronics9020336

Olan-Nuñez, K. N., Morales-Lovera, H. N., & Murphy-Arteaga, R. S. (2023, July). Mechanically Reconfigurable Multi-Band Antenna on a 3D-Printed Substrate for C/X Band Applications. In *2023 IEEE International Symposium on Antennas and Propagation and USNC-URSI Radio Science Meeting (USNC-URSI)* (pp. 1073-1074). IEEE. 10.1109/USNC-URSI52151.2023.10238192

Pandhare, R. A., Abegaonkar, M. P., & Dhote, C. (2023). A Reconfigurable Octa-Band Antenna with a High Gain for a Variety of Wireless Applications. *Journal of the Institution of Electronics and Telecommunication Engineers*, 1–11. doi:10.1080/03772063.2023.2176370

Pandhare, R. A., Abegaonkar, M. P., & Dhote, C. (2023). High gain wideband and multi-band on-demand reconfigurable antenna for modern wireless application. *International Journal of Microwave and Wireless Technologies*, *15*(4), 649–664. doi:10.1017/S1759078722000630

Pandhare, R. A., Abegaonkar, M. P., & Dhote, C. (2023). High Gain Compact Dual Band Reconfigurable Antenna Using Multilayer FSS for WLAN and Radar Applications. *Progress in Electromagnetics Research*, *C*, 130, 227–240.

Patel, S. K., Lavadiya, S. P., Parmar, J., Das, S., Ahmed, K., & Taya, S. A. (2023). Low-cost, compact, and reconfigurable antennas using complementary split-ring resonator metasurface for next-generation communication systems. *International Journal of Microwave and Wireless Technologies*, *15*(5), 860–870. doi:10.1017/S175907872200099X

Peroulis, D., Sarabandi, K., & Katehi, L. P. (2005). Design of reconfigurable slot antennas. *IEEE Transactions on Antennas and Propagation*, *53*(2), 645–654. doi:10.1109/TAP.2004.841339

Pokorny, M., Vertat, I., Panek, D., Hazdra, P., Svanda, M., Kracek, J., & Karban, P. (2024). Reconfigurable Antenna Array Testbed for Quantized Controlling. *IEEE Access : Practical Innovations, Open Solutions*, *12*, 11458–11469. doi:10.1109/ACCESS.2024.3356358

Raghavendra, C., Naveen, K., Pradeep, V., Reddy, K. V., & Manikiran, V. J. (2023). Investigation and Implementation of Inverted U-slot Reconfigurable Microstrip Patch for Monopulse Radar Applications. *Telecommunications and Radio Engineering*, *82*(6).

Raut, H. D., Shevada, L. K., Malekar, R. R., Dixit, A. S., & Kumar, S. (2021). Metamaterials in 5G antenna designs: A bibliometric survey. *Library Philosophy and Practice, 5401.*

Refaat, S. A., Mohamed, H. A., Abdelhady, A. M., & Mohra, A. S. (2023). A 28/38 GHz tuned reconfigurable antenna for 5G mobile communications. *Indonesian Journal of Electrical Engineering and Computer Science, 31*(1), 248. doi:10.11591/ijeecs.v31.i1.pp248-258

Row, J. S., Liu, W. L., & Chen, T. R. (2012). Circular polarization and polarization reconfigurable designs for annular slot antennas. *IEEE Transactions on Antennas and Propagation, 60*(12), 5998–6002. doi:10.1109/TAP.2012.2211556

Saraswat, R. K., & Kumar, M. (2023). Implementation of the Metamaterial Multiband Frequency Reconfigurable Antenna for IoT Wireless Standards. *Journal of the Institution of Electronics and Telecommunication Engineers*, 1–12. doi:10.1080/03772063.2023.2234863

Sathiaraj, S., Ravichandran, M., Balasubramaniyan, B., & Sivaraman, D. (2023, August). Versatile analysis of reconfigurable antenna arrays for MIMO applications. In AIP Conference Proceedings (Vol. 2790, No. 1). AIP Publishing. doi:10.1063/5.0152810

SathuluriM. R.SasikalaG. (2023). Electrostatic and Electromagnetic Parametric Analysis of RF MEMS Switches for Reconfigurable Antenna. Research Square. doi:10.21203/rs.3.rs-3323038/v1

Seid, A. A. (2023). *Performance Analysis of Software-Defined Radio Based Reconfigurable Antenna Array in The Millimeter Wave Wireless Communications* [Doctoral dissertation]. Haramaya University.

Selvam, Y. P., Sangamithira, N. M., Karuppasamy, T., Puvvadi, M., & Jayapalan, A. (2023, January). Optically Reconfigurable Slotted Waveguide Antenna Array for 5G Applications. In *2023 International Conference on Computer Communication and Informatics (ICCCI)* (pp. 1-4). IEEE. 10.1109/ICCCI56745.2023.10128601

Shankar, S., & Upadhyay, D. K. (2023). A compact maple leaf and triangular hybrid fractal shaped UWB frequency reconfigurable antenna with wide continuous tunable band notch. *Iranian Journal of Science and Technology. Transaction of Electrical Engineering, 47*(1), 301–316. doi:10.1007/s40998-022-00536-y

Sharma, A., Rathi, A., Al-Khafaji, H. M. R., Siddiqui, M. G., & Yadav, A. K. (2023). Hybrid mode reconfigurable antenna with V-shaped extrudes for cognitive radio applications. *Electronics (Basel), 12*(7), 1680. doi:10.3390/electronics12071680

Sharma, B. L., Mathur, D., & Sharma, M. K. (2023). Design of miniaturised circular polarisation-reconfigurable and frequency agile slow-wave SIW based microstrip antenna for wireless applications. *International Journal of Electronics*, 1–23. doi: 10.1080/00207217.2023.2192973

Shaw, M., & Choukiker, Y. K. (2023). Reconfigurable polarization and pattern microstrip antenna for IRNSS band applications. *AEÜ. International Journal of Electronics and Communications*, *160*, 154501. doi:10.1016/j.aeue.2022.154501

Shevada, L. K., Raut, H. D., Malekar, R. R., Dixit, A. S., & Kumar, S. (2021). A bibliometric survey on ultra-wideband multiple input multiple output antenna with improved isolation. *Library Philosophy and Practice*, 1-20.

Sievenpiper, D. F., Schmitz, A. E., Schaffner, J. H., Tangonan, G. L., Hsu, T. Y., Loo, R. Y., & Miles, R. S. (2003). *U.S. Patent No. 6,670,921*. Washington, DC: U.S. Patent and Trademark Office.

Sri Pravallika, N., Anitha, V. R., & Rama Naidu, K. (2023). X-band Miniaturized frequency reconfigurable travelling wave antenna. *International Journal of Electronics Letters*, *11*(2), 221–231. doi:10.1080/21681724.2022.2068194

Subashini, V., & Chitra, S. (2024). A Compact Honeycomb-Structured Reconfigurable Antenna with Coplanar Waveguide Feed for Multiband Wireless Applications. *Telecommunications and Radio Engineering*, *83*(2).

Subbaraj, S., & Thomas, S. B. (2023). Reconfigurable antennas and their practical applications-a review. *Radio Science*, *58*(9), 1–13. doi:10.1029/2023RS007656

Suman, P. N., Chandan, R. K., & Mishra, G. K. (2023). A Study of Different 5G Reconfigurable Antennas for Communication: Futuristic Glimpse for 6G. In Handbook of Research on Technological Advances of Library and Information Science in Industry 5.0 (pp. 178-195). IGI Global.

Tandel, T., & Trapasiya, S. (2023). Reconfigurable Antenna for Wireless Communication: Recent Developments, Challenges and Future. *Wireless Personal Communications*, *133*(2), 725–768. doi:10.1007/s11277-023-10785-7

Tathare, S. S., & Goswami, P. (2023). Squirrel Search Optimized FSS Based Filtered Frequency Reconfigurable Antenna for 5G Applications (Sub-6 GHz). *Wireless Personal Communications*, *131*(4), 2811–2839. doi:10.1007/s11277-023-10581-3

Tawk, Y., Costantine, J., Avery, K., & Christodoulou, C. G. (2011). Implementation of a cognitive radio front-end using rotatable controlled reconfigurable antennas. *IEEE Transactions on Antennas and Propagation*, *59*(5), 1773–1778. doi:10.1109/TAP.2011.2122239

Tawk, Y., Costantine, J., Hemmady, S., Balakrishnan, G., Avery, K., & Christodoulou, C. G. (2011). Demonstration of a cognitive radio front end using an optically pumped reconfigurable antenna system (OPRAS). *IEEE Transactions on Antennas and Propagation*, *60*(2), 1075–1083. doi:10.1109/TAP.2011.2173139

Tiwari, A., Yadav, D., Sharma, P., & Yadav, M. V. (2023). Design of Wide Notched-Band Circular Monopole Ultra-Wideband Reconfigurable Antenna Using PIN Diodes Switches. *Progress in Electromagnetics Research C*, *139*, 107–118.

Tran, H. H., Kim-Thi, P., Nguyen, T. T. L., & Thao, H. T. P. (2023). A flexible and bandwidth reconfigurable antenna for cognitive communication. *IEEE Access : Practical Innovations, Open Solutions*, *11*, 89561–89567. doi:10.1109/ACCESS.2023.3306451

Tripathi, S. K., & Tiwari, B. B. (2023, March). Design of a Reconfigurable Slot Antenna for WiMAX, C-band, and X-band Applications. In *2023 First International Conference on Microwave, Antenna and Communication (MAC)* (pp. 1-4). IEEE. 10.1109/MAC58191.2023.10189374

Tripathi, S. K., & Tiwari, B. B. (2024). Design and analysis of multiband reconfigurable slot antenna for S, C, and X-band applications. *Journal of Integrated Science and Technology*, *12*(3), 766–766. doi:10.62110/sciencein.jist.2024.v12.766

Uchendu, I., & Kelly, J. R. (2016). Survey of beam steering techniques available for millimeter wave applications. *Progress in Electromagnetics Research B. Pier B*, *68*, 35–54. doi:10.2528/PIERB16030703

Umamaheswari, G., Praveena, A., & Vignesh, S. B. (2023). Reconfigurable Antennas for Internet of Things (IoT) Devices. In *Internet of Things Enabled Antennas for Biomedical Devices and Systems: Impact, Challenges and Applications* (pp. 113–128). Springer Nature Singapore. doi:10.1007/978-981-99-0212-5_10

Varshney, A., Neebha, T. M., Sharma, V., & Andrushia, A. D. (2023). Low-Cost L-Band to Ku-Band Frequency Reconfigurable BAR64-02V Controlled Antenna for Satellite, Military, and Radar Applications. *Journal of the Institution of Electronics and Telecommunication Engineers*, 1–14. doi:10.1080/03772063.2023.2260343

Verma, K. N., & Asthana, R. (2023). Performance Evaluation and Analysis of Frequency Reconfigurable Dielectric Resonator Antenna: A Review. *International Research Journal of Engineering and Technology*, *10*(04), 374–379.

Vinayagam, K., & Rajesh, N. (2023, May). Design of Reconfigurable Patch Antenna with Polarization Agility. In *2023 2nd International Conference on Vision Towards Emerging Trends in Communication and Networking Technologies (ViTECoN)* (pp. 1-4). IEEE. 10.1109/ViTECoN58111.2023.10157248

Wang, P., Jia, Y., Liu, Z., & Liu, Y. (2023, May). Low-RCS Banyan Tree Antenna Array with Reconfigurable Scattering Patterns. In *2023 International Conference on Microwave and Millimeter Wave Technology (ICMMT)* (pp. 1-3). IEEE. 10.1109/ICMMT58241.2023.10277663

Wang, Y., Guo, Z., Li, C., & Yan, X. (2023, June). Design of graphene frequency reconfigurable antenna in terahertz band. In *2023 5th International Conference on Electronic Engineering and Informatics (EEI)* (pp. 521-525). IEEE. 10.1109/EEI59236.2023.10212546

Wang, Z., & Dong, Y. (2023, May). Low-Cost Metamaterial-Based Reconfigurable Antenna with Even-Odd Mode Pattern Control Method. In *2023 International Conference on Microwave and Millimeter Wave Technology (ICMMT)* (pp. 1-3). IEEE. 10.1109/ICMMT58241.2023.10276552

Wang, Z., & Dong, Y. (2024). Low-Cost 2-Bit Phase and 1-Bit Amplitude Reconfigurable Antenna for Beam Steering and Shaping. *IEEE Transactions on Antennas and Propagation*, *72*(3), 1–12. doi:10.1109/TAP.2023.3348868

Weedon, W. H., Payne, W. J., & Rebeiz, G. M. (2001, July). MEMS-switched reconfigurable antennas. In *IEEE Antennas and Propagation Society International Symposium. 2001 Digest. Held in conjunction with: USNC/URSI National Radio Science Meeting (Cat. No. 01CH37229)* (Vol. 3, pp. 654-657). IEEE. 10.1109/APS.2001.960181

Xia, Y., Yuan, M., Dobrea, A., Li, C., Heidari, H., Mottram, N., & Ghannam, R. (2023). Reconfigurable wearable antenna for 5G applications using nematic liquid crystals. *Nano Select*, *4*(8), 513–524. doi:10.1002/nano.202200209

Xiao, H., & Hao, S. (2021). Public participation in infrastructure projects: An integrative review and prospects for the future research. *Engineering, Construction, and Architectural Management*, *30*(2), 456–477. doi:10.1108/ECAM-06-2021-0495

Yan, L., Zou, W., Zheng, K., Yang, G., & Luo, Y. (2024). A Dual-Polarized CTS Array Antenna with Four Reconfigurable Beams for mm-Wave Wind Profile Radar. *Electronics (Basel)*, *13*(1), 238. doi:10.3390/electronics13010238

Yang, S., Zhang, C., Pan, H. K., Fathy, A. E., & Nair, V. K. (2009). Frequency-reconfigurable antennas for multiradio wireless platforms. *IEEE Microwave Magazine*, *10*(1), 66–83. doi:10.1109/MMM.2008.930677

Yang, S. L., & Luk, K. M. (2006). Design of a wide-band L-probe patch antenna for pattern reconfiguration or diversity applications. *IEEE Transactions on Antennas and Propagation*, *54*(2), 433–438. doi:10.1109/TAP.2005.863376

Yang, X. S., Wang, B. Z., Wu, W., & Xiao, S. (2007). Yagi patch antenna with dual-band and pattern reconfigurable characteristics. *IEEE Antennas and Wireless Propagation Letters*, *6*, 168–171. doi:10.1109/LAWP.2007.895292

Yu, F., & Hayes, B. E. (2018). Applying data analytics and visualization to assessing the research impact of the cancer cell biology (ccb) program at the university of north carolina at chapel hill. *Journal of Escience Librarianship*, *7*(1), e1123. doi:10.7191/jeslib.2018.1123

Zhao, W., & Wang, Y. (2023). A Reconfigurable Antenna with One Slit and A Small Clearance for Nona-Band Metal-Bezel Mobile Phones. *IEEE Transactions on Antennas and Propagation*, *71*(9), 7172–7183. doi:10.1109/TAP.2023.3291496

Zhu, Y., Yang, W., Chen, K., Zhao, J., & Feng, Y. (2023, November). A Low-Scattering Microwave Antenna with Reconfigurable Frequency Band. In *2023 IEEE MTT-S International Microwave Workshop Series on Advanced Materials and Processes for RF and THz Applications (IMWS-AMP)* (pp. 1-3). IEEE. 10.1109/IMWS-AMP57814.2023.10381336

Chapter 4
Reconfigurable Antenna System for 5G/6G

Smrity Dwivedi
IIT BHU, India

ABSTRACT

The timely upgradation in 5G technology standards has already been released by third generation partnership project, which enabled the researchers to refine the research to improve the communication and support in the development. After development of fifth generation technologies, researchers, scientists, and engineers are looking for wide bandwidth which should improve wireless systems and devices to provide better services and experiences. Because of that, beyond 5G, a sixth generation is the next step to be widely used in early 2028 or intensively used around 2030 as per demand. Parameters like gain, bandwidth, efficiency, size of the antenna, technique of fabrication, and other useful quantities can be modified and maximized, and also the cost of fabricating such antennas will be reduced and developed.

INTRODUCTION

As growing demand of wireless communications as well as use of the integration of many number of wireless protocol into a single place, it is seriously important that all the polarizations of antennas can be re change for the use. Antennas which is reconfigurable can change their working frequency, all bandwidth including other parameters as per the demand of applications and requirement of systems. These antennas radiates lots of plots at different frequencies and polarizations simultaneously. According to as per today's requirement, design of such antennas which should be cost effective, giving all properties with easy integration like frequency, radiation

DOI: 10.4018/979-8-3693-0916-2.ch004

pattern, polarization is a critical thing for design engineers. As moving from fifth generation to sixth generation, these procedure and derivations are good and nice to design these antennas for the lots of applications. Designing of any simple antenna and converting it a changing device by using other techniques to change structure by using physical and mathematical methods has been typical and challenging task. There are so many values as to be used such as getting all parameters throughout all the in fixed environment (Brown. 1998), (Erdil et al. 2007), (Grau, 2010). After that, based on these reviews there are several methods to use to find out changing properties and can be defined for these antennas that are classified as: (i) Electrical; (ii) Optical; (iii) Physical; (iii) antennas with smart materials such as graphene and gold, etc. Apart from that there are other techniques which is useful to defined the antenna characteristics as by using some semiconductor products like microwave switches, some high frequency diodes (Nikolaou et al. 2009). These diodes is widely used switch as it has acceptable performance and a low price compared to that switches and others. 1. Different methods to achieve reconfigurability:

1.1. Diode Such as PIN

The electrical quantity of the this such as capacitance and resistance can play role to turn ON and OFF the diode for getting reconfiguration as forward and reverse bias (Quin et al. 2010). PIN diodes are mostly used diode which utilizes the exchange methods in many remote frameworks as per demand. For reconfigurable transmission and reception, an arrangement has been done such supply as DC and bias to work with it as to use it with radio frequency. There are one or two PIN diodes can be used with antenna design with any materials with designed frequency. As per various conditions of the PIN diodes, these two groups for the long term evolution including many frequencies for different ranges for applications. It requires a high speed, large current in the start and also taking care of this which should be taken care off. So, in any case, it is useful by giving decision for reconfiguration with minimal and solid effort get reconfiguration property as shown in figure 1.

Figure 1. PIN diode fabrication, equivalent circuit diagram (b), positive bias, and (c) negative bias (Clemente, 2012)

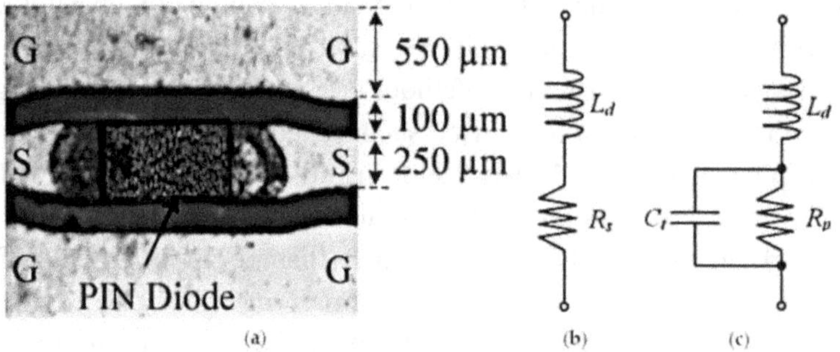

1.2 Diodes Such as Varactor

This is another diode which can be used with very high frequency can be easily embedded into design such as into patch or into feeding line to get reconfiguration property. It is voltage dependent diode where property can be changed by changing the voltage and due that capacitance can be changed which gives various properties. As by implementing the varactor diode into design to tune the frequency and easily frequency reconfiguration can be achieved (Tawk et al. 2012). According to Figure 2b, for many level of voltage for different capacitance, where the given frequency of the design is being tuned in GHz, without any other and unusal interferences due to proposed antenna in the design. This diode is not linear as like any diode with a low range. It has requires a crucial bias circuit including electrical parameters. Hence, as compared with other all electronic component which is not passive, such as a first diode and new, it has being used with small current and flow tuning which is less required compared to first diode.

Figure 2. Antenna design (a), return loss parameter (b) (Tawk et al., 2012)

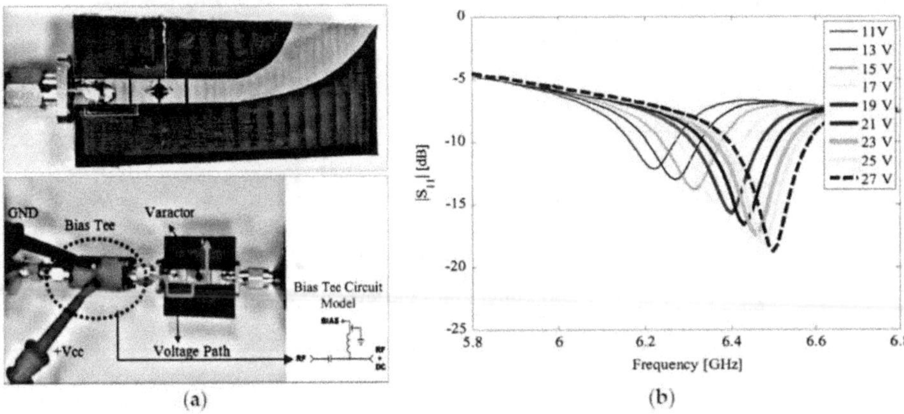

1.3 MEMS and Its Applications

MEMS with high frequency can be used to achieve reconfiguration properties in designed antenna. It is widely used switch but its design is complex as compared to first and second diodes. New switches are used in design to get the changing properties for given antenna. Although it is good but response id slow as compared to other methods as (Jaafar et al. 2013) (Kovitz et al. 2013).

Figure 3. MEMS implemented reconfigurable antenna (a) and big and clear view (b) (Kovitz et al., 2013)

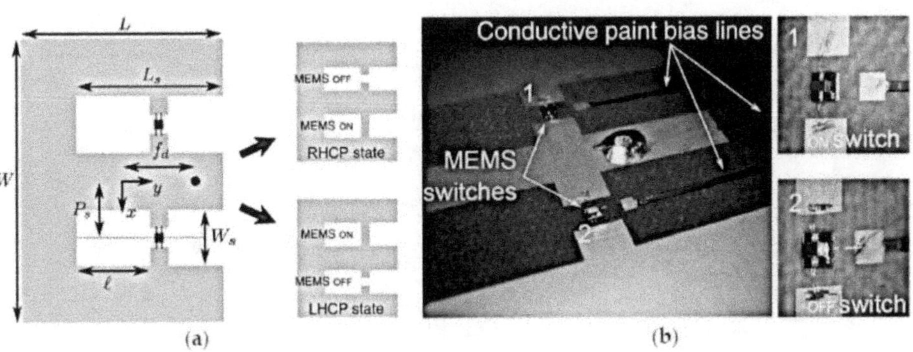

According to reviews and research work, a comparison table has been given below with proper explanation in Table 1. And (Cummings et al. 2013).

Figure 4. Reconfiguration technique advantages and disadvantages

Reconfiguration Technique	Advantages	Disadvantages
PIN-Diodes	• Very reliable • Extremely low-cost • Common choice for reconfiguration	• High tuning speed • High DC bias in ON-state • High power handling capacity
Varactors	• Small current flow • Continuous tuning • Ease of integration	• Nonlinear • Low dynamic range • Complex bias circuitry
RF MEMS	• High isolation and linearity • Wide impedance bandwidth • Low power losses and low noise figure	• High-control voltage • Slow switching speed • Limited life cycle

2. ANTENNAS DESIGN FOR 5G, BEYOND 5G, AND 6G SYSTEMS

2.1 Antenna Design for Fifth Generation Technologies

There are several techniques to get the properties of 5G antennas and these are easy to apply as fabrication is little bit tough. In many papers as per requirement of fifth generation, so many designs have been discussed. Here also many papers have been given for the help. These antennas which is having changing properties has designed and material characteristics has taken. As per high frequency operation, different materials can also be used or the design (H.-C. Huang, 2018), (W. Hong et al, 2017). (Sumit Kumar et al., 2020).

Figure 5. Specifications of 5G technology (Sumit Kumar et al., 2020)

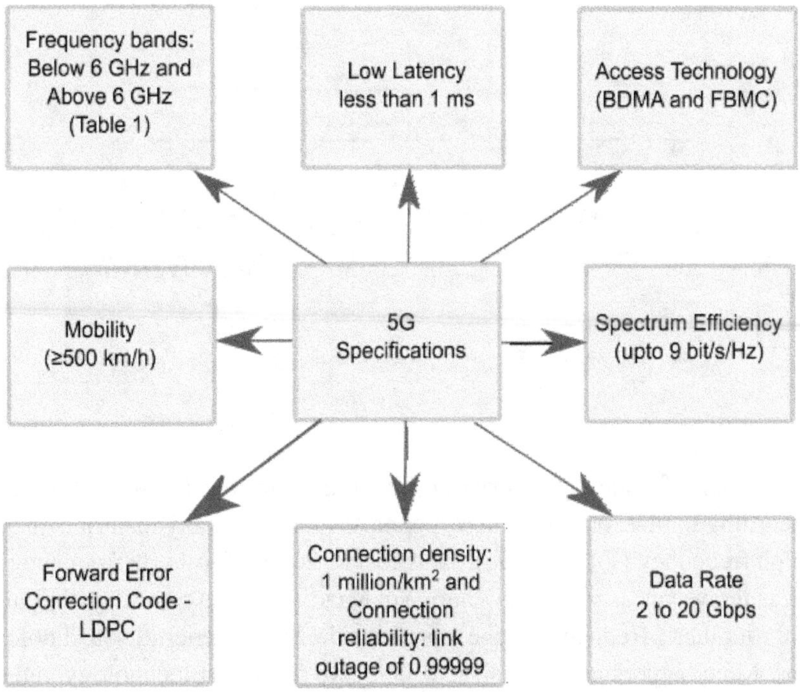

2.2 Antenna Design for Sixth Generation Technologies

proposed antenna with microstrip design and also having with high demand and their compact size as well as easy to fabricate are widely used as given (Sarthak Singhal, 2019), (Lin Guo et al., 2014). Many research institutes has done this things and published their items related to sixth generation systems, its main motive to take advantage and innovate design but during that era, it was not possible as mentioned (Samsung Research, 2020). After that many institutes and universities worked together including with industries and generates many products which is useful and giving better results (K. B. Letaief, 2019). As beyond 5G systems will design for terahertz frequency or the frequency which is in millimetre wave and that basically work and that ranges from 0.1 and 10 THz, and there after they divide into more part as subterahertz and pure terahertz having frequency 0.1–0.3 THz and 0.3–10 THz respectively as given in figure 5, (B. Duan, 2020), (L. Chi, et al., 2020), (M. K. Emara et al. 2020), (M. Kokkonen, 2020).

Figure 6. Range of frequency in beyond 5G and sixth generations (Fitri & Akbartx, 2017)

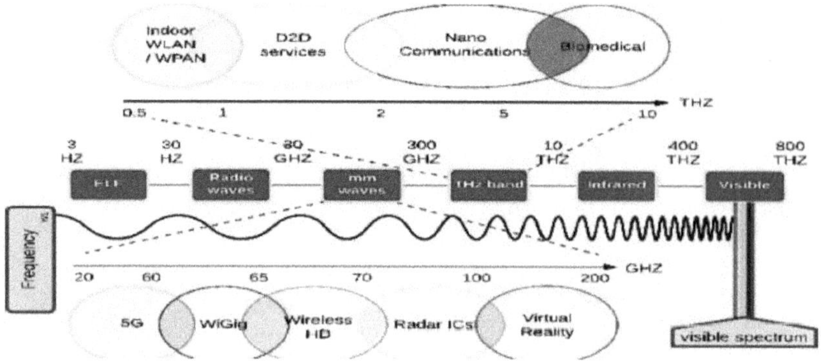

The terahertz frequency is very range having some gaps known as terahertz gap and according to that material can responds as there are many materials that can at such high frequency (I. Fitri and A. A. Akbartx, 2017). Terahertz frequency is also known as terawaves as given in (Zhuo-Wei Miao, 2018). According to figure 5, in which the terahertz frequency range represents the higher generation technology and systems. Many papers are published which give the smart technology and design for this frequency range and also already developed design for fifth generation systems which can be used for smart application and in intelligence. Now a days, in research area, different types of antenna are mentioned with different design and patterns which is also mentioned in this chapter for high frequency such as reflect array antenna given in (Zhuo-Wei Miao, 2018). In this design polymide has been taken as substrate material to achieve high speed and it can be used for medical field to check the inner working of body parts and bio-medical devices. There are few more design based on different shape has been mentioned for high gain and better directivity with wide bandwidth. They are having easy fabrication and radiate to get higher gain with less losses. Now, demand in change in frequency, radiation and position of electric field motivate the researchers to develop the new design which is known as reconfigurable system for antenna which can be developed using metamaterials, metasurfaces, gold, graphene and many mores just because of they can easily work with high frequency, even coppor and PEC both can be used for design of antenna for beyond 5G systems design. They are actually smart materials and can be fabricate easily but ofcourse with high frequency, there are so many precautions have taken. New antenna is also one option to achieve such high gain and wide bandwidth under this new generation as per increasing demand of users. There are several technologies have been developed with active components while designing the such high gain antenna with changing properties and also keep away from losses like isolation (I. Fitri and A. A. Akbartx, 2017), (Binbin Cheng, 2018), (Bellow Ghaida Muttashar Abdulsahib, 2023), (Xingsi Xue et al., 2023), (Raad Z. Homod et al., 2023), (Xingsi Xue et al., 2023), (S. K. Rana et al., 2023), (Oshamah Ibrahim Khalaf et al., 2023), (Shivani Chandra and Smrity Dwivedi, 2019), (Shivani Chandra and Smrity Dwivedi, 2021), (Smrity Dwivedi, 2022).

Therefore, for the future systems design for the antenna which can work with ground wave, sky wave as well as surface wave modes and can be fabricated in the micrometer range parameters, are now possible as per several materials like gold, graphene, polymide etc and many equipments that can be used with less ohmic losses and dielectric losses as base of design is same with ground, patch and substrate. There are high frequency network analyzers are available to measure all the important parameters. Also there are new technology developed whicg work for plasmon basically surface plasmon or spoof plasmon polaritron for the design of antennas (Shivani Chandra and Smrity Dwivedi, 2021), (Smrity Dwivedi, 2022), (Smrity Dwivedi, 2023).

3. DIFFERENT TECHNIQUES TO RECEIVE RECONFIGURATIONS

As there are several techniques to achieve reconfiguration in design of antenna with different methods and ways. Optical, electrical, material and physical reconfiguration are major process to change the properties of antennas. PIN diode, varactor diode and RF MEMS as well as smart materials can do these things very easily. As per discussion above about the PIN diode, varactor diode and RFMEMS these can be easily integrated with design and get the reconfiguration in frequency, radiation and polarization to use it in various applications. Hence, many applications like beam steering, control and command, directive beam etc. can be used with these antennas as shown in figure 6.

3.1 Electrically Reconfigurable Antenna

Under this techniques, various electronics switches have been used which is PIN diode, Varactor diode and RFMEMS. According to research (Dussopt et at. 2003), PIN diode is the best and easy to use as it gives high response and having low cost but Varactor diode although it can support high frequency compared to PIN, giving very high speed, but it fabrication is very costly and complicated. RFMEMS is kind of mechanical switch which require proper attention to fabricate and slow response as described.

Figure 7. Various techniques for reconfiguration

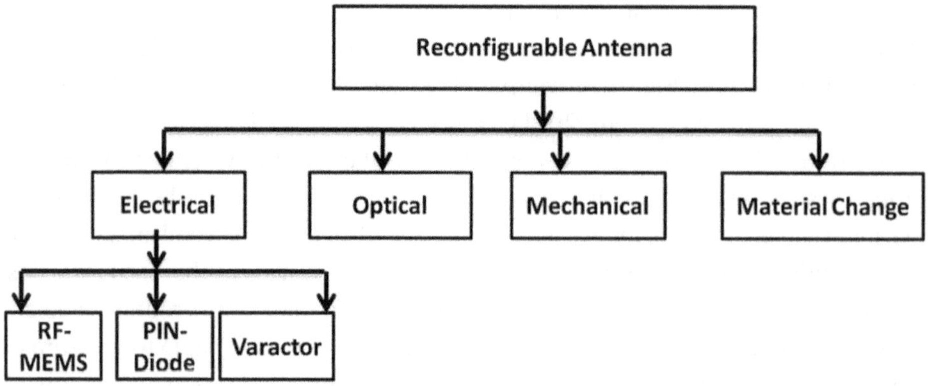

3.2 Optically Reconfigurable Antenna

It is more usable techniques as it does not require any bias line or any kind of wired connection to control DC potential and current associated with it. Here, it is being used as to change the radiating property of antenna by changing the radiation elements and also resonant frequency can be easily achieved. Actually, optical switches have used to get the property in antennas. Here, frequency range is high as compared to electrical techniques. Few materials like P3HT can be used to achieve reconfiguration (Rainville et al. 2003).

3.3 Mechanically/Physical Reconfigurable Antennas

Here, antennas properties can be changed physically by tuning it. Any probe or by any mean like stubs, irises that can implemented over it and by tuning it properties like inductance and capacitance can be changed to achieve reconfiguration as given in (Tan et al. 2013).

3.4 Smart Materials-Based Reconfigurable Antennas

Under this, there are several materials can be used to get the change in properties of antennas. There are several materials like metamaterials, metasurfaces, liquid crystal, graphene, gold etc. can be used to achieve this property. Depending on several properties like, chemical potential in graphene, negative permittivity and permeability in metamaterials and metasurfaces, domain formation as per diamagnetic properties in liquid crystal, reconfiguration is used (Huff et al. 2007).

4. REVIEW OF RECENTLY PUBLISHED PAPER

4.1 In the first paper review, Author has proposed for the design which is novel reconfigurable open slot antenna which is working in LTE systems for smart applications using wide bandwidth. Here, varactor diode is being used to achieve reconfiguration between the small spaces. There are three layer where on the top layer which open slot, diode placed. It is high coverage antenna and it is proposed as MIMO implementation. It is designed, fabricated and tested as well. Frequency range is high for this antenna. Thus, the proposed antenna is smart antenna for LTE smart phones and mobile devices (Rakhi Patil et al. 2022).

Figure 8. Design of for the frequency 0.6 GHz to 1.12 GHz, 1.04 GHz to 2.47 GHz, and 1.89 GHz to 2.70 GHz

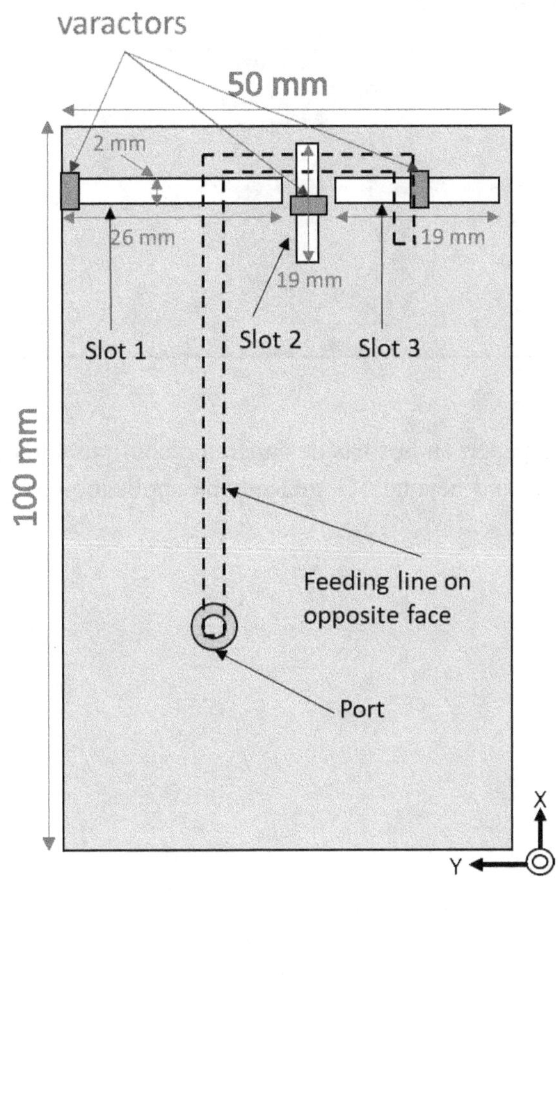

4.2 Under second paper review, again a very new microstrip circular patch antenna is proposed for reconfiguration. The antenna presented below in the figure is a wide bandwidth monopole frequency reconfigurable antenna and having switching between two resonating frequencies 24.9GHz and 25.3GHz due to the use of active diodes which are placed on one end to the feedline and the other end the PIN diodes are connected to the radiators through metal strip. RT Duroid is used as the substrate of this microstrip circular patch antenna. The monopole design helps to provide a wide bandwidth and the frequency reconfiguration which is achieved by the two PIN diodes as a electrical switches. The antenna structure given in the photo has been designed and simulated performed on HFSS software and results are obtained on the same (Urja Sudhir Ingle et al. 2022).

Figure 9. Proposed reconfigurable monopole micro strip patch antenna design (Urja Sudhir Ingle et al., 2022)

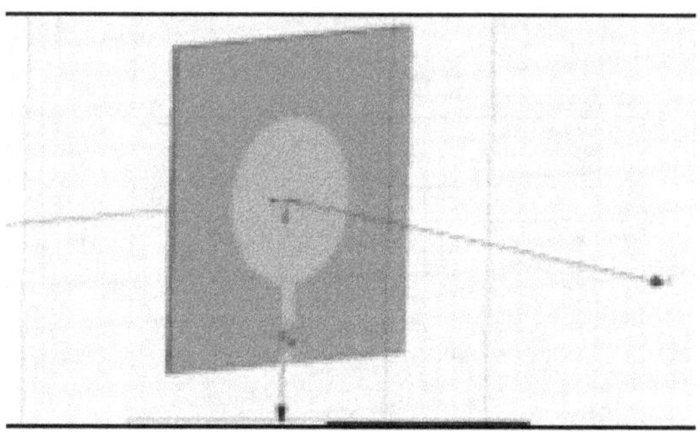

4.3 In the third paper, author has designed a reconfigurable three dimensional slot antenna for 4G and beyond 5G smartphone application as given in 9, (Peng Yang, 2020).

Figure 10. Structure implementation given in (a), (b)

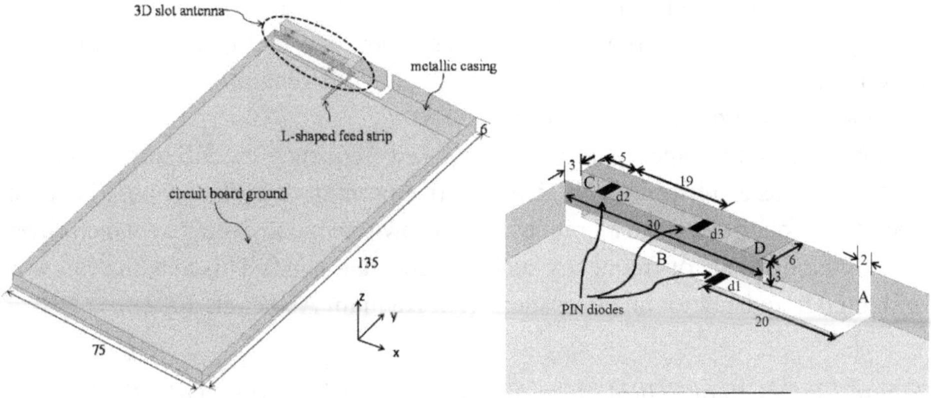

4.4 In the fourth paper, author has suggested the design antennas operating on multiple frequency bands due to the extensive use modern wireless technology and the demand for the multiple services in a single device as per requirement. The proposed design which is easy, can be changed and multiple frequency reconfigurable antenna uses a flexible roger RT duroid as a substrate with the miniature dimensions. The resonant frequency can be controlled by adjusting the mode of switches in connected or disconnected state. The proposed reconfigurable antenna is very much useful for different recent 5G\6G communication applications and this antenna is designed using CST tool. The simulated antenna has VSWR < 2 and a good radiation pattern (Dharavathu Krishna et al., 2022).

Figure 11. Fourth paper antenna design (top view) with yellow region represents metal and white represents substrate (Dharavathu Krishna et al., 2022)

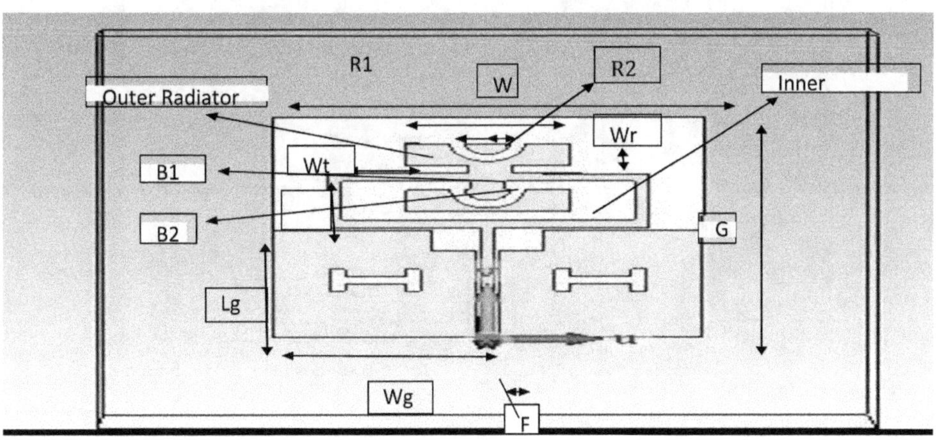

4.5 In the fifth paper, author proposes the FORMAT array, a reconfigurable millimeter-wave antenna array which is based on antenna tiles. This is giving 4.8 Gbps speed as per requirement of systems. There are multiplexing methods are used to achive this speed, which can work with fifth generation and beyond fifth generation (Eduardo V. P. Anjos et al. 2022).

4.6 In the sixth paper, author has proposed frequency reconfigurable antenna with Rogers as a substrate with given relative permittivity. The proposed antenna operates in four modes and all modes have its own frequency and gain which covers nine different bands of frequency such as three dual bands This antenna can work with high frequency for all generations (Ali Kadhum Abd et al. 2022).

Figure 12. Design of FORMAT array antenna on PCB (Anjos et al., 2022)

Figure 13. The suggested antenna geometry as given above according to paper

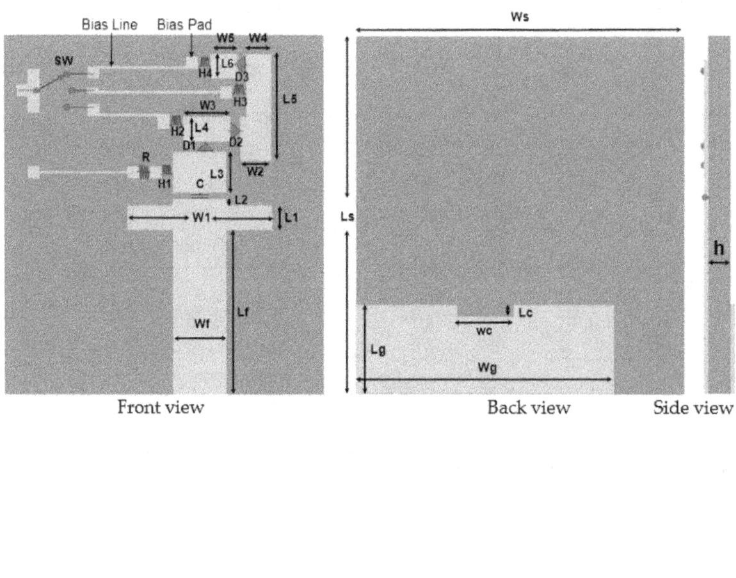

4.7 This is author own paper in which multilayer antenna is designed with different materials and air gaps. All designs are having same parameters and but the patch, ground and substrate are changed according to results. Also it is basically high frequency antenna have terahertz frequency and dimensions are in micrometer. This design has given 10.9 Db gain with 52.9 GHz bandwidth as given in figure. Under this design, frequency is increased by 0.3 THz to 0.6 and accordingly gain is improved 3.02 dB to 10.09 dB (Smrity Dwivedi, 2023).

Figure 14. Antenna design with different layering of materials and mentioned with different colors

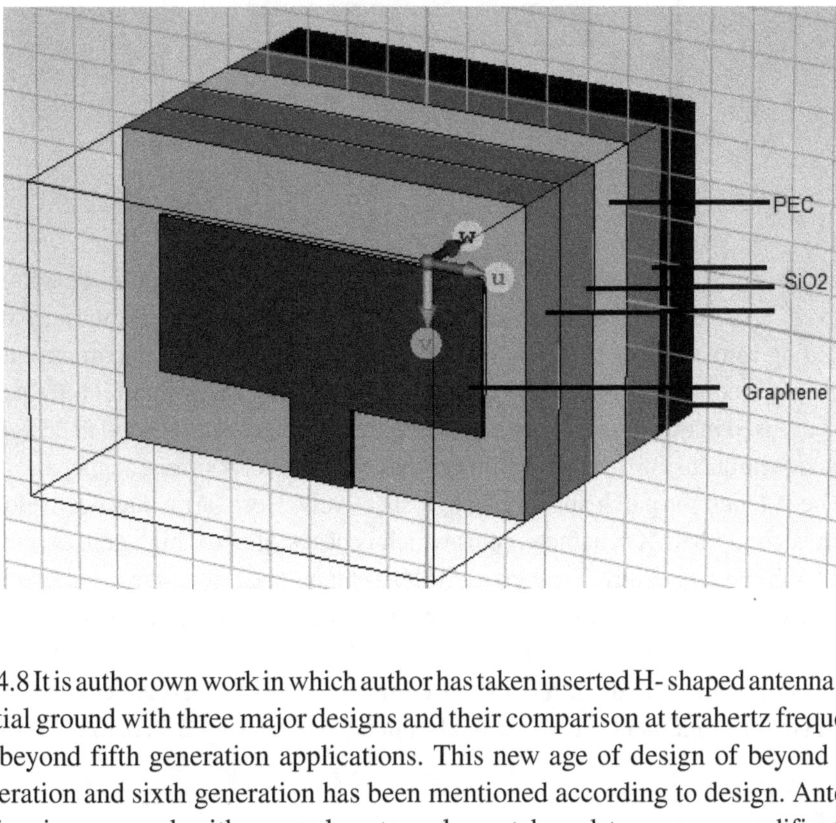

4.8 It is author own work in which author has taken inserted H- shaped antenna with partial ground with three major designs and their comparison at terahertz frequency for beyond fifth generation applications. This new age of design of beyond fifth generation and sixth generation has been mentioned according to design. Antenna design is proposed with normal rectangular patch and two more modifications have been done in same patch and it is simulated with the help of CST MWS for its important parameters such as bandwidth and gain. For the three structures first design, second design and third design bandwidth is 100 GHz, 250 GHz, 400 GHz with gain 13.5 dBi, 14.4 dBi, 14.8 dBi respectively. A simple PEC material as a radiating element and polyamide as substrate have been taken. Partial ground has been taken with microstrip line feedline (Smrity Dwivedi, 2023).

Figure 15. Final structure antenna with inserted h- shaped

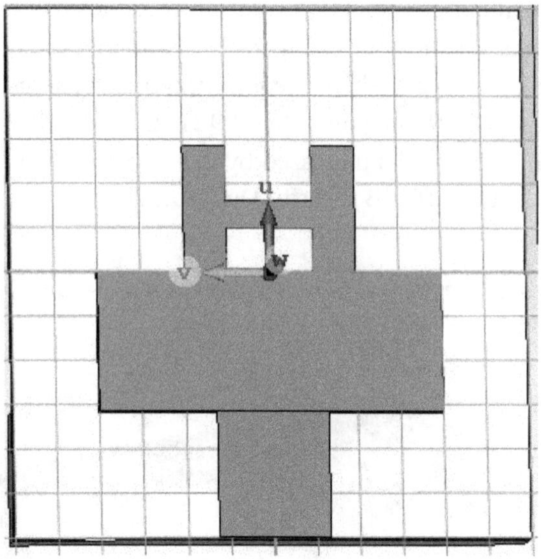

4.9 This is again author own work in which author has proposed full parameters of design an antenna as given in photograph for beyond fifth generation systems. According to the purpose, requirement of high bandwidth with maximum gain is one primary constraint for improving performance because of that high frequency regime is used in this paper. Hence, a high frequency has been selected to design the antenna, without using the material like graphene, gold etc. A simple copper and roger have been taken for patch and substrate respectively. Here, array has been adopted which is giving MIMO characteristics which is best suited for high gain extra wide band terahertz frequency. Terahertz MIMO antenna used here, having 2-elements and 3- elements arrays. Antenna structures have been simulated using commercial software. The gain of three element arrays is 24.5 dBi and the gain of two element arrays is 22.1 dBi. Bandwidth is increased from 0.2 THz for 2-element arrays and 0.45 THz for 3- element arrays for the frequency range from 2 THz to 6 THz.

Figure 16. The patch and substrate are mentioned in above figure with front view of 3-elements MIMO antenna

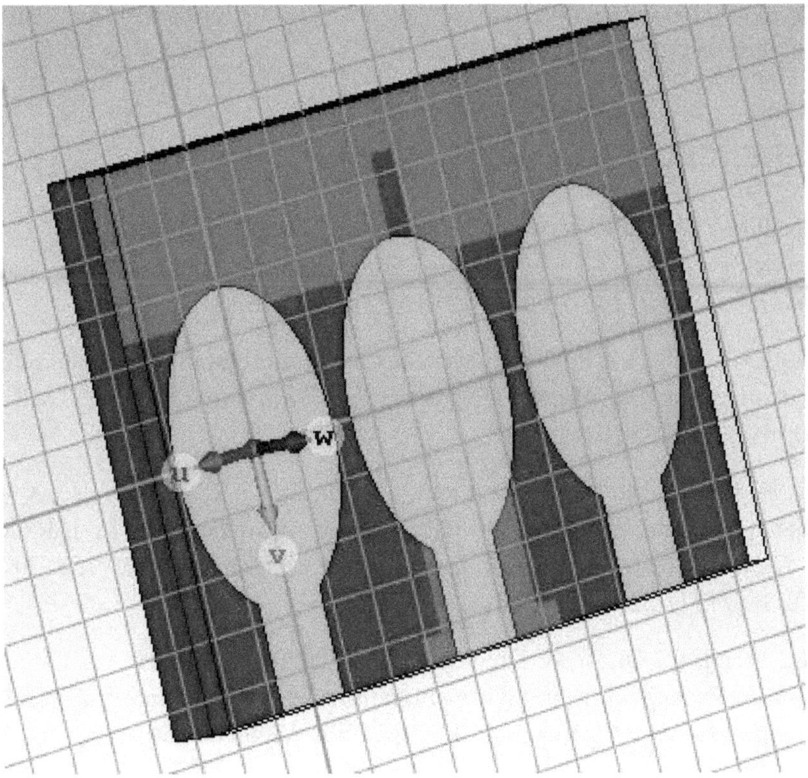

5. CONCLUSION

As in above chapter, there are various topics have covered related to history of fifth and sixth generation, design of antennas in 5G and 6G generation in which many recent papers including author own work have been discussed. Also, advantages for this new era and impact on our society have been discussed. Reconfiguration which is covered and that is main topic of this chapter under which designs have been taken and elaborated with figures and explanation. Different way to achieve reconfiguration and types of reconfiguration has been given with impact on materials and uses. Reconfigurable antenna is very important to achieve high frequency applications to apply for 5G and 6G applications. As there are three reconfiguration properties have been discussed frequency, radiation and polarization for many applications. Electrical, optical, physical/mechanical and material based reconfiguration techniques have proposed for which electronic switches, optical switches, smart materials, and probe like methods are being used. Many review papers have been discussed with its results. Also author of this chapter has discussed own paper for 5G/6G systems. Overall, different types of antennas have been proposed for research under this new generation systems. Further, there are many scope for this work remaining as technologies is growing day by day.

REFERENCES

Abdulsahib. (2023). Reverse polarity optical Orthogonal frequency Division Multiplexing for High-Speed visible light communications system. *Egyptian Informatics Journal, 24*(4). . doi:10.1016/j.eij.2023.100407

Brown, E. R. (1998). RF-MEMS switches for reconfigurable integrated circuits. *IEEE Transactions on Microwave Theory and Techniques*, *46*(11), 1868–1880. doi:10.1109/22.734501

Cheng, B., Cui, Z., Lu, B., Qin, Y., Liu, Q., Chen, P., He, Y., Jiang, J., He, X., Deng, X., Zhang, J., & Zhu, L. (2018). 340-GHz 3-D imaging radar with 4Tx-16Rx MIMO array. *IEEE Transactions on Terahertz Science and Technology*, *8*(5), 509–519. doi:10.1109/TTHZ.2018.2853551

Chi, L., Weng, Z., Qi, Y., & Drewniak, J. L. (2020). A 60 GHz PCB Wideband Antenna-in-Package for 5G/6G Applications. *IEEE Antennas and Wireless Propagation Letters*, *1225*(c), 1–1. doi:10.1109/LAWP.2020.3006873

Cummings, N. P. (2003). *Active Antenna Bandwidth Control Using Reconfigurable Antenna Elements* [Ph.D. Thesis]. Virginia Tech, Bradley Department of Electrical and Computer Engineering, Blacksburg, VA, USA.

Duan. (2020). Evolution and innovation of antenna systems for beyond 5G and 6G. *Frontiers of Information Technology & Electronic Engineering, 21*(1), 1–3. doi:10.1631/FITEE.2010000

Dussopt, L., & Rebeiz, G. M. (2003, April). Intermodulation distortion and power handling in RF MEMS switches, varactors, and tunable filters. *IEEE Transactions on Antennas and Propagation*, *51*(4), 1247–1256.

Dwivedi, S. (2022). *ReconfigurableArray antenna*. Intech Open. doi:10.5772/intechopen.106343

Dwivedi. (2022). Design and Analysis of Metamaterial Waveguide Antenna for broadband applications. *ICT4SD*. . doi:10.1007/978-981-19-5224-1_42

Dwivedi, S. (2023). *Metamaterials-Based Antenna for 5G and Beyond*. IGI Global. doi:10.4018/978-1-6684-8287-2.ch001

Eduardo, V. P. (2022). FORMAT: A Reconfigurable Tile-Based Antenna Array System for 5G and 6G Millimeter-Wave Testbeds. *IEEE Systems Journal*, *16*(3), 4489–4500. Advance online publication. doi:10.1109/JSYST.2022.3146360

Emara, M. K., Stuhec-Leonard, S. K., Tomura, T., Hirokawa, J., & Gupta, S. (2020). Laser-Drilled All- Dielectric Huygens' Transmit-Arrays as 120 GHz Band Beamformers. *IEEE Access : Practical Innovations, Open Solutions*, *8*, 153815–153825. doi:10.1109/ACCESS.2020.3018297

Erdil, E., Topalli, K., Unlu, M., Civi, O. A., & Akin, T. (2007, April). Frequency tunable patch antenna using RF MEMs technology. *IEEE Transactions on Antennas and Propagation*, *55*(4), 1193–1196. doi:10.1109/TAP.2007.893426

Fitri, I., & Akbart, A. A. (2018). A new gridded parasitic patch stacked microstrip antenna for enhanced wide bandwidth in 60 GHz Band. *2017 International Conference on Broadband Communication, Wireless Sensors and Powering, BCWSP 2017*, 1–4. 10.1109/BCWSP.2017.8272571

Grau, A., Romeu, J., Ming-Jer Lee, Blanch, S., Jofre, L., & De Flaviis, F. (2010, January). A dual linearly polarized MEMS-reconfigurable antenna for narrow band MIMO communication systems. *IEEE Transactions on Antennas and Propagation*, *58*(1), 4–16. doi:10.1109/TAP.2009.2036197

Guo, L., Huang, F., & Tang, X. (2014). A novel integrated MEMS helix antenna for terahertz applications. *Optik (Stuttgart)*, *125*(1), 101–103. doi:10.1016/j.ijleo.2013.06.016

Homod, R. Z., Mohammed, H. I., Abderrahmane, A., Alawi, O. A., Khalaf, O. I., Mahdi, J. M., Guedri, K., Dhaidan, N. S., Albahri, A. S., Sadeq, A. M., & Yaseen, Z. M. (2023, December). Deep clustering of Lagrangian trajectory for multi-task learning to energy saving in intelligent buildings using cooperative multi-agent. *Applied Energy*, *351*(1), 121843. doi:10.1016/j.apenergy.2023.121843

Hong, W., Jiang, Z. H., Yu, C., Zhou, J., Chen, P., Yu, Z., Zhang, H., Yang, B., Pang, X., Jiang, M., Cheng, Y., Al-Nuaimi, M. K. T., Zhang, Y., Chen, J., & He, S. (2017, December). Multibeam antenna technologies for 5G wireless communications. *IEEE Transactions on Antennas and Propagation*, *65*(12), 6231–6249. doi:10.1109/TAP.2017.2712819

Huang, H.-C. (2018). Overview of antenna designs and considerations in 5G cellular phones. *Proc. Int. Workshop Antenna Technol. (iWAT),* 1–4. 10.1109/IWAT.2018.8379253

Huff, G. H., Bahukudumbi, P. B., Everett, W. N., Beskok, A., Bevan, M. A., Lagoudas, D., & Ounaies, Z. (2007). Microfluidic reconfiguration of antennas. *Proc. Antenna Appl. Symp.,* 241-258.

Jaafar, H., Sidek, O., Miskam, A., & Abdul-Rahman, R. (2013). High-isolation and low-loss RF MEMS shunt switches. ACES J.-. *Appl. Comput. Electromagnet. Soc.*, *25*, 780–786.

Kovitz, J. M., Rajagopalan, H., & Rahmat-Samii, Y. (2015). Design and implementation of broadband MEMS RHCP/LHCP reconfigurable arrays using rotated E-shaped patch elements. *IEEE Transactions on Antennas and Propagation*, *63*(6), 2497–2507. doi:10.1109/TAP.2015.2417892

Letaief, K. B., Chen, W., Shi, Y., Zhang, J., & Zhang, Y. J. A. (2019, August). The Roadmap to 6G: AI Empowered Wireless Networks. *IEEE Communications Magazine*, *57*(8), 84–90. doi:10.1109/MCOM.2019.1900271

Miao, Z.-W., Hao, Z.-C., Wang, Y., Jin, B.-B., Wu, J.-B., & Hong, W. (2018). A 400-GHz High-Gain Quartz-Based Single Layered Folded Reflectarray Antenna for Terahertz Applications. *IEEE Transactions on Terahertz Science and Technology*, *9*(1), 78–88. doi:10.1109/TTHZ.2018.2883215

Nikolaou, S., Kingsley, N. D., Ponchak, G. E., Papapolymerou, J., & Tentzeris, M. M. (2009, August). UWB elliptical monopoles with reconfigurable band notch using MEMS switches actuated without bias lines. *IEEE Transactions on Antennas and Propagation*, *57*(8), 2242–2251. doi:10.1109/TAP.2009.2024450

Qin, P. Y., Weily, A. R., Guo, Y. J., Bird, T. S., & Liang, C. H. (2010, August). Frequency reconfigurable quasi-yagi folded dipole antenna. *IEEE Transactions on Antennas and Propagation*, *58*(8), 2742–2747. doi:10.1109/TAP.2010.2050455

Rainville, P. J., & Harackewiez, F. J. (1992, December). Magnetic tuning of a microstrip antenna fabricated on a ferrite film. *IEEE Microw. Guided Wave Lett.*, *2*(12), 483–485.

Rana, S. K., Rana, A. K., Rana, S. K., Sharma, V., Lilhore, U. K., Khalaf, O. I., & Galletta, A. (2023). Decentralized Model to Protect Digital Evidence via Smart Contracts Using Layer 2 Polygon Blockchain. *IEEE Access : Practical Innovations, Open Solutions*, *11*, 83289–83300. doi:10.1109/ACCESS.2023.3302771

Samsung Research. (2020). *6G: The Next Hyper Connected Experience for All.* Available: https://research.samsung.com/next-generation-communication

Shivani & Dwivedi. (2019). Comparative analysis of Reconfigurable Patch Antenna Array for different Liquid Crystal Substrates. *URSI AP-RASC*. doi: RASC.2019.8738497 doi:10.23919/URSIAP

Shivani & Dwivedi. (2020). Simulation analysis of High Gain and Low Loss Antenna with Metallic Ring and spoof SPP Transmission Line. *7th International Conference on Signal Processing & Integrated Networks.* 10.1109/SPIN48934.2020.9071379

Shivani & Dwivedi. (2021). Graphene based Radiation Pattern reconfigurable antenna. *International Conference for convergence in Technology*. 10.1109/I2CT51068.2021.9417938

Singhal, S. (2019). Elliptical ring terahertz fractal antenna. *Optik (Stuttgart)*, *194*, 163129. Advance online publication. doi:10.1016/j.ijleo.2019.163129

Smrity. (2023). *Metamaterials-Based Antenna for 5G and Beyond.* doi:10.4018/978-1-6684-8287-2.ch001

Sumit Kumar, Dixt, Malekar, Raut, & Shevada. (2020). Fifth Generation Antennas: A Comprehensive Review of Design and Performance Enhancement Techniques. *IEEE Access*. . doi:10.1109/ACCESS.2020.3020952

Tan, L. R., Wu, R. X., Wang, C. Y., & Poo, Y. (2013, March). Magnetically Tunable Ferrite Loaded SIW Antenna. *IEEE Antennas and Wireless Propagation Letters*, *12*, 273–275. doi:10.1109/LAWP.2013.2248113

Tawk, Y., Costantine, J., & Christodoulou, C. G. (2012). A varactor-based reconfigurable filtenna. *IEEE Antennas and Wireless Propagation Letters*, *11*, 716–719. doi:10.1109/LAWP.2012.2204850

Xue, X. (2023). A Novel partial sequence technique based Chaotic biogeography optimization for PAPR reduction in eneralized frequency division multiplexing waveform. Heliyon, 9. doi:10.1016/j.heliyon.2023.e19451

Xue, X., Abdulsahib, G. M., Khalaf, O. I., Jagan, J., Loganathan, K., Makota, C., & Ponraj, B. (2023, October). Soft computing approach on estimating the lateral confinement coefficient of CFRP veiled circular columns. *Alexandria Engineering Journal*, *81*(15), 599–619. doi:10.1016/j.aej.2023.09.053

Chapter 5
Integration of Reconfigurable Intelligent Surfaces With Antennas for 6G Wireless Communications

Pallavi Sapkale
https://orcid.org/0000-0002-8403-8908
Ramrao Adik Institute of Technology, India

Shilpa Mehta
https://orcid.org/0000-0002-4691-3185
Auckland University of Technology, New Zealand

ABSTRACT

The evolution of wireless communication networks towards 6G demands innovative solutions to address the challenges of increasing data rates, network capacity, and energy efficiency. Reconfigurable intelligent surfaces (RIS) emerge as a promising technology that revolutionizes how wireless signals propagate and interact with the environment. In this chapter, the authors explore the integration of RIS with antennas for 6G wireless communications. RIS, comprised of programmable meta-surfaces, can dynamically manipulate electromagnetic waves to optimize signal quality, mitigate interference, and enhance network performance. By providing adaptive beam forming, beam steering, and signal enhancement capabilities, RIS-enabled antennas offer unprecedented flexibility and efficiency in communication systems. This chapter overviews RIS principles, discusses their integration with antennas, and highlights their potential to shape the future of 6G wireless communications.

DOI: 10.4018/979-8-3693-0916-2.ch005

1. INTRODUCTION

Reconfigurable intelligent surfaces (RISs) are a potential system for improving seamless connectivity system's spectrum and environment friendly. Fulfilling the user's requirements for future 6G wireless networks is hampered by two major issues: power consumption and hardware expense discussed by Q. Wu (2020). Currently, the emerging reconfigurable intelligent surface (RIS) is being considered a feasible technique for increasing the transmission parameters with the spectrum performance of next-generation networks by exerting affordable active reflecting portions Pan, Cunhua, et al. (2021) shown in their work. These devices can reconfigure the wireless networks by precisely tweaking the Phase transformation of a large number of inexpensive passives reflecting components. However, Current cellular communication theories constrain. The advancement of RIS-assisted 6G networks, posing considerable challenges. In this chapter, we focused on the integration of RIS antenna for 6G with an improvement of network performance and reducing the co-channel interference.

1.1 RIS Background

Meta surfaces, the two-dimensional in-nature equivalents of metamaterials, are artificial structures made from tiny meta-atoms that are often organized in a periodic or aperiodic pattern on a plane. Meta surfaces have advanced rapidly over the last ten years and are now widely used in the creation of novel gadgets and services in the microwave, terahertz, and visible light spectrums. Snell's law describes both the radio wave reflecting and refracting over the point of interaction between two isotropic similar media. A revised version of The theory put forward by Snell asserts that: the presence of the phase differential causes irregular absorption or reflections, was put forth in 2011. At the meta surface, which acts as the interface between the two mediums explained by Hong, (2022).

1.2: Reconfigurable Intelligent Surfaces Significance

RISs are a demanding system in the realm of future communication and signal processing. In Basar E. et al. (2019) they are composed of collections of passive elements, such as antennas or reflectors, which can Dynamically modify electromagnetic radiation that travels across devices. These surfaces can be controlled in real-time to alter the propagation environment of wireless signals. The primary characteristic of RISs is their capacity to adaptively modify the Phase and frequency of the transmission. By doing so, G.C. Trichopoulos, (2022) can optimize the wireless communication links between transmitters and receivers by directing signals, reducing interference, expanding coverage, and enhancing the overall efficiency of wireless networks. The fundamental physical layer technology used in fifth-generation (5G) wireless networks, which have already been implemented internationally, is huge multiple-input multiple-output (MIMO) within the below-6 GHz range. Which has been intended to be a single for the primary systems of future communication, but has not gained widespread adoption. Because severe path loss problems and advanced applications of 6G such as intelligent systems, robotics, and virtual reality require a high quality of service (QoS). Conventional systems fail to solve the blockage problem during high frequencies. Deploying base stations can solve such critical issues but such systems are too expensive. So, to overcome this issue Reconfigurable intelligent surfaces (RISs) are the accurate solution. RISs are often mounted across wide planes (e.g., indoor and outdoor) for reflection of radio-frequency (RF) radiation across barriers and also establish a virtualized traveling pathway from a millimeter-wave source and its intended location discussed in Basharat, Sarah, et al. (2022). The Reflecting Intelligent Surface (RIS) is made up of a flat surface with an array of passive reflecting components. Every component can individually apply a needed phase shift for transmitting the network. These reflecting elements are classified into two categories according to the elements that are used: structures based on antenna arrays and Metasurfaces. By properly regulating these elements' phase changes, the RIS may intelligently rearrange the reflected signals, thereby directing them in the desired direction.

2. LITERATURE SURVEY

Table 1. Previous work insights and techniques for RIS in 6G

Ref.	Main Objectives	Technique	Insights
Scott, A., Gerber. (2023)	the state-of-the-art hardware architectures designed for channel estimation in RIS-	model-based and model-free techniques	- Overview of RIS hardware architectures for channel estimation in wireless systems. - Categorization of channel estimation approaches into model-based and model-free techniques.
M., H., Almekhlafi et. al.(2023)	To propose a low-complexity solution for optimizing the configuration of RIS elements to minimize the total transmit power in a single base station (BS) serving multiple single-antenna users in 6G networks	a linear transformation approach and element-wise Karush-Kuhn-Tucker (KKT) approach to obtain a closed-form expression for the RIS phase shifts	Research has focused on studying the gains and optimal configuration of RIS elements. - Existing literature decomposes the problem into power control and RIS phase shift sub-problems.
Jaafar et. al (2023)	To present a model for accurately estimating the reflection coefficient amplitude and phase of reflective intelligent surfaces	Incorporating varactor diodes into the surface of the cell unit to achieve reconfigurability.	The provided paper does not discuss the integration of reconfigurable intelligent surfaces with antennas for 6G wireless communications
Moazzam et al. (2023)	to highlight the benefits and applications of RISs in programming to improve connectivity, spectrum efficiency, and energy efficiency in wireless networks	orthogonal frequency-division multiplexing (OFDM) and multi-user multiple-input multiple-output (MIMO) systems.	- Overview of principles and applications of RISs. - Discussion of advantages and challenges in wireless communications
Giacomo et al. (2023)	To explore and propose dynamic control strategies for reconfigurable intelligent surfaces (RISs) in the context of smart electromagnetic (EM) environments for 6G networks.	enable efficient and adaptive control of RISs to optimize wireless communication performance in smart EM environments.	- The paper discusses the challenges and strategies for implementing dynamic control algorithms for Reconfigurable Intelligent Surfaces (RISs) in 6G networks. - The objective is to illustrate recent advances and emerging solution strategies for RIS configuration and adaptation in next-generation networks.

3. THE ARCHITECTURE OF A TYPICAL RIS

Figure 1. The architecture of a typical RIS

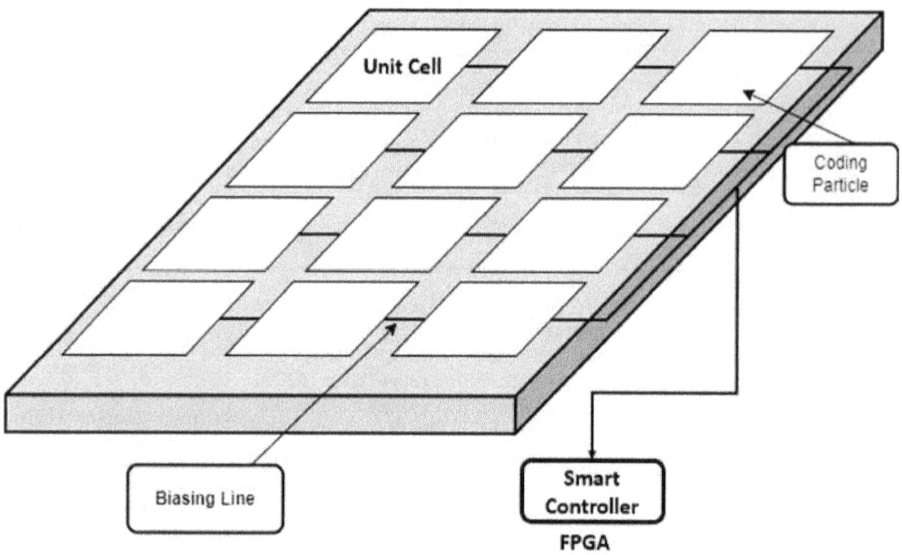

The structure of a common RIS, fabricated using metamaterials, primarily comprises a flat surface and a controller, as illustrated in Figure 1. The initial layer of this design displays a substantial quantity of reflective elements imprinted on a dielectric substrate known as unit cells. These components possess a pivotal function in directly impacting the incoming signals. The subsequent layer consists of a copper panel, strategically positioned to avert any undesired signal and energy seepage. Finally, the third layer includes a circuit board that allows for adjusting the reflection coefficients of the RIS elements. RISs also offer significant benefits for practical deployments. The RIS-reflecting components passively reflect incoming signals, without requiring RF transceiver gear for complex signal processing (SP). As a result, as compared to traditional active transmitters, RISs may run at significantly cheaper hardware and power consumption expenditures discussed by Hong, I.-P.(2023).

Reconfigurable intelligent surfaces (RISs) are gaining popularity as a critical technology for achieving smart radio environments (SREs) beyond 5G (B5G)/6G described by Z. Chen (2022), Figure 2 illustrates the future of the 6G environment. B5G/6G aims to enhance connectivity and satisfy varied QoS needs. The transmission performance of wireless networks has been proven using practical testing and simulations that make use of RIS's reconfigurability.

Figure 2. Anticipated 6G technologies (Hong, 2023)

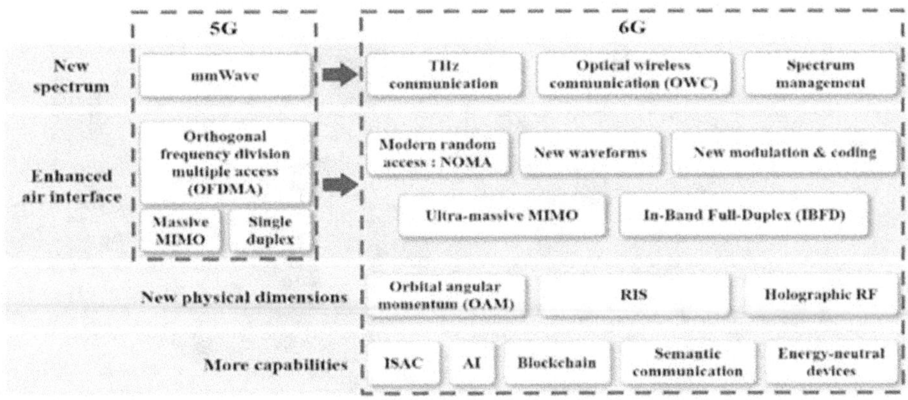

Currently, the three-gigahertz (3GPP) spectrum bands used for 5G networks are divided between FR1 (410-7125 MHz) as well as FR2 (greater than 24 and less than 52 GHz) . As introduced in the period from 2018 the RIS concept has received widespread attention for its advantages and practical application in wireless communication networks shown in Björnson, E (2020). Reliable electromagnetic modeling of the RIS is the first step, and several studies on the subject have concluded that the metasurface. Technical parameters of phase shifting, like Regulate solutions, RIS substrate properties, incident radio wave transmission, reflection, and polarization characteristics, and RIS size should be taken into account, however. For instance, It won't be feasible to acquire acceptable SNR characteristics when the RIS cannot exceed a specific size, hence estimating the RIS's size becomes a critical difficulty.

4. ADVANCED RIS USES ENABLING 6G WIRELESS SYSTEMS

In this section, the emerging applications of RIS for 6G wireless networks are discussed, and Multiplicative fading must be taken into consideration in an actual time spectrum monitoring and estimate approach. Figure 3 depicts the emerging applications of RIS for 6G. RIS is modified to address channel optimization problems by the need for the performance metric of interest for suitable application in every use circumstance. To enhance the expected signal strength at the receiving end, the signals reflected from the RIS should be superimposed constructively with those from the direct path, or combined destructively to lessen the negative effects of multiuser interference with appropriate adjustment of the phase shift of the unit's cell reflects the RIS.

Figure 3. Various applications of ris for 6G wireless networks from Trichopoulos et al. (2022)

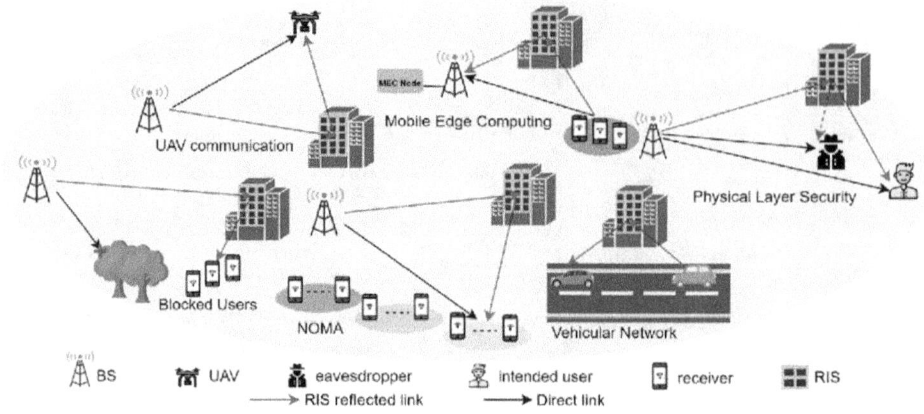

Reconfigurable Intelligent Surfaces (RIS) are going to play a major role in the development of 6G wireless networks, offering various applications and advantages. Some emerging uses of RIS for 6G wireless networks are:

4.1 RIS for Non-Orthogonal Multiple Access (NOMA)

New communication technology, can be paired with RIS to boost coverage region and connectivity. The beamforming design and power allocation play a crucial role in achieving improved performance in Farshad et al (2023). Increasing the number of passive elements in the RIS and shortening the distance between the base station and the RIS are effective ways to further enhance the performance by Adam (2023).

4.2 RIS for Mobile Edge Computing (MEC) Networks

In mobile edge computing the integration of RIS technology can improve the latency performance. In the future Virtual reality and image and video computation technology will work in real time. However, to fulfill the requirements extra power supply and hardware will be used. Due to limited traditional devices such tasks can't be fulfilled, to overcome such issues. These computationally difficult jobs can be offloaded to powerful computing nodes, which are often located near the network's edge. However, in some circumstances where these devices are remote from the mobile edge computing (MEC) node, they may experience a poor data offloading rate due to high route loss, resulting in excessive offloading delays. To address this issue, a unique RIS-aided MEC framework was presented in T. Bai et al. (2020).

4.3 RIS for Physical Layer Security Networks

Due to the electromagnetic nature of wireless networks, there is a major chance of leaking the data through the transmission link. Physical layer security (PLS) approaches have currently obtained significant R&D attention because they can avoid cumbersome key exchange protocols and are suited for latency-sensitive applications. To maximize the rate of a secure communication link, both artificial noise and additional antennas have been proposed by Nematullo Rahmatov (2023). However, when both actual users and hackers use correlated channels, or when the eavesdroppers are closer to the BS than the actual users, the feasible secure rate is limited.

5. THE INTEGRATION OF RIS ANTENNA FOR 6G WITH AN IMPROVEMENT IN NETWORK PERFORMANCE

RIS technology introduces an innovative approach to wireless communication systems by enabling dynamic control over signal propagation environments. Here's how the integration of RIS antennas can improve network performance in the context of 6G:

5.1 Improved Signal Strength and Coverage: RIS antennas can manipulate the propagation environment by reflecting, refracting, or diffracting electromagnetic waves. By strategically adjusting the phase and amplitude of reflected signals, RIS antennas can enhance signal strength and coverage, especially in areas with poor signal quality or coverage gaps.

5.2 Mitigation of Multipath Interference: Multipath interference occurs when signals take multiple paths to reach the receiver, causing signal distortion and degradation. RIS antennas can help mitigate multipath interference by optimizing signal paths and reducing signal attenuation, resulting in improved signal quality and reliability.

5.3 Increased Spectral Efficiency: RIS technology can enhance spectral efficiency by optimizing signal transmission and reception parameters. By dynamically adjusting the phase and amplitude of reflected signals, RIS antennas can minimize interference and maximize the utilization of the available frequency spectrum, thereby increasing data transmission rates and network capacity.

5.4 Enhanced Security and Privacy: RIS antennas can also contribute to enhancing security and privacy in 6G networks by providing additional layers of encryption and authentication mechanisms. By controlling signal propagation environments, RIS antennas can help prevent unauthorized access and eavesdropping, ensuring secure and private communication channels.

5.5 Energy Efficiency: By optimizing signal propagation paths and reducing signal attenuation, RIS antennas can help minimize energy consumption in 6G networks. By enhancing signal quality and coverage, RIS technology enables more efficient use of resources, leading to reduced power consumption and extended battery life for mobile devices.

5.6 Support for Massive MIMO (Multiple Input Multiple Output): RIS technology can complement Massive MIMO systems by improving signal propagation and reducing interference. By optimizing signal paths and enhancing signal-to-noise ratios, RIS antennas can enable more efficient utilization of MIMO techniques, leading to higher data rates and improved network performance.

Overall, the integration of RIS antennas represents a promising technology for enhancing the performance of 6G networks by optimizing signal propagation environments, improving spectral efficiency, enhancing security and privacy, and promoting energy efficiency. As 6G networks continue to evolve, RIS technology is expected to play a significant role in shaping the future of wireless communications.

6. MAJOR CHALLENGES IN RIS FOR FUTURE 6G NETWORKS

Using RIS could lead to the development of an entirely novel networking framework and new opportunities for the next generation but it also presents numerous new technological and practical difficulties. Currently, RIS used in engineering technology research is still in its early stages, with more technical issues and limited space. This study will only analyze and discuss the key technical issues of channel rank reduction, network cohabitation, network deployment, and other key technical challenges, as well as suggest potential answers or research ideas. We will analyze and discuss other technological challenges that were not addressed in this study at a later time. Here, we shall provide a brief list for researchers' reference. The introduction of RIS may create a new network. While RISs are intriguing for the previously mentioned purposes, they also provide practical issues. We have highlighted a few major issues as follows:

6.1 Impacts on Wireless Channel Characteristics

The enactment of RIS transforms the natural, uncontrollable electromagnetic transmission context into an artificially managed wireless propagated environment. The active manipulation of the wireless propagation environment may result in a novel channel characterization paradigm. It optimizes the quality of the signal but limits the spatial mobility and channel ranking. Large-scale changes, such as the multiplicative link between NB-RIS-UE path loss and segmented channel distance.

Significant near-field and spatial no stationary features. Compact construction with basic unit spacing of less than half wavelength leads to mutual coupling issues.

6.2 RIS Channel Estimation

The quantity of reflecting elements in the RIS is a crucial factor in achieving accurate acquisition. Since RIS does not have baseband processing capabilities, the system is designed to have a large RIS dimension, which is very disadvantageous for an RIS implementation. Therefore, the implementation bottleneck of RIS-enabled MIMO systems is how many reflective RIS elements must be satisfied for accurate acquisition.

Channel estimation in RIS-enabled systems is challenging due to the large number of reflecting elements and the dynamic nature of the propagation environment. Traditional channel estimation techniques, such as pilot-based estimation or channel state information (CSI) feedback, may not be directly applicable in RIS scenarios. Therefore, novel channel estimation algorithms tailored to the unique characteristics of RIS, such as sparse channel representations, spatial correlations, and channel sparsity, need to be developed to accurately capture the channel state and enable effective beamforming.

6.3 Beamforming Model

One of the most difficult aspects of RIS-assisted beamforming is acquiring the necessary ratio between entering along with leaving signals. However, customers are mobile, and climatic qualities fluctuate, potentially leading to an inaccuracy in the beamforming design. The other essential challenge is to obtain the ideal beamforming design while keeping the inaccuracy of cascade channel estimates at a reasonable level.

Passive beamforming refers to the manipulation of electromagnetic waves using passive elements, such as reflective surfaces or metamaterials, to achieve desired signal shaping and radiation patterns. In RIS technology, passive beamforming relies on controlling the phase shifts of the reflecting elements to steer or focus the transmitted signals towards the intended receiver, thereby enhancing signal strength, coverage, and capacity.

Passive beamforming with RIS offers several advantages over traditional active beamforming techniques, such as reduced hardware complexity, lower power consumption, and greater flexibility in adapting to changing environmental conditions. By intelligently adjusting the phase shifts of the reflecting elements based on channel state information, passive beamforming enables dynamic optimization of signal propagation paths, interference suppression, and spatial multiplexing, leading to improved spectral efficiency and communication reliability.

Overall, channel estimation and passive beamforming are fundamental techniques in RIS technology for optimizing signal transmission, enhancing coverage, and mitigating interference, thereby unlocking the full potential of RIS-enabled communication systems in future wireless networks. Continued research and innovation in these areas are essential for realizing the promise of RIS in enabling transformative wireless communication capabilities in 6G and beyond.

6.4 Co-Channel Interference

To increase the throughput of RIS-assisted systems, there is a need for proper transmission and reception of beamforming. During the mobility of the user, the various base stations are enabled, and co-channel interference will become a critical issue for future 6G communication networks. As a result, B5G/6G communication necessitates strict interference coordination methods to counteract co-channel interference. Fig 4 (a) shows RIS for coverage extension, whereas Fig 4 (b) shows RIS for channel rank improvement.

Figure 4. RIS-assisted coverage extension

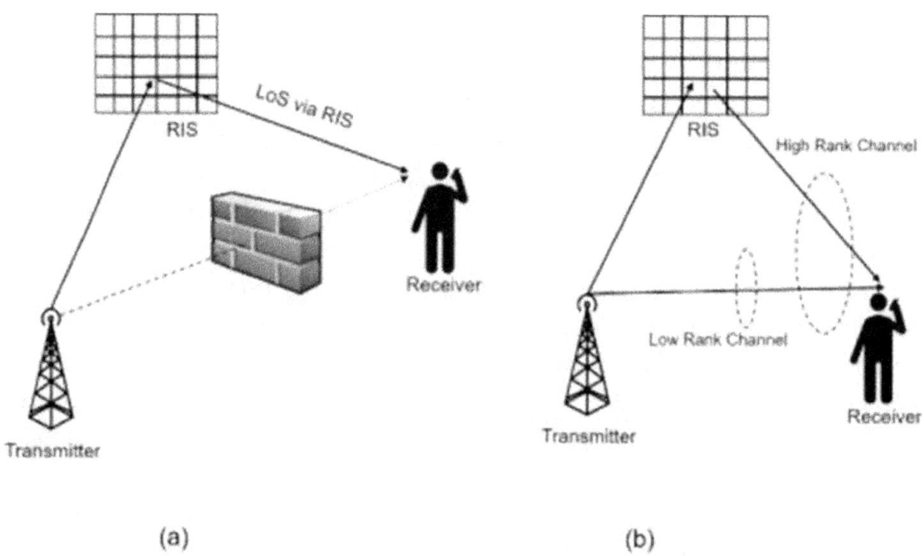

Research X. Tang et al. (2022) suggest several techniques to enhance the data confidentiality ratio while taking into account CSI channel state information. Due to the comprehensive channel state information and jamming power strategy, researchers used bipartite search and presented a data-driven deep neural network (DNN) to estimate effective UAV localization.

Figure 5. A Dual-hop RF-RIS-aided system

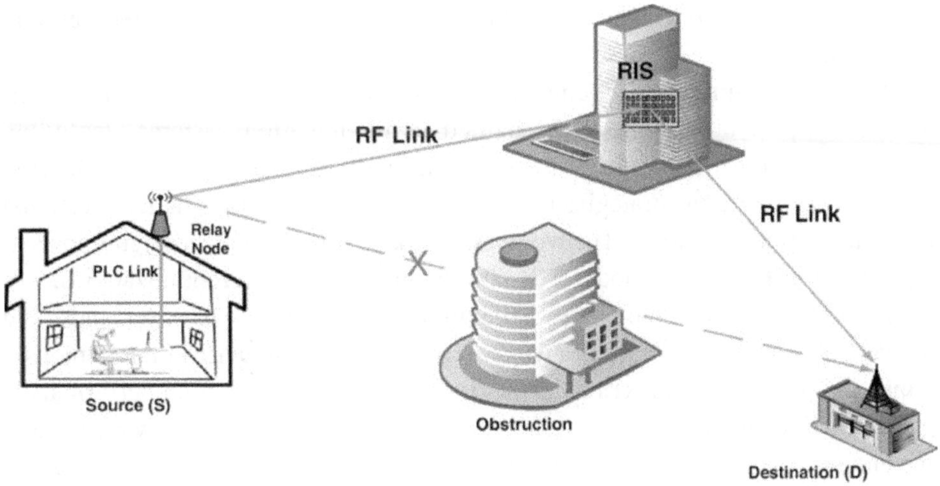

We analyze a RIS-based dual-hop mixed FSO-RF system with a source node (S) communicating with a destination node (D) which is shown in fig 5. By using decode-and-forward (DF) relay node (R) and RIS. We assume that the direct links between S-D and R-D do not exist due to obstruction caused by some buildings and obstacles shown in Zhou (2023). Here, D is corrupted by CCI and R has RF and FSO capabilities. To mitigate the impact of AT and PEs on the FSO link, we employ multiple FSO apertures at the transmitter end explained in M. R. Bhatnagar et al.(2016). Hence, S is considered to be connected with multiple FSO apertures and R and D are connected with a single antenna. Here, the RIS can obtain full channel state information (CSI). In this chapter, we review the integration with various antennas. For high path loss, sub-THz links are very sensible because of hurdles. Due to such sensible properties, it's difficult to track the movable users in such cases the beam steerable antennas can be used for accurate propagation. In terahertz antenna Lens Antennas are used to fulfill the high-gain values for focusing the lenses and improving the directivity shown by Jiang, R. et al. (2023). Another is a phased array antenna; such antennas are made up of several elements and they are fed using phase shifters at every element. It helps them to electrically focus their beam in particular directions shown in C. A. Fernandes (2016).

7. FUTURE DIRECTIONS AND RESEARCH OPPORTUNITIES FOR RECONFIGURABLE INTELLIGENT SURFACES (RIS) IN 6G COMMUNICATION:

Fure 6G encompasses a wide range of areas aimed at advancing the understanding, development, and deployment of RIS technology to realize its full potential in future wireless networks. Here's a detailed exploration of potential directions and opportunities:

Advanced Beamforming Techniques: Investigate novel beamforming algorithms and techniques tailored to RIS-enabled communication systems, including intelligent beam steering, beam tracking, and multi-user beamforming. Develop adaptive beamforming strategies that can dynamically adjust to changing channel conditions, user mobility, and network requirements. Dynamic Reconfiguration and Optimization: Explore methods for dynamic reconfiguration and optimization of RIS configurations in real-time based on environmental feedback, user demands, and network objectives [26]. Develop adaptive control algorithms that can autonomously adjust the phase shifts of RIS elements to optimize communication performance, enhance coverage, and mitigate interference. Integrated RIS-Enabled Networks: Study the integration of RIS technology with other emerging technologies such as massive MIMO, millimeter-wave communication, terahertz communication, and visible light communication given by X. Fu (2020), Ouamri (2023). Investigate the synergies and potential benefits of combining RIS with these technologies to achieve higher data rates, lower latency, and improved reliability in 6G networks. Machine Learning and AI Applications: Explore the use of machine learning and artificial intelligence (AI) techniques to enhance RIS-enabled communication systems. Investigate how machine learning algorithms can optimize RIS configurations, predict channel conditions, and adaptively adjust system parameters to maximize performance and efficiency. Security and Privacy Enhancements: Address security and privacy concerns related to RIS-enabled communication systems by developing robust authentication, encryption, and intrusion detection mechanisms. Investigate potential vulnerabilities and threats specific to RIS technology and develop countermeasures to protect against cyberattacks and unauthorized access. Energy-Efficient Designs: Focus on designing energy-efficient RIS architectures and control mechanisms to minimize power consumption and extend battery life in 6G networks explain L. Ge et al (2020). Investigate energy harvesting techniques and self-powered RIS solutions to enable autonomous operation and sustainability in resource-constrained environments. Standardization and Interoperability: Contribute to standardization efforts for RIS technology in 6G communication standards bodies and organizations. Define interoperability standards, protocols, and interfaces to ensure seamless integration of RIS-enabled communication systems with existing and future wireless networks. Real-World Deployment and Testbeds: Conduct field trials and testbed experiments to validate the performance of RIS technology in real-world deployment scenarios. Collaborate with industry partners and network operators to deploy RIS-enabled communication systems in urban, suburban, and rural environments and assess their practical feasibility, scalability, and performance. Cross-Disciplinary Collaboration: Foster collaboration between researchers from diverse disciplines such as electrical engineering, computer science, physics, and materials science to address interdisciplinary challenges and explore new avenues for innovation in RIS technology shown in Sharma, T. et. al.(2021). Encourage knowledge exchange and collaboration with industry stakeholders, policymakers, and regulatory bodies to accelerate the adoption of RIS technology in 6G communication networks.

7.1 Standard Timeline for RIS

By releasing the 2023 framework document of the International Telecommunication Union Radiocommunication Sector (ITU-R) IMT-2030, RIS is a major developing technology in 6G communication by Viswanathan (2021). Figure 6 illustrates the standardization for future networks with RIS.

Figure 6. Standard timeline for RIS

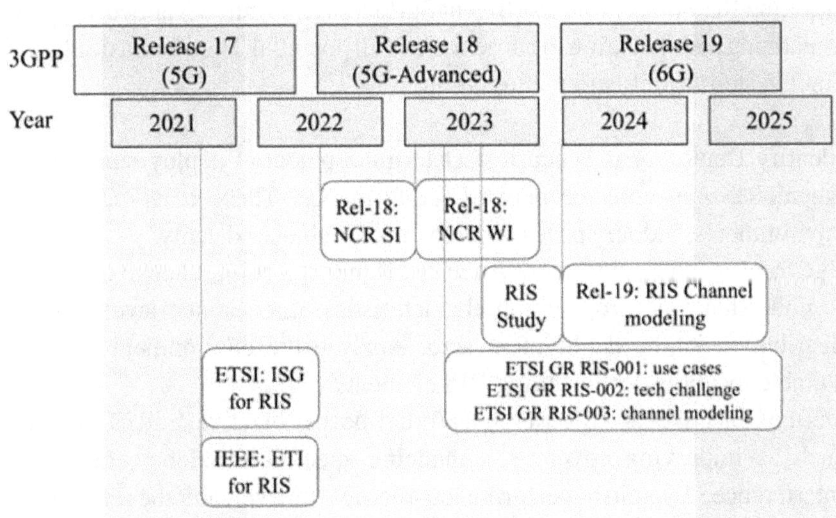

As a result, multiple studies in the literature have examined RIS-assisted communications, covering the theoretical foundation, physical design of metasurfaces, physical architecture, and associated channel modeling at various operational frequencies shown in JungSook Bae (2024), S.Arzykulov (2023). This chapter reviews the key features useful to meet the expectations of a 6G communications network in terms of antennas and beam reconfiguration techniques.

8. FUTURE TRENDS IN RIS

New future trends for RIS are Algorithms with Low Overhead swapping, the use of RIS in Terahertz Communications, Mobility Management, Deployment challenges, AI-based models, and its upgradation. Also, major issues like security and privacy.

Privacy and Security Considerations: As RIS becomes more pervasive, there will be increasing attention to privacy and security considerations. Measures such as encryption, authentication, and access control will be essential to protect RIS-enabled networks and prevent unauthorized access or tampering.

Regulatory and Policy Implications: The deployment of RIS may raise regulatory and policy considerations related to spectrum allocation, licensing, and environmental impact. Policymakers and regulatory bodies will need to address these issues to facilitate the responsible deployment and operation of RIS technology.

Integrating Reconfigurable Intelligent Surfaces (RIS) into 6G networks requires careful planning and execution to harness their full potential. Here's a basic algorithm outlining the steps for effective integration:

1. Identify Deployment Scenarios: Determine potential deployment scenarios where RIS can enhance network performance. These may include urban environments, indoor spaces, or areas with high user density.
2. Site Survey and Environment Analysis: Conduct a comprehensive site survey to understand the propagation characteristics, interference levels, and user distribution in the deployment area. Analyze the environment to identify suitable locations for installing RIS elements.
3. Define Objectives and Metrics: Clearly define the objectives of RIS integration, such as improving coverage, enhancing spectral efficiency, or reducing interference. Establish performance metrics to evaluate the effectiveness of RIS deployment, such as signal-to-interference-plus-noise ratio (SINR), throughput, and latency.
4. Modeling and Simulation: Utilize propagation models and simulation tools to evaluate the impact of RIS deployment on network performance. Simulate

various deployment scenarios and configurations to optimize RIS placement and configuration parameters given in H. Li (2023) .

8.1 Algorithm for Effective Integration of RIS in 6G

a. Initialization: Initialize RIS configuration parameters such as phase shifts, element positions, and activation patterns.
b. Optimization Iterations:
   i. Evaluate the current RIS configuration using performance metrics.
   ii. Generate new configurations by adjusting RIS parameters based on optimization techniques such as genetic algorithms, simulated annealing, or reinforcement learning.
   iii. Assess the performance of the new configurations and select the one that maximizes the predefined objectives.
   iv. Repeat the optimization iterations until convergence or a stopping criterion is met.
c. Final Configuration Selection: Choose the optimal RIS configuration based on the performance achieved during optimization.
d. Integration with Network Planning Tools: Develop interfaces to integrate RIS configurations with existing network planning tools. This enables seamless incorporation of RIS into 6G network design processes.

Figure 7. A basic flowchart for effective integration of RIS in 6G

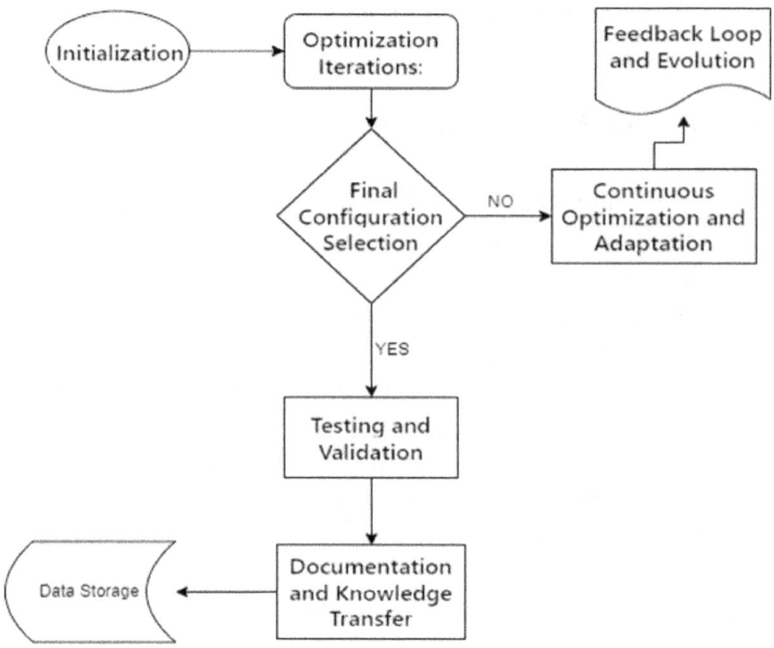

e. Testing and Validation: Conduct extensive testing and validation of the deployed RIS system in real-world conditions. Measure performance metrics and compare them against simulation results to validate the effectiveness of RIS integration.

f. Continuous Optimization and Adaptation: Monitor network performance continuously and adapt RIS configurations dynamically to accommodate changes in user behavior, environmental conditions, and network traffic patterns.

g. Documentation and Knowledge Transfer: Document the RIS integration process, including deployment strategies, configuration parameters, optimization techniques, and performance evaluation results. Transfer knowledge to network operators and maintenance teams to ensure efficient operation and troubleshooting.

h. Feedback Loop and Evolution: Establish a feedback loop to collect data on RIS performance and user satisfaction. Use this feedback to refine RIS deployment strategies and algorithms, enabling continuous evolution and improvement of the integrated system. Figure 7 indicates the basic Flowchart for the effective integration of RIS in 6G.

By following this algorithm, network operators can effectively integrate Reconfigurable Intelligent Surfaces into 6G networks, unlocking their potential to enhance coverage, capacity, and quality of service

9. CONCLUSION

In conclusion, the development of Reconfigurable Intelligent Surfaces (RIS) represents a breakthrough in the field of wireless technology, particularly in the context of 6G technology. Through harnessing the power of intelligent surfaces to manipulate electromagnetic waves, RIS offers unparalleled potential to enhance network performance, efficiency, and coverage. As discussed, RIS enables dynamic control over signal propagation, effectively mitigating issues such as path loss, interference, and fading. Its adaptability and reconfigurability pave the way for innovative approaches to spectrum utilization and resource allocation, promising to revolutionize wireless networks' operation. Furthermore, the integration of artificial intelligence and machine learning techniques with RIS opens up new avenues for optimization and self-configuration, leading to autonomous and intelligent network management systems. However, while the potential benefits of RIS are vast, challenges such as hardware complexity, deployment cost, and regulatory considerations must be addressed for widespread adoption. In summary, the RIS technique holds immense promise for shaping the future of 6G wireless communication networks, offering unprecedented opportunities for enhancing connectivity, capacity, and user experience. As research and development in this field continue to progress, RIS is poised to play a pivotal role in realizing the vision of a truly connected and intelligent world.

REFERENCES

Adam, F. (2023). Selected Aspects of Nonorthogonal Multiple Access for Future Wireless Communications. Mathematics in Computer Science. doi:10.1007/s11786-023-00561-y

Almekhlafi, M. H., Arfaoui, M. A., Assi, C., & Ghrayeb, A. (2023, May). A Low Complexity Passive Beamforming Design for Reconfigurable Intelligent Surface (RIS) in 6G Networks. *IEEE Transactions on Vehicular Technology*, *72*(5), 6309–6321. Advance online publication. doi:10.1109/TVT.2022.3233469

Arzykulov, S., Celik, A., Nauryzbayev, G. M., & Eltawil, A. (2023). Artificial noise and RIS-aided physical layer security: Optimal RIS partitioning and power control. *IEEE Wireless Communications Letters*, *12*(6), 992–996. doi:10.1109/LWC.2023.3256001

Bai, T., Pan, C., Deng, Y., Elkashlan, M., Nallanathan, A., & Hanzo, L. (2020). Latency Minimization for Intelligent Reflecting Surface Aided Mobile Edge Computing. *IEEE Journal on Selected Areas in Communications*, *38*(11), 2666–2682. doi:10.1109/JSAC.2020.3007035

Basar E., Di Renzo M., De Rosny J., Debbah M., Alouini M.-S., & Zhang R., (2019). Wireless communications through reconfigurable intelligent surfaces. *IEEE Access, 7*.

Basharat, S., Khan, M., Iqbal, M., Hashmi, U. S., Zaidi, S. A. R., & Robertson, I. (2022). Exploring reconfigurable intelligent surfaces for 6G: State-of-the-art and the road ahead. *IET Communications*, *16*(13), 1458–1474. doi:10.1049/cmu2.12364

Bhatnagar, M. R., & Ghassemlooy, Z. (2016). Performance analysis of Gamma–Gamma fading FSO MIMO links with pointing errors. *Journal of Lightwave Technology*, *34*(9), 2158–2169. doi:10.1109/JLT.2016.2526053

Bjornson, E., Ozdogan, O., & Larsson, E. G. (2020). Reconfigurable intelligent surfaces: Three myths and two critical questions. *IEEE Communications Magazine*, *58*(12), 90–96. doi:10.1109/MCOM.001.2000407

Chen, Z., Tang, J., Zhang, X. Y., Wu, Q., Wang, Y., Jin, S., & Wong, K.-K. (2022). Offset learning-based channel estimation for intelligent reflecting surface-assisted indoor communication. *IEEE Journal of Selected Topics in Signal Processing*, *16*(1), 41–55. doi:10.1109/JSTSP.2021.3129350

Farshad, R. (2023). *Analytical Characterization of Coverage Regions for STAR-RIS-aided NOMA/OMA Communication Systems*. ArXiv.org, doi: /arXiv.2306.00404. doi:10.48550

Fernandes, C. A., Lima, E. B., & Costa, J. R. (2016). Dielectric lens antennas in Handbook of Antenna Technologies. Springer.

Fu, X., Yang, F., Liu, C., Wu, X., & Cui, T. J. (2020). Terahertz beam steering technologies: From phased arrays to field-programmable metasurfaces. *Advanced Optical Materials*, *8*(3), 1900628. doi:10.1002/adom.201900628

Ge, L., Dong, P., Zhang, H., Wang, J. B., & You, X. (2020). Joint beamforming and trajectory optimization for intelligent reflecting surfaces-assisted UAV communication. *IEEE Access : Practical Innovations, Open Solutions*, *8*, 78702–78712. doi:10.1109/ACCESS.2020.2990166

Giacomo, Benoni, Rocca, Salucci, & Massa. (2023). *RIS Dynamic Control Strategies in Smart EM Environments for 6G Networks*. . doi:10.23919/EuCAP57121.2023.10133521

Hong, I.-P. (2023). Reviews Based on the Reconfigurable Intelligent Surface Technical Issues. *Electronics (Basel)*, *12*(21), 4489. doi:10.3390/electronics12214489

Hong. (2022). Review of antenna technology using metasurfaces. *Electromagnet. Tech. KIEES*, *33*, 20–53.

Jaafar, Kadhim, & Hasan. (2023). Design and Implementation Unit Cell for 6G Reconfigurable Intelligent Surface Application. *International Journal of Online Engineering*. doi:10.3991/ijoe.v19i05.37585

Jiang, M., Cezanne, J., Sampath, A., Shental, O., Wu, Q., Koymen, O., Bedewy, A., & Li, J. (2022). Wireless front haul for 5G and future radio access networks: Challenges and enabling technologies. *IEEE Wireless Communications*, *29*(2), 108–114. doi:10.1109/MWC.003.2100482

Jiang, R., Fei, Z., Huang, S., Wang, X., Wu, Q., & Ren, S. (2023). Bivariate Pilot Optimization for Compressed Channel Estimation in RIS-Assisted Multiuser MISO-OFDM Systems. *IEEE Transactions on Vehicular Technology*, *72*(7), 9115–9130. doi:10.1109/TVT.2023.3250252

JungSook, Khalid, Lee, Lee, Noh, & Yu. (2024). Overview of RIS-enabled secure transmission in 6G wireless networks. *Digital Communications and Networks*. doi:10.1016/j.dcan.2024.02.005

Li, H., Shen, S., & Clerckx, B. (2023). Beyond diagonal reconfigurable intelligent surfaces: From transmitting and reflecting modes to single-, group-, and fully-connected architectures. *IEEE Transactions on Wireless Communications*, *22*(4), 2311–2324. doi:10.1109/TWC.2022.3210706

MoazzamBukhari, Sundus, & Muhammad. (2023). *Wireless Communications Beyond Antennas: The Role of Reconfigurable Intelligent Surfaces*. . doi:10.3390/engproc2023032010

Ouamri, M. A., Alkanhel, R., Gueguen, C., Alohali, M. A., & Ghoneim, S. S. (2023). Modeling and analysis of UAV-assisted mobile network with imperfect beam alignment. *CMC-Comput.*, *74*, 453–467. doi:10.32604/cmc.2023.031450

Pan, C., Ren, H., Wang, K., Kolb, J. F., Elkashlan, M., Chen, M., Di Renzo, M., Hao, Y., Wang, J., Swindlehurst, A. L., You, X., & Hanzo, L. (2021). Reconfigurable intelligent surfaces for 6G systems: Principles, applications, and research directions. *IEEE Communications Magazine*, *59*(6), 14–20. doi:10.1109/MCOM.001.2001076

Rahmatov, N., & Baek, H. (2023). RIS-carried UAV communication: Current research, challenges, and future trends. *ICT Express*, *9*(5), 961–973. doi:10.1016/j.icte.2023.03.004

Scott, A. (2023_. *Wireless Communications Empowered by Reconfigurable Intelligent Surfaces: Model-Based vs Model-Free Channel Estimation*. doi:10.1016/j.jiixd.2023.06.010

Sharma, T., Chehri, A., & Fortier, P. (2021). Reconfigurable Intelligent Surfaces for 5G and beyond Wireless Communications: A Comprehensive Survey. *Energies*, *14*(24), 8219. doi:10.3390/en14248219

Tang, X., Liu, N., Zhang, R., & Han, Z. (2022). *Deep learning-assisted secure UAV-relaying*. Academic Press.

Trichopoulos, G. C., Theofanopoulos, P., Kashyap, B., Shekhawat, A., Modi, A., Osman, T., Kumar, S., Sengar, A., Chang, A., & Alkhateeb, A. (2022). Design and evaluation of reconfigurable intelligent surfaces in a real-world environment. *IEEE Open Journal of the Communications Society*, *3*, 462–474. doi:10.1109/OJCOMS.2022.3158310

Viswanathan, H., & Mogensen, P. E. (2020). Communications in the 6G era. *IEEE Access : Practical Innovations, Open Solutions*, *8*, 57063–57074. doi:10.1109/ACCESS.2020.2981745

Wu, Q., & Zhang, R. (2020). Towards smart and reconfigurable environment: Intelligent reflecting surface aided wireless network. *IEEE Communications Magazine*, *58*(1), 106–112. doi:10.1109/MCOM.001.1900107

Zhou, E., Liu, Z., Lan, P., Xiao, W., Yang, W., & Niu, X. (2023). Interference Avoidance through Periodic UAV Scheduling in RIS-Aided UAV Cluster Communications. *Electronics (Basel)*, *12*(21), 4539. doi:10.3390/electronics12214539

Chapter 6
Revolutionizing Wireless Communication:
A Comprehensive Study on Modern Antenna Technologies

Pardhu Thottempudi
https://orcid.org/0000-0002-9653-1951
BVRIT HYDERABAD College of Engineering for Women, India

Vijay Kumar
Vellore Institute of Technology, India

ABSTRACT

In the rapidly evolving wireless communication landscape, antenna technology is indispensable. This chapter provides an in-depth review of recent advancements and trends in antenna technology transforming modern wireless communication systems. The authors delve into vital technological innovations, including massive MIMO, beamforming, metamaterial antennas, and reconfigurable intelligent surfaces, shedding light on their functions, potential benefits, and implications. The study underscores that the ongoing evolution in antenna technology holds immense potential to revolutionize wireless communication systems, enabling more efficient, high-speed, and sustainable networks. These insights will benefit communication engineers, researchers, and academicians, offering a broad understanding of antenna technology's current state and future trajectories in wireless communication systems.

DOI: 10.4018/979-8-3693-0916-2.ch006

INTRODUCTION

The rapid evolution of wireless communication technologies has significantly transformed our social, professional, educational, recreational, and commercial interactions, fostering a level of interdependence previously unimaginable. The antennas are central to this technological revolution, crucial for transmitting and receiving wireless signals in everything from compact wearable devices to extensive data networks. As the number of connected devices skyrockets and demands for higher data rates soar, antenna technology advances, embracing multifaceted disciplines, including electromagnetics, materials science, and signal processing.

Today's antennas are not limited to basic signal transmission and reception; they are integral to sophisticated next-generation wireless systems equipped for complex functions like multi-band operation, beam steering, and multiple input and output (MIMO) capabilities. Modern challenges demand efficient, cost-effective, compact, and lightweight antennas.

This research explores the forefront of antenna technology and its implications for evolving wireless communication networks. Comprehending these breakthroughs is essential for understanding the enhanced functionalities of future-generation wireless frameworks in our progressively digitized environment. Among these innovations, reconfigurable antennas are noteworthy for their ability to support operations across multiple bands, shape directional beams, and adjust polarization, thereby significantly broadening their utility. Such attributes play a pivotal role in minimizing the size and complexity of antenna setups, consequently elevating the performance of radio frequency (RF) systems (Nadeem et al., 2018), (Bhartia & Bahl, 1982). Characterized by their flexibility to alter operating frequencies, radiation patterns, and polarization, reconfigurable antennas adapt seamlessly to various communication services, protocols, and specific use-case demands. The degree of their adaptability often correlates with including active elements like switches or capacitors. Adopting RF microelectromechanical systems (MEMS), varactor diodes, and PIN diodes is essential for enabling a wide range of adjustability and rapid switching capabilities (Grau & Lee, 2010), (Tawk et al., 2012), (Nikolaou et al., 2006).

This segment delves into diverse designs of reconfigurable antennas suited for mobile and wireless contexts, underscoring their ability to perform singular or multiple reconfiguration tasks. Through this examination, the document furnishes readers with a deeper understanding of cutting-edge antenna technologies, steering them toward promising avenues for future investigation in antenna development. This is crucial for developing next-generation wireless communication infrastructures (Bernhard, 2007),(Peroulis et al., 2005),(Christodoulou et al., 2012),(Balanis, 1998),(Bernhard, 2005).

EVOLUTION OF ANTENNA TECHNOLOGY AND ITS EFFECT ON MODERN WIRELESS COMMUNICATION SYSTEMS

A. Evolution of Antenna Technologies

- **The Dawn of Wireless Communication: Marconi's Early Experiments**

The inception of wireless communication traces back to the late 19th century with Guglielmo Marconi's invention of the radio. Utilizing simple dipole antennas, which were essentially straight rods divided in the middle, Marconi pioneered the first steps into wireless transmissions. Despite their simplicity, these early antennas laid the groundwork for the future of wireless communication.

- **Microwave Era: The Rise of Parabolic and Horn Antennas**

The mid-20th century brought about the era of microwave communication, necessitating the development of new antenna designs. Parabolic (dish) and horn antennas emerged as frontrunners for their exceptional performance in point-to-point and satellite communications, significantly enhancing the efficiency and reach of wireless communication systems.

- **The Digital Explosion: Log-periodic and Yagi-Uda Antennas**

The digital revolution saw a boom in broadcasting media such as television and radio, propelling the adoption of multi-element antennas like log-periodic and Yagi-Uda designs. Known for their excellent directionality and wide frequency range, these antennas enabled high-quality, reliable broadcasts, setting the stage for modern mass communication technologies.

- **Wireless Revolution: The Advent of Microstrip Antennas**

The rapid spread of mobile phones and Wi-Fi technology called for compact, efficient antennas. Microstrip or patch antennas became ubiquitous, notable for their small size and compatibility with printed circuit board manufacturing. These antennas are integral to billions of devices worldwide and are critical to the widespread access to wireless internet and telephony services.

- **Contemporary Advances: Smart Antennas**

Wireless communication demands high data throughput, extensive coverage, and minimal interference. Meeting these needs, smart antennas like MIMO and phased array systems offer dynamic radiation pattern adjustments to enhance signal quality, supporting advanced technologies such as 4G and 5G.

B. Impact on Modern Wireless Communication

The progression of antenna technology has fundamentally transformed modern wireless communication systems. The miniaturization and increased gain of antennas have been pivotal for the widespread adoption of personal wireless devices, such as smartphones and laptops. Smart antennas have been instrumental in developing high-speed, high-capacity networks, accommodating the exponential growth of data traffic.

Technologies like MIMO and beamforming have significantly boosted the efficiency of wireless networks, allowing for higher data rates without additional spectrum allocation. These advancements are crucial for contemporary wireless protocols, including LTE and 5G, facilitating applications from real-time video streaming to the Internet of Things.

Ongoing research into innovative antenna technologies, such as metamaterials and reconfigurable antennas, promises even more sophisticated, flexible, and resilient wireless communication systems. As connectivity becomes increasingly integral to our daily lives, the evolution of antenna technology remains at the heart of future wireless communication developments.

MIMO AND BEAMFORMING FOR MODERN WIRELESS COMMUNICATION SYSTEMS

1. Multiple-Input Multiple-Output (MIMO) Technology

MIMO technology has revolutionized wireless networking by employing multiple antennas to transmit and receive data. This setup not only multiplies the data transmission rate by channeling multiple data streams through distinct antennas but also bolsters the robustness of wireless connections through diversity gain, which provides several paths for data to traverse.

MIMO manifests in three primary forms: single-user (SU-MIMO), multi-user (MU-MIMO), and massive MIMO. SU-MIMO enhances data throughput for individual users by dispatching multiple data streams concurrently, while MU-MIMO extends this capability across several users simultaneously, thus amplifying the overall system capacity. Massive MIMO, a cornerstone of 5G technology, employs an extensive array of antennas at the base station to simultaneously cater to numerous users, significantly advancing efficiency, system capacity, and reliability.

2. Beamforming Technology

Beamforming refines signal transmission and reception by precisely focusing signals through sensor arrays, thereby improving wireless system range and data rates. This technique enhances the signal-to-interference-plus-noise ratio (SINR) by amplifying signal strength in a specific direction. In a beamforming setup, each antenna array element transmits or receives identical signals at slightly staggered intervals, fostering constructive interference in targeted directions while minimizing it elsewhere. This strategic concentration of signal energy gives beamforming its name, as it effectively shapes a focused beam of communication.

3. Synergy of MIMO and Beamforming

Integrating MIMO with beamforming yields a formidable synergy for wireless networks. MIMO's capability to transmit multiple data streams simultaneously complements beamforming's precision in directing these streams to their destinations. This combination is instrumental in enhancing the performance of contemporary wireless systems by supporting higher data speeds, extending coverage, increasing network capacity, and minimizing interference. As the backbone of modern and future wireless communications, including the deployment of 5G networks, the fusion of MIMO and beamforming technologies is pivotal.

PHASED ARRAY ANTENNAS FOR STATE-OF-THE-ART WIRELESS NETWORKS

In today's state-of-the-art wireless communication systems, phased array antennas are indispensable. They are antenna arrays in which the effective radiation pattern of the array is strengthened in a desired direction while suppressed in undesired ones by adjusting the relative phases of the corresponding signals feeding the antennas.

1. The Theory Behind Phased Array Antennas

A phased array antenna consists of numerous smaller antennas, each receiving a uniquely phase-shifted signal. This setup allows for the precise steering of the antenna's beam in desired directions through electronic adjustments of the signal's phase at each element, eliminating the need for mechanical movement and enabling swift and accurate control over the radiation pattern.

Key Advantage of Phased Array Antennas

The foremost advantage of phased array antennas is their capability to precisely direct signals. This targeted approach minimizes interference, boosts signal quality, and extends coverage area. The agility of phased arrays in beam steering makes them invaluable across various applications, including radar, satellite, and mobile communications.

An additional advantage is their capacity for generating multiple beams simultaneously, significantly enhancing the efficiency and reach of wireless networks through multi-beam operations.

Phased Array Antennas and the Evolution of Wireless Networks

Phased array antennas are pivotal in modern wireless communication systems, underpinning high-speed data transmission, low latency, and extensive network coverage. They form the backbone of advanced technologies such as multiple input and multiple output (MIMO) and beamforming, which are critical for the infrastructure of 4G and 5G networks.

Phased array antennas are essential for 5G communications, particularly millimeter-wave frequencies. These frequencies experience notable attenuation, necessitating phased arrays to focus beams and ensure reliable coverage effectively.

For satellite communications, the ability of phased array antennas to dynamically steer beams is crucial, maintaining uninterrupted links with satellites in motion. This feature underscores the vital role of phased array antennas in ensuring seamless global connectivity.

2. Future Perspectives

Phased array antennas will continue to progress due to the persistent demand for faster data rates and lower latencies in wireless communication. Topics of study include: • creating smaller and more powerful array designs.

Combining with existing technology; for example, incorporating metamaterials.

Novel uses, such as terahertz communication and beyond 5G networks, are being investigated.

METAMATERIALANTENNASINMODERNWIRELESSCOMMUNICATION SYSTEMS

Metamaterials have catalyzed a significant transformation in antenna technology, distinguished by their artificially engineered attributes that diverge from natural occurrences. These materials possess extraordinary electromagnetic properties, such as a negative refractive index, positioning them as prime candidates for cutting-edge antenna applications.

A. Metamaterials: A Primer

Metamaterials are artificially fabricated structures that showcase properties rarely, if ever, seen in nature. Constructed from assemblies of units composed of conventional microscale materials like metals or plastics, these structures are arrayed in precise patterns. The distinctiveness of metamaterials arises not from the constituent materials themselves but from their deliberate architectural design. The shape, geometry, size, orientation, and arrangement of these structures dictate their interaction with electromagnetic waves, ushering in capabilities unachievable with conventional materials.

B. Advancements in Antenna Design through Metamaterials

Leveraging metamaterials' unique attributes, including negative permittivity and permeability, has birthed a novel category of antennas boasting superior performance features. These include miniaturization, bandwidth enhancement, increased directivity, and augmented gain. For example, designs for patch antennas inspired by metamaterials, where traditional dielectric substrates are substituted with metamaterial frameworks, facilitate the development of compact antennas with broader bandwidths compared to standard microstrip patch antennas.

C. The Role of Metamaterial Antennas in Contemporary Wireless Networks

Metamaterial antennas emerge as a strategic solution in today's wireless communication landscapes, characterized by a relentless quest for compact, high-efficiency, and high-capacity antennas. They are particularly pivotal in navigating the challenges of 5G and future network generations, such as the scarcity of spectrum resources at lower frequencies. By enabling efficient operation at higher frequency bands, including the millimeter-wave spectrum, metamaterial antennas can significantly contribute to overcoming these hurdles, thanks to their superior gain and directivity.

D. Prospects and Research Directions

The potential of metamaterial antennas in reshaping wireless communication is immense. Research efforts are geared towards addressing the production and deployment challenges of metamaterial antennas for widespread commercial use. Fields poised to reap substantial benefits from advancements in this technology include the Internet of Things (IoT), 6G wireless communications, satellite connectivity, and wireless power transmission.

Despite existing challenges, the unique properties of metamaterials open up exciting possibilities for redefining antenna technology and enhancing wireless communication systems.

INTERNET OF THINGS (IOT) AND ANTENNAS

The Internet of Things (IoT) has rapidly evolved into a transformative force across various sectors, heralding significant changes in everyday life and industrial practices. It encompasses a vast network of physical entities—called 'things'—embedded with sensors, software, and connectivity capabilities, facilitating their interaction and data exchange over the internet.

A. Antennas' Crucial Role in IoT

Antennas are foundational to IoT's success, given the reliance on wireless communication to interconnect devices. These components are essential for efficient transmission and reception of signals within IoT systems, necessitating dependable, efficient, and adaptable designs to diverse operational contexts.

B. Key Antenna Specifications and Challenges

- **Compact Design:** IoT gadgets often require small, portable designs, demanding antennas that match these physical constraints without sacrificing functionality.
- **Energy Efficiency:** Given many IoT devices' battery-operated and sometimes inaccessible nature, their antennas must minimize power consumption.
- **Frequency Versatility:** IoT applications vary widely, with devices operating across different spectral bands. Antennas must either support multiband functionality or allow for straightforward reconfiguration.

- **Environmental Robustness:** Deployment across varied settings, from indoor to extreme outdoor environments, requires antennas that consistently perform well under any condition.

C. Innovative Antenna Solutions for IoT

To address these demands, several antenna technologies have been deployed in IoT devices, including:

- **Chip Antennas:** Ideal for IoT due to their integration ease on circuit boards, offering robust performance across multiple bands.
- **Printed Antennas:** Favored for their cost-effectiveness and manufacturing simplicity, they can be tailored for various frequencies and directly printed onto devices.
- **Reconfigurable Antennas:** These adapt their operational parameters to meet specific communication needs, supporting diverse standards within IoT.
- **Metamaterial Antennas:** They provide benefits such as device miniaturization and performance optimization, enhancing IoT applications.

D. The Future of Antenna Technology in IoT

As IoT expands, the demand for advanced antenna technologies will escalate, necessitating innovations that promise further miniaturization, improved energy efficiency, and greater adaptability. Future developments will likely leverage sophisticated technologies like MIMO, beamforming, and phased arrays, especially as IoT integrates with 5G and subsequent generations for superior connectivity and throughput.

Antennas in the 6G Era and Beyond

The advent of 5G and prospective 6G networks introduces challenges and opportunities for antenna technology, driven by the quest for higher data rates, reduced latency, energy savings, and extensive device connectivity. Innovations in antenna design will be critical to fulfilling these emerging demands, signifying a vibrant and evolving area of research and development in wireless communication.

ANTENNA TECHNOLOGIES FOR 5G

5G networks are expanding into the millimeter-wave (mm-wave) frequency bands to support the burgeoning demand for high-speed data transmission. This leap into higher frequencies necessitates advancements in antenna technology to manage these signals effectively.

Massive MIMO in 5G

A cornerstone of 5G technology is Massive Multiple-Input Multiple-Output (MIMO), which significantly increases network capacity and spectral efficiency. Massive MIMO can substantially boost the data throughput and efficiency of wireless networks by deploying numerous antennas at a single base station to serve multiple users simultaneously.

Beamforming and Its Role

In conjunction with massive MIMO, beamforming is a technique that concentrates a signal's power towards a specific user, enhancing the signal quality for the recipient while reducing overall network interference. This targeted approach is essential for optimizing individual user experiences in dense network environments.

Phased Array Antennas for mmWave

Due to the inherently high propagation losses at mmWave frequencies, phased array antennas are critical. They generate focused beams of radiation to overcome these losses, making efficient use of the mmWave spectrum by directing signal power precisely where needed.

Advancements for Beyond 5G (B5G) and 6G Networks

As the deployment of 5G progresses, research into Beyond 5G (B5G) and 6G networks is already underway. These future networks aim to deliver even higher data speeds, reduced latency, and more pervasive connectivity, likely extending into the sub-terahertz (sub-THz) and terahertz (THz) bands. This shift presents both challenges and opportunities for antenna technology:

- **Metamaterial Antennas:** Utilizing the unique properties of metamaterials, such as negative refractive indices and enhanced directivity, can lead to compact, efficient antenna designs suitable for B5G/6G frequencies.

- **Reconfigurable Intelligent Surfaces (RIS):** This innovative technology employs surfaces embedded with numerous small antennas to manipulate incoming signals, potentially enhancing network coverage and efficiency while reducing energy consumption.
- **Holographic Beamforming:** Offering precise control over beam shaping and direction, holographic beamforming could significantly improve network performance in densely populated and dynamically changing environments.

These advancements in antenna technology are pivotal for navigating the complexities of future wireless networks, ensuring that they can meet the ever-increasing demands for data speed, reliability, and connectivity energy-efficiently.

USE OF RECONFIGURABLE ANTENNAS IN STATE-OF-THE-ART WIRELESS NETWORKS

Reconfigurable antennas stand out for their ability to dynamically adjust to changes in the communication environment or system requirements by altering their frequency, radiation pattern, and polarization. This adaptability is achieved by modifying the antenna's electrical properties or physical structure.

A. The Essence of Reconfigurable Antennas

The adaptability of reconfigurable antennas is often facilitated by micro-electrical mechanical systems (MEMS), RF switches, varactors, and PIN diodes. These technologies enable antennas to vary their operational characteristics, optimizing performance across signal conditions and communication standards.

B. Application in Contemporary Wireless Systems

The ability to support many bands, modes, and beam directions is crucial in the current wireless communication landscape due to the diversity of wireless services and applications. With their versatile nature, reconfigurable antennas can single-handedly fulfill these multifaceted requirements, eliminating the need for multiple fixed antennas. Frequency reconfigurability allows these antennas to operate simultaneously across various networks (e.g., Wi-Fi, Bluetooth, cellular). Radiation pattern reconfigurability enhances signal quality by focusing the beam toward the intended receiver and minimizing interference. Additionally, switching polarization can optimize communication efficiency, catering to different transmission requirements like circular polarization for satellite communications and linear polarization for terrestrial links.

C. Practical Applications

Reconfigurable antennas find significant applications in several areas of wireless communication:

- Mobile and computing devices leverage them to support diverse wireless protocols.
- In MIMO systems, dynamic adjustment of radiation patterns through beamforming techniques enhances signal quality and data throughput.
- Cognitive radio systems benefit from the frequency agility of reconfigurable antennas, allowing for the efficient utilization of available spectrum bands.

D. Future Prospects

The future development of reconfigurable antennas is driven by the evolving demands of wireless communication technologies, including 5G and beyond, the Internet of Things (IoT), and vehicle-to-everything (V2X) communications. Despite challenges in design and fabrication, the potential benefits of reconfigurable antennas—such as enhanced performance, flexibility, and compactness—make them a pivotal component in advancing wireless systems.

The trajectory of reconfigurable antenna technology underscores its critical role in shaping the adaptability and efficiency of future wireless networks, promising significant advancements in connectivity and communication capabilities.

Antennas With a Single Reconfiguration Feature

Reconfigurable antennas can execute various actions, including but not limited to frequency switching, bandwidth adjustment, emission pattern change, and polarization modification. The features above are accomplished by actively manipulating the dimensions and configuration of the antenna's radiating elements using active components. This section examines many types of reconfigurable antennas, each specifically engineered to fulfill a singular reconfigurable function.

Frequency Reconfigurable Antennas

Frequency-reconfigurable antennas are crucial for contemporary wireless devices like smartphones and tablets, where space-saving is crucial. It is crucial to comprehend the basic principles of these antennas. The devices are engineered to switch between various frequency bands dynamically or adapt constantly within a given frequency range, as (Smith et al., 2020) described. The antennas play a crucial role in multi-band/multi-service systems by enabling the effective reallocation of frequencies for dynamic spectrum access. One typical method is to change the electrical length of the antenna's radiator by utilizing PIN diodes (Parchin et al., 2018), (Ojaroudi et al., 2014), (Ojaroudi et al., 2014).

Varactor diode switches alter the operating frequency, allowing a transition from wideband or ultra-wideband (UWB) modes to specific narrowband modes. Matching networks are involved in determining the resonance frequency (Hinsz & Braaten, 2014), (Hannula et al., 2017).

Studies have investigated alternative methods, such as adding a shorting post or modifying the input impedance, to improve the antenna's frequency response.

Figure 1, depicted in reference (Sam et al., 2014), offers a visual aid for comprehending the design and operation of frequency-reconfigurable antennas. The design utilizes a composite right/left-handed transmission line (CRLH-TL) with parallel and series capacitance/inductance components. The addition of a varactor to the meander-line slot enables the antenna to operate across frequencies from 4.13 GHz to 4.50 GHz by applying voltages between 0 and 36 V. This design showcases the capability of these antennas for precise frequency tuning, rendering them appropriate for phased array systems in 5G base stations.

Figure 1. (a) Constructed model of the reconfigurable antenna and (b) frequency response across various voltage levels (Sam et al., 2014)

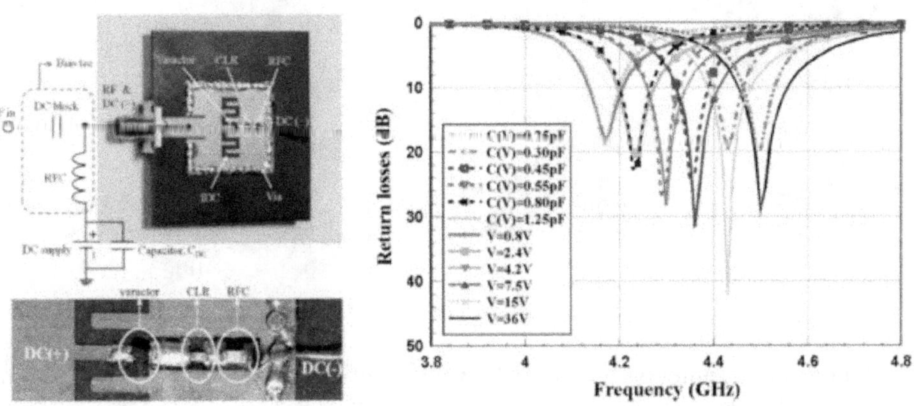

(a) (b)

Bandwidth Reconfigurable Antennas

The bandwidth of an antenna, which delineates the range of frequencies it can send or receive, is adaptable from narrow to broad spectrums. This adaptability is often achieved by integrating a matching network or employing reconfigurable filter designs within the antenna's feed system (Anand et al., 2016),(Anagnostou et al., 2017),(Meng et al., 2017).

For the antenna to deliver peak performance across its entire frequency range, it must sustain superior performance levels across the total bandwidth of interest. A prevalent strategy involves modifying the ground plane, particularly for patch or monopole antenna types. Research (Chen et al., 2016) introduced an innovative antenna designed to adjust its bandwidth to meet the demands of WLAN and WiMAX applications. This antenna incorporates a rectangular radiating patch with a unique, rotated F-shaped slot connected to the microstrip feed line, as illustrated in Figure 2a. The antenna's underside hosts a truncated ground plane on the FR-4 substrate. Central to the slot, two PIN diodes are precisely placed to regulate current flow, allowing for the bandwidth adjustment from narrow to wide ranges, as shown in Figure 2b. This capability grants the antenna an impressive impedance bandwidth range from 22% to 78%, enhancing its utility across various wireless communication platforms.

Figure 2. (a) Configurable bandwidth antenna design and (b) performance across various PIN diode states (Chen et al., 2016)

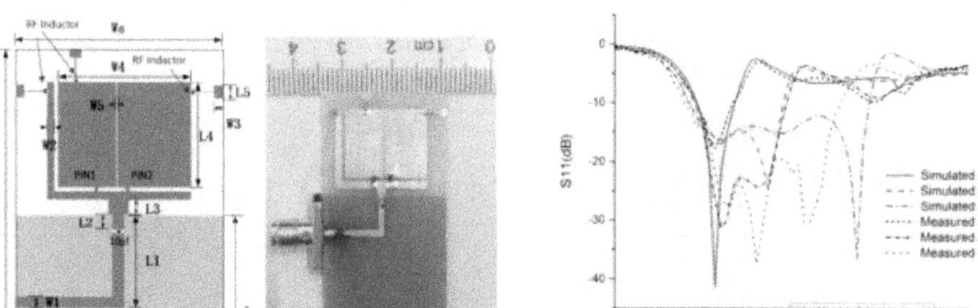

(a) (b)

Moreover, the antenna can toggle between multi-resonance and wideband/Ultra-Wideband (UWB) operational modes. This versatility mitigates potential interference with UWB systems and other wireless technologies, including 4G, WLAN, WiMAX, and satellite communications (Horestani et al., 2016). Its ability to accommodate multiple communication services within a single device enhances its efficiency in multi-mode wireless communication environments.

Pattern Reconfigurable Antennas

Antennas with pattern reconfigurability, which can produce diverse radiation patterns at a consistent frequency, are exceptionally suited for surveillance and tracking tasks (Abdulraheem et al., 2014),(Zhong-Liang et al., 2014). Such antennas are typically designed by crafting specific patterns or utilizing adaptive feeding mechanisms, a principle borrowed from array theory (Chen et al., 2018), (Tang et al., 2018). A notable design (Jin et al., 2018), depicted in Figure 3a, features four identical arc-shaped dipoles positioned around a Rogers 4350B substrate linked to a reconfigurable feed network. Modifying the diodes allows the antenna to shift its end-fire radiation pattern by 90 degrees within the azimuthal plane, enabling omnidirectional coverage, as demonstrated in Figure 3b. The design ensures a steady reflection coefficient (S11) across the 2.3–3.2 GHz spectrum, making it compatible with 4G and 5G communications.

Techniques involving shorting pins, capacitors, and adaptable ground planes have been applied to modify the electric field distribution on the antenna substrate, thereby altering the direction of radiation patterns (Zhong, Hong, & Zhou, 2016), (Lu, Yang, & Tan, 2017), (Tran, Nguyen-Trong, Le, Abbosh, & Park, 2018). A particular study (Zhang, Zhang, Lin, Fan, & Pedersen, 2018) unveiled a pattern-reconfigurable phased array, instrumental for pinpointing transmission angles in the 28 GHz band of 5G mobile networks. Moreover, pattern reconfigurable antennas are lauded for their contribution to the efficiency of multiple-input multiple-output (MIMO) wireless communication systems by diminishing interference and boosting overall performance. This aspect and its distinct advantages are comprehensively examined in the study referenced (Chamok, Yılmaz, Arslan, & Ali, 2016).

Polarization Reconfigurable Antennas

Polarization-reconfigurable antennas have emerged as a pivotal solution in mitigating multipath fading impacts, a phenomenon well-documented in prior studies. These antennas are capable of transitioning between different polarization modes (Wu & Luk, 2017), with significant research focus on the ability to switch between right-hand circular polarization (RHCP) and left-hand circular polarization (LHCP) at designated frequencies (Lin, Chen, Ziolkowski, & Guo, 2018). Interest also extends to linear polarization, especially in the vertical and horizontal planes (Tong & Huang, 2008), (Wang & Wong, 2017). The methodologies for attaining polarization flexibility encompass the integration of slots, slits, parasitic elements, and truncated corners alongside active components (Cai, Gao, Yin, Li, & Luo, 2016), (Wu, Okoniewski, & Hayden, 2014). Furthermore, meta-surfaces and sophisticated feeding mechanisms are acknowledged for enabling polarization variability (Zhu, Cheung, Liu, & Yuk, 2014), while programmable polarizers and phase shifters are cited as viable alternatives for achieving this reconfigurability (Row & Hou, 2014).

A particular investigation (Mak, Lai, Luk, & Ho, 2017) revealed an innovative design for a polarization-reconfigurable antenna tailored for the 2.2-2.8 GHz range, pertinent to 4G technologies. This design, illustrated in Figure 4a, includes a circular radiating element with a C-shaped slot and dual diodes over a concentric circular slot. As demonstrated in Figure 4b, this antenna is adept at oscillating between vertical and horizontal linear polarizations (VP/HP) and between LHCP and RHCP, rendering it advantageous for future 5G wireless networks. Additionally, two distinct polarization reconfigurable patch antennas suitable for 5G use were detailed in studies (Al-Yasir et al., 2018a), (Al-Yasir et al., 2018b), with an extensive exploration of their operational characteristics.

Figure 3. (a) S11 measurement results for antenna configuration and (b) radiation patterns corresponding to varied diode states (Jin et al., 2018)

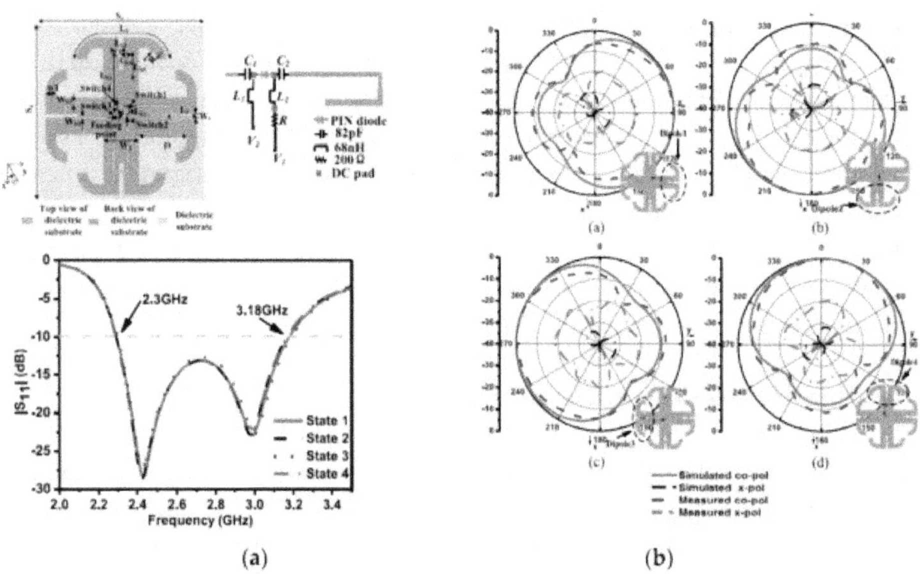

Figure 4. (a) Antenna design schematic and (b) analysis of standing wave ratio (SWR), gain, and axial ratio attributes of the antenna (Mak, Lai, Luk, & Ho, 2017)

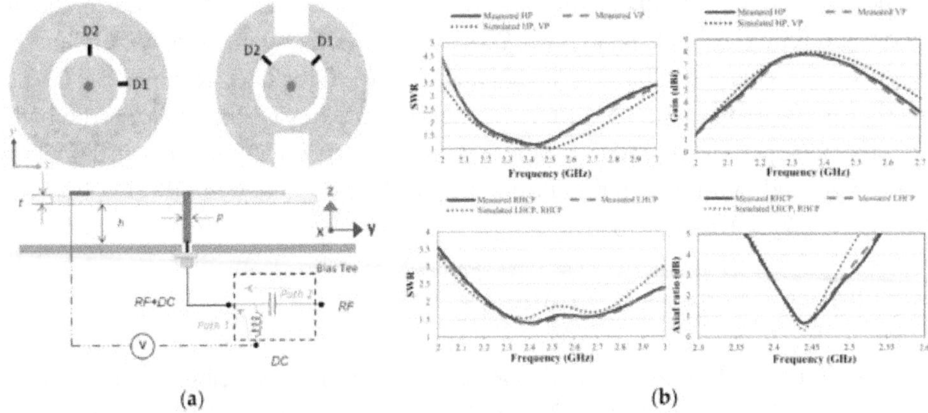

Antennas With Multi Reconfigurable Features

Multi-reconfigurable antennas possess the capability to independently modify two or more features, including frequency and bandwidth, frequency and radiation pattern, frequency and polarisation, radiation pattern and polarisation, as well as frequency, radiation pattern, and polarisation.

Antennas With Frequency and Bandwidth Configuration

The rising need for multi-service radios that support diverse spectrum standards has led to an increased focus on developing antennas capable of transitioning between single-band, multi-band, or wideband operations. This multifunctional capability has become a key area of research, as highlighted in various studies (Qin, Wei, & Guo, 2015), (Kingsly et al., 2018), (Hussain & Sharawi, 2017), (Bitchikh, Rili, & Mokhtar, 2018), (Hussain & Sharawi, 2015a), (Hussain & Sharawi, 2015b), (Chacko, Augustin, & Denidni, 2015), (Nachouane, Najid, Tribak, & Riouch, 2016a), (Nachouane, Najid, Tribak, & Riouch, 2016b). There are several design approaches for antennas to achieve this versatile functionality.

Adapting Antenna Operation Through Design Innovations

One approach to achieving broad operational flexibility involves integrating reconfigurable band-pass filters into the antenna's feed line (Qin, Wei, & Guo, 2015), (Kingsly et al., 2018), (Hussain & Sharawi, 2017), (Bitchikh, Rili, & Mokhtar,

2018). This allows for a seamless shift from narrowband to wideband functionalities. Another strategy involves designing a specialized ground plane featuring parasitic elements, slits, and the inclusion of active components like PIN diodes to enable reconfigurability (Hussain & Sharawi, 2015a), (Hussain & Sharawi, 2015b), (Chacko, Augustin, & Denidni, 2015), (Nachouane, Najid, Tribak, & Riouch, 2016a),). Moreover, the combination of different antenna types to support either narrowband or wideband operation is under exploration, particularly for applications in cognitive radio (CR) and multiple-input multiple-output (MIMO) technologies (Nachouane, Najid, Tribak, & Riouch, 2016b).

Highlighted Research Examples

- **Switchable Filtering for Broadband Antennas:** Research (Qin, Wei, & Guo, 2015), (Kingsly et al., 2018), (Hussain & Sharawi, 2017), (Bitchikh, Rili, & Mokhtar, 2018) delves into incorporating switchable filtering components within the feed line of broadband antennas. A study (Qin, Wei, & Guo, 2015) showcased a microstrip-fed monopole antenna with circular disc geometry capable of altering between wideband and narrowband configurations. The integration of a reconfigurable band-pass filter in the feed line, as illustrated, enables the antenna's impedance bandwidth to be modified, shifting from wideband to narrowband settings. This adjustment is facilitated by varactor diodes, tuning the antenna's frequency range from 3.9 to 4.7 GHz while maintaining isolation in narrowband mode.
- **Elliptical Monopole Antenna with Custom Ground Plane:** In (Kingsly et al., 2018), an elliptical monopole antenna featuring a rectangular slit in its partial ground plane is introduced. The design includes four resonators linked through a split transmission line and a reconfigurable band-pass filter on the feed line. Utilizing four PIN diodes as switches allows the antenna to operate at various frequencies, covering 1.8 GHz, 2.4 GHz, 3.5 GHz (WiMAX/5G), and 5.2 GHz (WLAN). The antenna spans a wide frequency range with all diodes deactivated, encapsulating all mentioned bands.
- **Integration of Narrowband and Ultra-Wideband Antennas:** Reference (Hussain & Sharawi, 2017) presents a unique design that integrates a narrowband antenna within a broader ultra-wideband (UWB) antenna. This configuration pairs a printed UWB monopole antenna atop the substrate with a printed reconfigurable feeding structure beneath. This setup offers a wide impedance bandwidth within the UWB spectrum and can be fine-tuned across six discrete narrow bands, making it particularly suitable for cognitive radio (CR) applications.

Figure 5. (a) Design layout and (b) operational performance of the frequency and bandwidth reconfigurable antenna (Qin, Wei, & Guo, 2015)

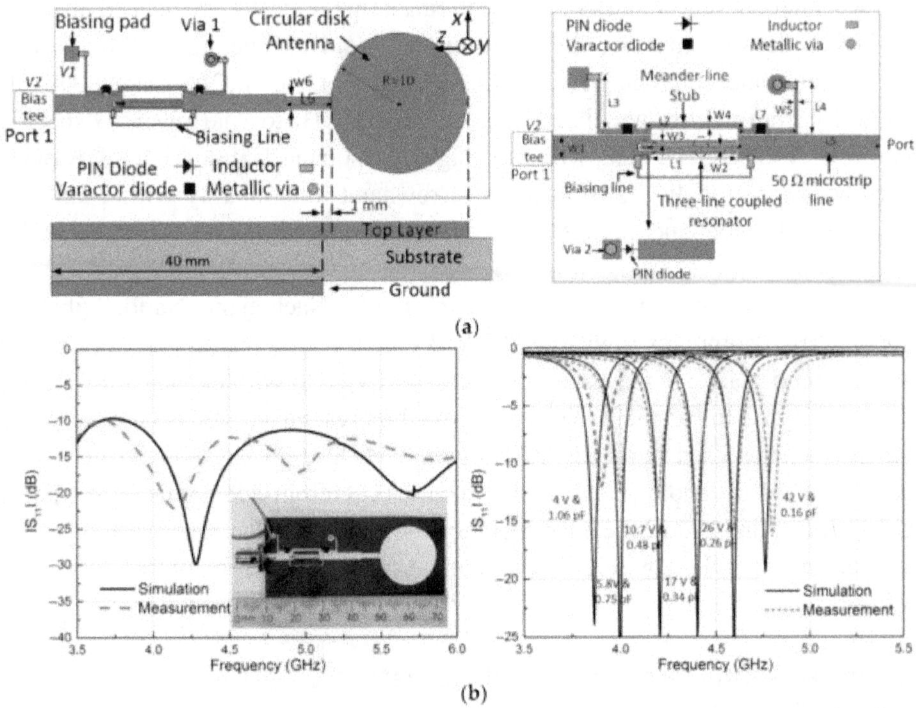

In a study parallel to the one mentioned (Bitchikh, Rili, & Mokhtar, 2018), an innovative antenna design incorporates a microstrip feed line connected to two resonators, furnishing the antenna with ultra-wideband (UWB) capabilities while enabling focused narrowband frequencies through each resonator. When both resonators are activated, the antenna operates in a bi-band mode. This research also highlights two integrated reconfigurable multiple-input multiple-output (MIMO) antenna designs optimized for 3G/4G/5G mobile devices, merging UWB sensing antennas with reconfigurable MIMO functionality on a unified substrate. The design detailed in (Hussain & Sharawi, 2015) features a pair of reconfigurable meander-line antennas on the circuit board's upper layer, with a UWB antenna on the reverse side acting as the ground plane, spanning a bandwidth from 720 MHz to 3440 MHz. Conversely, the reconfigurable MIMO antennas can switch frequencies from 573 MHz to 2550 MHz.

Another design (Hussain & Sharawi, 2015) employs customized printed inverted-F antennas (PIFA), equipped with digital biasing circuits and four PIN diodes, supporting four distinct operational modes within a 755 to 3450 MHz frequency spectrum.

Further developments (Chacko, Augustin, & Denidni, 2015), (Nachouane, Najid, Tribak, & Riouch, 2016) introduce uni-planar designs for comprehensive to narrowband monopole antennas, incorporating a ring-slot filter in the feed line with two PIN diodes to seamlessly shift between wideband and specific communication bands. For instance, one design (Chacko, Augustin, & Denidni, 2015) can modify its bandwidth from an ultra-wideband range of 1.6–6.0 GHz to a narrower 5G-specific range of 3.39–3.80 GHz. Meanwhile, a design (Nachouane, Najid, Tribak, & Riouch, 2016) uses varactor diodes to achieve narrowband functionality within 2.55-3.2 GHz, while its wideband mode extends over approximately 5 GHz, from 1.35 GHz to 6.2 GHz.

A unique integrated antenna system explored in (Nachouane, Najid, Tribak, & Riouch, 2016) combines a wideband UWB antenna for spectrum sensing with a narrowband dipole antenna, creating a dual-port system with extensive frequency coverage from 2 to 5.5 GHz. This setup exhibits resonance at 2.8 GHz, aligning with frequencies used by IEEE 802.11ac/n, UMTS 2000, and 4G/5G networks, making it an ideal candidate for cognitive radio (CR) front-end applications. These advancements underscore the ongoing innovation in antenna design, which is crucial for supporting modern wireless communication systems' versatile and dynamic requirements.

Antennas With Frequency and Radiation Pattern Reconfiguration

This category of antennas is distinguished by its capability to adjust operational frequencies or radiation pattern parameters, significantly enhancing wireless system performance. These antennas can switch radiation patterns to select different operational modes, such as broadside, end-fire, and omnidirectional, optimizing signal direction, reducing energy consumption, and mitigating interference.

Enhancing Wireless Communications With Adaptive Antennas

The adoption of frequency tuning in antennas offers considerable advantages, including minimizing interference from other wireless systems and reducing the necessity for multiple antennas, thereby streamlining communication setups. The literature reveals a variety of antenna designs that embody these reconfigurable characteristics, employing innovative methods to achieve adaptability in frequency and radiation patterns (Zainarry, Nguyen-Trong, & Fumeaux, 2018), (Li, Shao, Wang, & Cheng, 2015), (Sahu & Sharma, 2017), (Han, Wang, Zhang, Ma, & Zeng, 2018), (Majid, Rahim, Hamid, & Ismail, 2014), (Zhu, Wang, You, & Gao, 2018), (Ye & Gao, 2015), (Pan, Ma, Xiong, Hou, & Zeng, 2015), (Li, Bao, Zhai, Li, & Li, 2015).

Technological Approaches to Reconfigurability

Active components like varactors and PIN diodes are frequently utilized to enable the switching of operational frequencies in many designs (Zainarry, Nguyen-Trong, & Fumeaux, 2018), (Li, Shao, Wang, & Cheng, 2015), (Sahu & Sharma, 2017), (Han, Wang, Zhang, Ma, & Zeng, 2018), (Majid, Rahim, Hamid, & Ismail, 2014). Pattern reconfiguration often involves modifications to the ground plane, such as integrating slots or slits and altering the current distribution within the antenna substrate for controlled radiation direction (Zhu, Wang, You, & Gao, 2018),(Ye & Gao, 2015). Alternatively, switchable directors around the primary antenna radiator offer another method for adjusting radiation patterns (Pan, Ma, Xiong, Hou, & Zeng, 2015), (Li, Bao, Zhai, Li, & Li, 2015).

Highlighted Research Contributions

- **Dual-Pattern and Frequency-Reconfigurable Patch Antenna:** A study (Trong, Hall, & Fumeaux, 2017) showcases a patch antenna capable of shifting between monopolar and broadside modes using four varactors, exemplifying dual-pattern reconfigurability.
- **Single-Fed, Double-Element Array:** Research (Zainarry, Nguyen-Trong, & Fumeaux, 2018) presents an array that adjusts both frequency (2.15 to 2.38 GHz) and radiation pattern (beam scanning of ±23°), demonstrating versatile functionality.
- **Monopole and Patch Antenna Integration:** An innovative design (Li, Shao, Wang, & Cheng, 2015) combines monopole and patch antennas for dual-mode operation, enabled by five PIN diodes, showcasing frequency and pattern reconfigurability.
- **Slot Antennas with Redesigned Ground Planes:** Studies (Sahu & Sharma, 2017), (Han, Wang, Zhang, Ma, & Zeng, 2018), (Majid, Rahim, Hamid, & Ismail, 2014) introduce slot antennas with configurable radiation modes and frequencies facilitated by PIN diodes and strategic ground plane designs.
- **Hexagonal Split Rings and PIN Diodes:** A hexagonal split ring design (Zhu, Wang, You, & Gao, 2018) achieves variable frequency and directional patterns, tuning within 1.9-2.4 GHz.
- **Back-to-Back F Semi-Circular Antenna Configuration:** This configuration (Ye & Gao, 2015) utilizes four patch elements for frequency band adjustments, indicating the potential for multi-frequency operation.
- **Circular Antenna with Central and Parasitic Radiators:** A unique circular design (Pan, Ma, Xiong, Hou, & Zeng, 2015) features adjustable radiators supporting various communication protocols and pattern alterations.

- **Dipole-Based Component with Pixelation:** A dipole component (Li, Bao, Zhai, Li, & Li, 2015) showcases frequency and pattern versatility, resonating at three distinct frequencies while offering directed and omnidirectional patterns.
- **Planar Antenna with Twin Symmetrical Radiators:** This antenna (Pan, Ma, Xiong, Hou, & Zeng, 2017) operates at three frequencies, demonstrating two unique emission patterns through its innovative design.

These advancements underscore the dynamic evolution of antenna technology, highlighting the importance of reconfigurability in meeting the diverse and growing demands of modern wireless communication systems.

Antennas With Frequency and Polarization Reconfiguration

The referenced studies highlight the crucial role of antennas in various applications, including imaging, sensing, tracking, and radar, where their ability to change frequency settings adaptively and utilize available spectrum efficiently is of significant value. Polarization diversity, a vital feature of these antennas, effectively mitigates the effects of multipath propagation, enhancing channel capacity.

Advancements in Reconfigurable Antennas

The trend towards integrating sophisticated features has propelled innovative antenna design, particularly in frequency and polarization reconfigurability. A notable strategy involves using active electromagnetic band gap (EBG) structures or metasurface topologies with embedded switches to alter the antenna's resonant behavior. However, a more common method relies on integrating active components, such as varactor and PIN diodes, along with structural modifications, to attain the desired flexibility in frequency and polarization (Nguyen-Trong, Hall, & Fumeaux, 2015), (Liu, Li, & Xu, 2018), (Rahman, Nishiyama, & Toyoda, 2017)
(Kumar & Choukiker, 2017).

Key Design Innovations and Their Implications

- **Electromagnetic Band Gap (EBG) Structures:** Designs from studies (Liang, Sanz-Izquierdo, Parker, & Batchelor, 2015), (Huang, Ma, Pan, & Luo, 2014) utilize EBG arrangements for circular polarization over a frequency range of 1.1 to 1.7 GHz, enabling switches between LHCP and RHCP. This capability demonstrates the potential for advanced communication systems requiring versatile polarization states.

- **Reconfigurable Antenna with Integrated Metasurface:** A planar slot antenna incorporating a metallic reflector and a metasurface (Chen, Zhang, & Wu, 2018) offers the ability to select frequencies between 8.0 and 11.2 GHz and switch polarization states, showcasing the antenna's adaptability to different communication needs.
- **Square-Ring Antenna Patch:** The antenna design (Niture, Govind, & Mahajan, 2016) includes a square-ring patch with horizontal gaps, achieving LP at 2.4 GHz and 4.18 GHz and both LHCP and RHCP at 3.4 GHz, highlighting the antenna's multi-polarization capabilities.
- **Stub-Loaded Reconfigurable Patch Antenna:** Featuring twelve loading stubs around the core radiator (Nguyen-Trong, Hall, & Fumeaux, 2015), this antenna design supports frequency range adjustments between 2.4 and 3.4 GHz and accommodates various polarizations for multi-mode communication systems.
- **Monopole Antenna with Reflector and Modified Ground Plane:** This antenna (Liu, Li, & Xu, 2018) showcases versatility in frequency and polarization states, underlining the potential for reconfigurable antennas in supporting diverse communication protocols.
- **Patch Antenna with Corner Segment:** A novel design (Rahman, Nishiyama, & Toyoda, 2017) facilitates seamless switching between 5.6-5.8 GHz and different polarization modes, leveraging a PIN diode for dynamic reconfigurability.
- **Reconfigurable Patch Antenna with Dual-Probe Coupler:** This design (Kumar & Choukiker, 2017) incorporates a varactor diode in a dual-layer structure, allowing for frequency adjustments within 2.05 to 3.13 GHz and enabling polarization diversity, exemplifying the advanced capabilities of modern reconfigurable antennas.

These advancements underscore the ongoing evolution in antenna technology, focusing on creating highly adaptable antennas that can meet the complex demands of modern wireless communication systems, ensuring optimal performance across a range of applications.

Figure 6. (a) Design scheme and (b) functional efficacy of the frequency and pattern reconfigurable antenna (Ye & Gao, 2015)

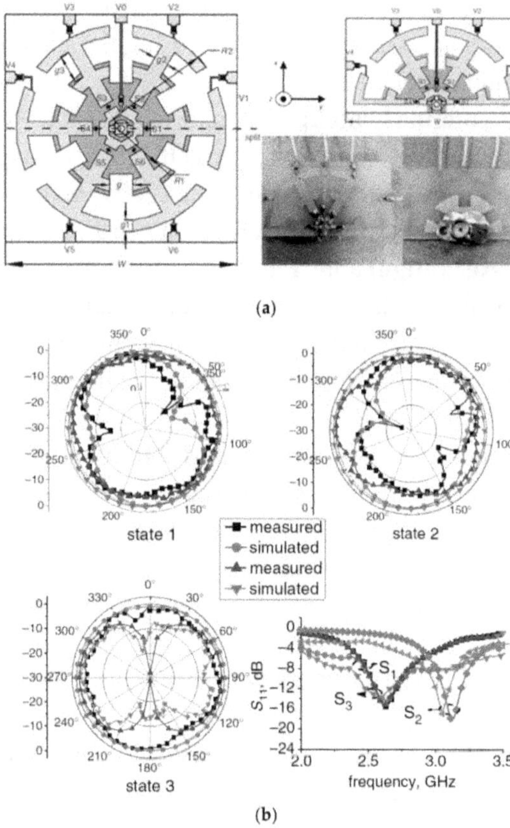

Figure 7. (a) Structural design and (b) efficacy in frequency and polarization reconfiguration of the antenna (Nguyen-Trong, Hall, & Fumeaux, 2015)

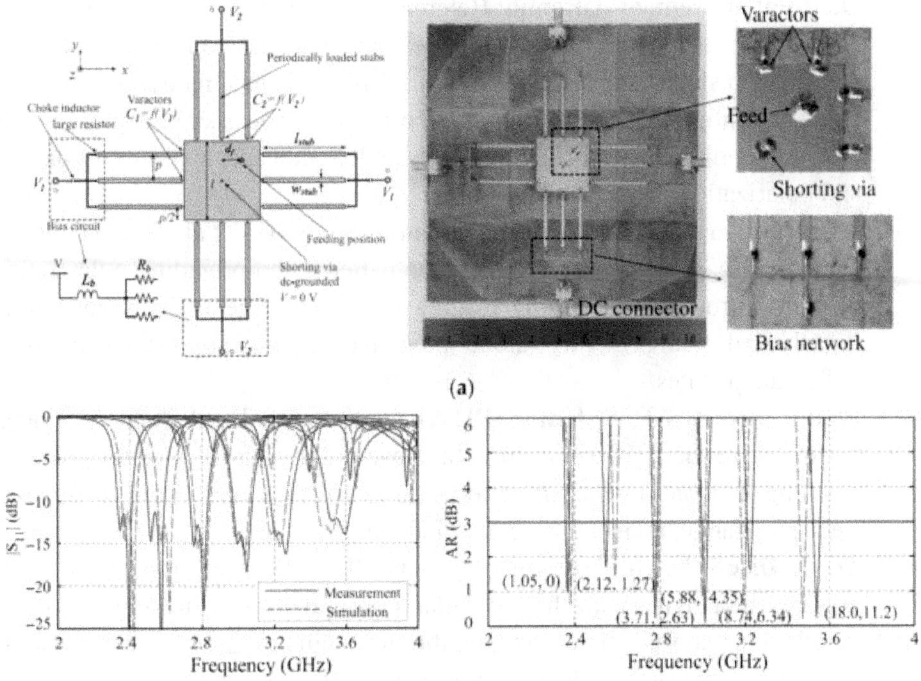

(a)

(b)

Antennas With Radiation Pattern and Polarization Reconfiguration

Radiation pattern and polarization reconfigurable antennas are pivotal in enhancing communication systems, offering the versatility required for diverse applications by enabling the adjustment of radiation patterns and polarization. This adaptability improves signal strength, optimizes coverage, and addresses polarization mismatches, thereby boosting channel capacity without enlarging the physical antenna structure. Research in this field has produced a range of innovative antenna designs (Narbudowicz, Bao, & Ammann, 2014), (Gu et al., 2015), (Yang, Loutridis, & Bao, 2016), (Lin, Wong, & Ziolkowski, 2018), (Chen, Ning, Wang, & Zhang, 2017), (Yi, Huitema, & Wong, 2018), (Trong, Mobashsher, & Abbosh, 2018), (Lin & Wong, 2015), demonstrating the breadth of potential applications and configurations.

Innovative Antenna Designs for Enhanced Communication

- **Dual-Patch Element Antenna:** Reference (Narbudowicz, Bao, & Ammann, 2014) discusses a microstrip-fed patch antenna that can switch between polarization states and radiation patterns. This antenna employs a dual-patch element configuration with a unique corner feed and shared ground plane, offering omnidirectional characteristics and significant flexibility in pattern and polarization control.
- **Versatile Parasitic Component Antenna:** Study (Gu et al., 2015) explores an antenna capable of altering radiation patterns and providing polarization diversity through strategically placing parasitic elements. This adaptation is achieved using a PIN diode-integrated dipole, supporting multiple polarization states.
- **Planar Inverted-F Antenna (PIFA):** The design highlighted in (Yang, Loutridis, & Bao, 2016) features an inverted F-shaped parasitic element enabling tunable radiation patterns and polarization within a specific frequency range, thanks to a single diode mechanism.
- **Triple-Disc Substrate Antenna:** Reference (Lin, Wong, & Ziolkowski, 2018) presents an antenna with three circular substrate discs, achieving versatility in radiation pattern and polarization through a unique assembly that supports transitions between broadside and conical circular polarized modes.
- **Metasurface-Integrated Antenna:** The antenna introduced in (Chen, Ning, Wang, & Zhang, 2017) incorporates a three-layer metasurface structure, enhancing gain, particularly at 5 GHz. PIN diodes and metasurface tuning provide control over polarization and radiation pattern directionality.
- **Reconfigurable Quadrifilar Helical Antenna:** Study (Yi, Huitema, & Wong, 2018) details a quadrifilar helical antenna with a switchable feeding network at 0.9 GHz, offering four polarization and pattern reconfiguration modes.
- **Square Radiator with Parasitic Structures:** Reference (Trong, Mobashsher, & Abbosh, 2018) describes a square-shaped radiator with interconnected parasitic elements, enabling three forms of polarization and pattern variety through PIN diode switching.
- **Wideband Circular Polarized Antenna:** The design in (Lin & Wong, 2015) features a wideband circular polarized antenna with four arms and a reconfigurable feeding network suitable for applications like GPS, RFID, and mobile communications. This antenna offers an axial ratio bandwidth from 1.5 to 1.9 GHz and can switch between bidirectional and broadside modes, with a metallic reflector enhancing its gain.

These advancements underscore the dynamic progress in antenna technology, focusing on creating versatile antennas that can meet modern wireless communication systems' complex and evolving demands, ensuring optimal performance across a wide range of applications.

Antennas With Frequency, Radiation Pattern, and Polarization Reconfiguration

While most antenna designs previously explored are limited in their reconfigurability, typically adjusting only one or two aspects, a groundbreaking design presented in reference (Rodrigo, Cetiner, & Jofre, 2014) showcases the ability to reconfigure three key parameters: frequency, radiation pattern, and polarization autonomously. A subsequent study (Selvam et al., 2017) also explores an antenna design offering adaptability in these three areas. Still, it highlights the need for enhanced tuning capabilities and the ability to achieve simultaneous and independent reconfiguration across these parameters. This gap in antenna design emphasizes the need for comprehensive solutions capable of adaptively modifying essential characteristics such as frequency, radiation pattern, and polarization, a challenge attributed to the complexity of such a multifaceted design.

Innovative Antenna Design Examination

The antenna introduced in (Rodrigo, Cetiner, & Jofre, 2014) is notable for its advanced reconfigurability. Illustrated in Figure 9, the antenna combines a patch antenna radiator with a parasitic pixel surface consisting of 66 pixels and 60 PIN diodes. This switched-grid pixel surface enables the simultaneous adjustment of frequency, radiation pattern, and polarization. The antenna's operational frequency range can be altered from 2.4 to 3 GHz, its radiation pattern can be tuned within a range of ± 30 degrees, and its polarization can be switched among left-hand circular polarization (LHCP), right-hand circular polarization (RHCP), or linear polarization (LP). This versatility makes the antenna highly suitable for various mobile and wireless communication applications.

Comparative Analysis of Reconfigurable Antennas

A comprehensive table (Table 1) provides an in-depth summary of various reconfigurable antennas documented in recent academic research. This analysis covers essential aspects such as antenna type, reconfiguration capabilities, the number of switches used, operating frequency range, and dimensions. By comparing and evaluating these factors, the table serves as a valuable resource for selecting the optimal reconfigurable antenna design for specific applications, addressing the need for antennas that adeptly navigate the complexities of modern wireless communication systems. This effort to catalog and analyze reconfigurable antennas marks a significant step toward identifying and addressing the gaps in current antenna technologies, aiming to propel the development of more sophisticated, adaptable antenna solutions.

Figure 8. Design and construction of the (a) planned and (b) realized pattern and polarization reconfigurable antenna (Yi, Huitema, & Wong, 2018)

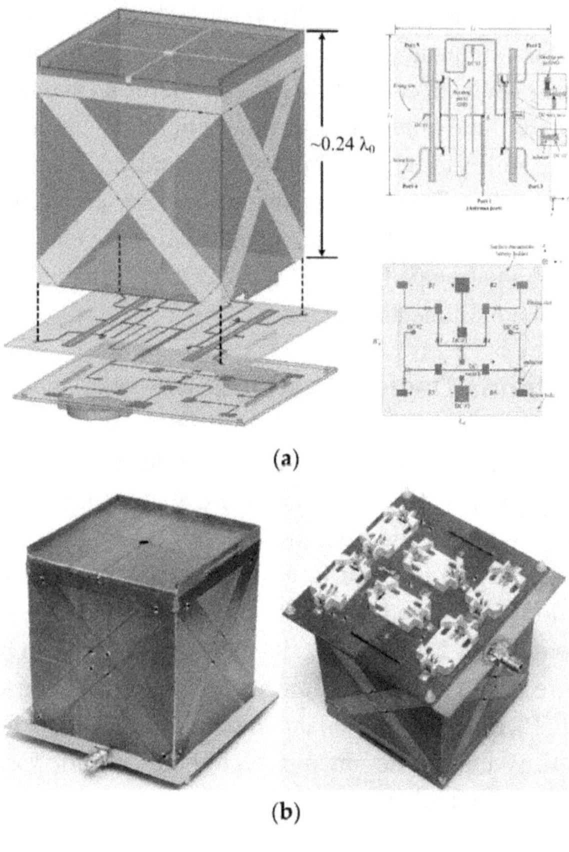

(a)

(b)

Figure 9. Multi-purpose reconfigurable antenna (Rodrigo, Cetiner, & Jofre, 2014)

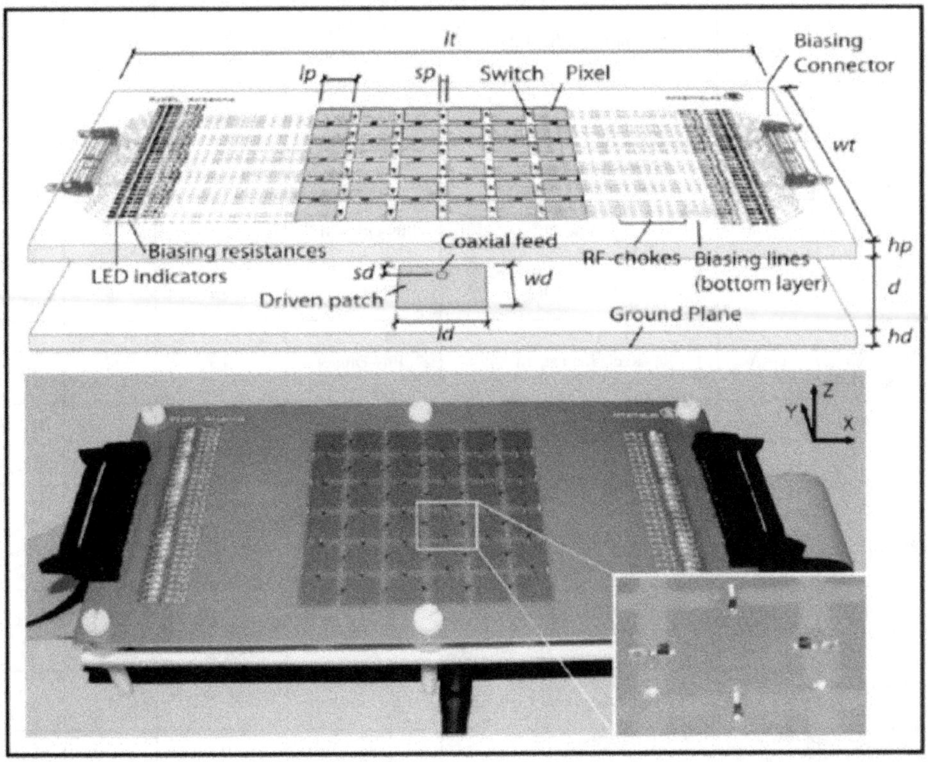

Table 1. Summary of various reconfigurable antennas

Reference	Antenna Type	Type of Reconfiguration	Means of Reconfiguration	Frequency Range (GHz)	Number of Switches	Size (mm)
(Hussain et al., 2018)	Monopole	Frequency	PIN-Diode	1.77–2.51	4	50 × 120
(Khan et al., 2018)	Slot	Frequency	PIN-Diode	1.73–2.28	2	50 × 80
(Anand et al., 2016)	Patch	Bandwidth	PIN-D/MEMS	2–4	8	12 × 12
(Meng et al., 2017)	Slot	Bandwidth	PIN-Diode	1.64–2.68	6	10 × 64
(Zhong, Hong, & Zhou, 2016)	DRA	Radiation Pattern	PIN-Diode	5.8	8	50 × 50
(Lu, Yang, & Tan, 2017)	Patch	Radiation Pattern	PIN-Diode	2.45	8	95.5 × 100
(Lin, Chen, Ziolkowski, & Guo, 2018)	Bow-Tie	Polarization	PIN-Diode	1.3–1.85	4	58 × 58
(Cai, Gao, Yin, Li, & Luo, 2016)	Patch	Polarization	PIN-Diode	2.1–2.6	24	60 × 60
(Kingsly et al., 2018)	Monopole	Frequency/Bandwidth	PIN/varactor	3.9–4.82	3	25 × 75
(Hussain & Sharawi, 2015)	Monopole	Frequency/Bandwidth	PIN/varactor	0.72–3.44	4	65 × 120
(Chacko, Augustin, & Denidni, 2015)	Monopole	Frequency/Bandwidth	PIN-Diode	3.4–8/4.7–5.4	4	70 × 70
(Zainarry, Nguyen-Trong, & Fumeaux, 2018)	Patch	Frequency/Radiation	varactor Diode	2.15/2.25/2.38	2	151.5 × 160.9
(Han, Wang, Zhang, Ma, & Zeng, 2018)	Slot	Frequency/Radiation	PIN-Diode	3.6/3.95	2	30 × 40
(Majid, Rahim, Hamid, & Ismail, 2014)	Slot	Frequency/Radiation	PIN-Diode	1.8/1.9/2.05	14	130 × 160

Continued on following page

Table 1. Continued

Reference	Antenna Type	Type of Reconfiguration	Means of Reconfiguration	Frequency Range (GHz)	Number of Switches	Size (mm)
(Liang, Sanz-Izquierdo, Parker, & Batchelor, 2015)	Monopole	Frequency/Polarization	varactor Diode	1–1.6	112	88 × 114
(Niture, Govind, & Mahajan, 2016)	Patch-Slot	Frequency/Polarization	PIN-Diode	2.4/3.4/4.1	2	60 × 65
(Liu, Li, & Xu, 2018)	Monopole	Frequency/Polarization	PIN-Diode	1.9–2.7	2	40 × 70
(Kumar & Choukiker, 2017)	Patch	Frequency/Polarization	varactor Diode	2–3.1	1	100 × 100
(Yang, Loutridis, & Bao, 2016)	PIFA	Radiation/Polarization	PIN-Diode	2.5	1	50 × 50
(Chen, Ning, Wang, & Zhang, 2017)	Slot	Radiation/Polarization	PIN-Diode	5	36	78 × 78
(Yi, Huitema, & Wong, 2018)	Helical	Radiation/Polarization	PIN-Diode	0.9	32	81 × 81
(Rodrigo, Cetiner, & Jofre, 2014)	Patch	Frequency/Radiation/Polarization	PIN-Diode	2.4–3	60	31 × 31

CONCLUSION

The evolution of antenna technology is essential to keep up with the advancements in wireless communication systems. The field has seen significant transformation from simple monopole designs to complex MIMO arrays and metamaterial-based antennas. Modern antennas must excel in efficiency, multi-band operation, and adaptability to varying communication scenarios. Innovations like MIMO and beamforming have revolutionized antenna usage in wireless systems, enhancing capacity and signal quality. Phased array antennas and metamaterials offer improved performance and miniaturization, while reconfigurable antennas provide adaptability to various communication standards and environments. The rise of the Internet of Things (IoT) and the advent of 5G networks present new challenges and opportunities, necessitating efficient, multi-band antennas. This study concludes that the continuous evolution in antenna design, particularly in achieving simultaneous and independent reconfigurability of frequency, radiation patterns, and polarization, is crucial and warrants further research to harness future wireless technologies' potential fully.

REFERENCES

Abdulraheem, Y. I., Abdullah, A. S., Mohammed, H. J., Mohammed, B. A., & Abd-Alhameed, R. A. (2014). Design of radiation pattern-reconfigurable 60-GHz antenna for 5G applications. *Journal of Telecommunications*, *52*(1), 1–5.

Al-Mahdi, A., & Al-Hassan, O. (2022). Reconfigurable Intelligent Surfaces: A game changer for future wireless systems. *Journal of Electromagnetic Waves and Applications*, *36*(13), 1789–1803.

Al-Yasir, Y., Abdullah, A., Ojaroudi Parchin, N., Abd-Alhameed, R., & Noras, J. (2018). A new polarization-reconfigurable antenna for 5G wireless communications. *Proceedings of the 9th International Conference on Broadband Communications, Networks, and Systems.*

Al-Yasir, Y. I. A., Abdullah, A. S., Ojaroudi Parchin, N., Abd-Alhameed, R. A., & Noras, J. M. (2018). A new polarization-reconfigurable antenna for 5G applications. *Electronics (Basel)*, *7*(11), 293. doi:10.3390/electronics7110293

Anagnostou, E., Torres, D., & Sepulveda, N. (2017). Vanadium dioxide switch for a reconfigurable bandwidth antenna. *Proceedings of the Loughborough Antennas & Propagation Conference (LAPC 2017).*

Anand, S., Raj, R. K., Sinha, S., Upadhyay, D., & Mishra, G. K. (2016). Bandwidth reconfigurable patch antenna for next generation wireless communication system applications. *Proceedings of the International Conference on Emerging Trends in Communication Technologies (ETCT).* 10.1109/ETCT.2016.7882991

Balanis, C. A. (1998). *Antenna Theory Analysis and Design* (4th ed.). John Wiley & Sons.

Bernhard, J. T. (2005). *Reconfigurable Antennas.* Wiley.

Bernhard, J. T. (2007). *Reconfigurable Antennas.* Morgan & Claypool Publishers. doi:10.1007/978-3-031-01535-9

Bhartia, P., & Bahl, I. J. (1982). Frequency agile microstrip antennas. *Microwave Journal*, *25*(1), 67–70.

Bitchikh, M., Rili, W., & Mokhtar, M. (2018). An UWB to narrow band and BI-bands reconfigurable octagonal antenna. *Progress in Electromagnetics Research Letters*, *74*, 69–75. doi:10.2528/PIERL18011023

Boukarkar, A., Lin, X. Q., Jiang, Y., Chen, Y. J., Nie, L. Y., & Mei, P. (2018). Compact mechanically frequency and pattern reconfigurable patch antenna. *IET Microwaves, Antennas & Propagation*, *12*(12), 1864–1869. doi:10.1049/iet-map.2017.0917

Cai, Y.-M., Gao, S., Yin, Y., Li, W., & Luo, Q. (2016). Compact-size low-profile wideband circularly polarized omnidirectional patch antenna with reconfigurable polarizations. *IEEE Transactions on Antennas and Propagation*, *64*(5), 2016–2021. doi:10.1109/TAP.2016.2535502

Chacko, P., Augustin, G., & Denidni, T. A. (2015). Electronically reconfigurable uniplanar antenna for cognitive radio applications. *IET Microwaves, Antennas & Propagation*, *9*(3), 213–216.

Chamok, N. H., Yılmaz, M. H., Arslan, A., & Ali, M. (2016). High-gain pattern reconfigurable MIMO antenna array for wireless handheld terminals. *IEEE Transactions on Antennas and Propagation*, *64*(10), 4306–4315. doi:10.1109/TAP.2016.2598201

Chen, A., Ning, X., Wang, L., & Zhang, Z. (2017). A design of radiation pattern and polarization reconfigurable antenna using metasurface. *Proceedings of the 2017 IEEE Asia Pacific Microwave Conference (APMC)*. 10.1109/APMC.2017.8251389

Chen, M. S., Zhang, Z. X., & Wu, X. L. (2018). Design of frequency-and polarization-reconfigurable antenna based on the polarization conversion metasurface. *IEEE Antennas and Wireless Propagation Letters*, *17*(1), 78–81. doi:10.1109/LAWP.2020.3040509

Chen, S. L., Qin, P. Y., Lin, W., & Guo, Y. J. (2018). Pattern-reconfigurable antenna with five switchable beams in elevation plane. *IEEE Antennas and Wireless Propagation Letters*, *17*(3), 454–457. doi:10.1109/LAWP.2018.2794990

Chen, S.-Y., Chu, Q.-X., & Shinohara, N. (2016). A bandwidth reconfigurable planar antenna for WLAN/WiMAX applications. *Proceedings of the Asia-Pacific Microwave Conference (APMC)*. 10.1109/APMC.2016.7931467

Chen, W., & Lu, Y. (2023). Reconfigurable intelligent surfaces: Unleashing potential for energy-efficient communication. *Journal of Wireless Systems*, *44*(3), 299–312.

Christodoulou, C. G., Tawk, Y., Lane, S. A., & Erwin, S. R. (2012). Reconfigurable antennas for wireless and space applications. *Proceedings of the IEEE, 100*(7), 2250–2261. doi:10.1109/JPROC.2012.2188249

Deshmukh, R., & Patel, S. (2021). A critical review of Reconfigurable Intelligent Surfaces in modern communication. *Journal of Electromagnetic Research and Applications*, *13*(2), 60–73.

Fernandez, G. L., & Montes, R. H. (2022). Integrating advanced antenna technologies for sustainable wireless networks. *IEEE Transactions on Sustainable Computing*, *7*(2), 230–241.

Fujimoto, K., & Nakamura, T. (2023). Emerging trends in antenna technology for 6G and beyond. *Future Wireless Systems Journal*, *6*(1), 9–24.

Ghani, U., & Rahman, F. (2022). A comparative study of traditional vs. Metamaterial Antennas for IoT devices. *IoT Systems and Applications*, *7*(3), 210–224.

Grau, J. R., & Lee, M. J. (2010). A dual-linearly-polarized MEMS-reconfigurable antenna for NB MIMO communication systems. *IEEE Transactions on Antennas and Propagation*, *58*(1), 4–17. doi:10.1109/TAP.2009.2036197

Gu, C., Gao, S., Liu, H., Luo, Q., Loh, T.-H., Sobhy, M., ... Wei, G. (2015). Compact smart antenna with electronic beam-switching and reconfigurable polarizations. *IEEE Transactions on Antennas and Propagation*, *63*(12), 5325–5333. doi:10.1109/TAP.2015.2490239

Han, L., Wang, C., Zhang, W., Ma, R., & Zeng, Q. (2018). Design of frequency- and pattern-reconfigurable wideband slot antenna. *International Journal of Antennas and Propagation*, *2018*, 1–7. doi:10.1155/2018/3678018

Hannula, J.-M., Saarinen, T., Holopainen, J., & Viikari, V. (2017). Frequency reconfigurable multiband handset antenna based on a multichannel transceiver. *IEEE Transactions on Antennas and Propagation*, *65*(9), 65. doi:10.1109/TAP.2017.2725384

Hinsz, L., & Braaten, B. D. (2014). A frequency reconfigurable transmitter antenna with autonomous switching capabilities. *IEEE Transactions on Antennas and Propagation*, *62*(7), 3809–3813. doi:10.1109/TAP.2014.2316298

Horestani, A. K., Shaterian, Z., Naqui, J., Martín, F., & Fumeaux, C. (2016). Reconfigurable and tunable S-shaped split-ring resonators and application in band-notched UWB antennas. *IEEE Transactions on Antennas and Propagation*, *64*(9), 3766–3776. doi:10.1109/TAP.2016.2585183

Huang, Y. C., Ma, X., Pan, W., & Luo, X. (2014). A low profile polarization reconfigurable dipole antenna using tunable electromagnetic band-gap surface. *Microwave and Optical Technology Letters*, *56*(6), 1281–1285. doi:10.1002/mop.29074

Hussain, R., Khan, M. U., & Sharawi, M. S. (2018). An integrated dual MIMO antenna system with dual-function GND-plane frequency-agile antenna. *IEEE Antennas and Wireless Propagation Letters*, *17*(1), 142–145. doi:10.1109/LAWP.2017.2778182

Hussain, R., & Sharawi, M. S. (2015). A cognitive radio reconfigurable MIMO and sensing antenna system. *IEEE Antennas and Wireless Propagation Letters*, *14*, 257–260. doi:10.1109/LAWP.2014.2361450

Hussain, R., & Sharawi, M. S. (2015). Integrated reconfigurable multiple-input–multiple-output antenna system with an ultra-wideband sensing antenna for cognitive radio platforms. *IET Microwaves, Antennas & Propagation*, *9*(9), 940–947. doi:10.1049/iet-map.2014.0605

Hussain, R., & Sharawi, M. S. (2017). An integrated frequency reconfigurable antenna for cognitive radio application. *Wuxiandian Gongcheng*, *26*(3), 746–754.

Hussain, R., Sharawi, M. S., & Shamim, A. (2018). An integrated four-element slot-based MIMO and a UWB sensing antenna system for CR platforms. *IEEE Transactions on Antennas and Propagation*, *66*(2), 978–983. doi:10.1109/TAP.2017.2781220

Iqbal, A., & Khan, Z. (2021). Sustainability in wireless networks: Role of advanced antenna systems. *Green Communications Review*, *12*(4), 245–260.

Ito, H., & Tanaka, S. (2020). Challenges and breakthroughs in next-gen wireless communication: A focus on antenna technologies. *Communications Research Letters*, *15*(4), 657–665.

Jin, G., Li, M., Liu, D., & Zeng, G. (2018). A simple planar pattern reconfigurable antenna based on arc dipoles. *IEEE Antennas and Wireless Propagation Letters*, *17*(9), 1664–1668. doi:10.1109/LAWP.2018.2862624

Kingsly, S., Thangarasu, D., Kanagasabai, M., Alsath, M. G., Thipparaju, R. R., Palaniswamy, S. K., & Sambandam, P. (2018). Multiband reconfigurable filtering monopole antenna for cognitive radio applications. *IEEE Antennas and Wireless Propagation Letters*, *17*(8), 1416–1420. doi:10.1109/LAWP.2018.2848702

Kothapudi, V. K., & Kumar, V. (2017). A Single Layer S/X-Band Series-Fed Shared Aperture Antenna for SAR Applications. *Progress In Electromagnetics Research (PIER-C), 76*, 207–219.

Kothapudi, V. K., & Kumar, V. (2018). A 6-port Two Dimensional 3x3 Series-fed Planar Array Antenna for Dual Polarized X-band Airborne Synthetic Aperture Radar Applications. *IEEE Access, 6,* 12001-12007.

Kothapudi, V. K., & Kumar, V. (2018). Compact 1 × 2 and 2 × 2 Dual Polarized Series-Fed Antenna Array for X-Band Airborne Synthetic Aperture Radar Applications. *Journal of Electromagnetic Engineering and Science, 18*(2), 117–128. doi:10.26866/jees.2018.18.2.117

. Kothapudi, V. K., & Kumar, V. (2019). Vertical polarized 1 × 3 series-fed linear array with gain and front-to-back ratio enhancement for airborne X-band SAR applications. *Progress in Electromagnetic Research (PIER-M), 80,* 169-179.

Kothapudi, V. K., & Kumar, V. (2019). SFCFOS Uniform and Chebyshev amplitude distribution linear array antenna for K-band Radar applications. *Journal of Electromagnetic Engineering and Science, 19*(1), 64–70. doi:10.26866/jees.2019.19.1.64

Kothapudi, V. K., & Kumar, V. (2021). Hybrid-fed shared aperture antenna array for X/K-band airborne synthetic aperture radar applications. *IET Microwaves, Antennas & Propagation, 15*(1), 93–103. doi:10.1049/mia2.12024

Kumar, M. S., & Choukiker, Y. K. (2017). Frequency and polarization reconfigurable antenna using BLC feed network. *Proceedings of the 2017 IEEE International Conference on Antenna Innovations & Modern Technologies for Ground, Aircraft and Satellite Applications (iAIM).* 10.1109/IAIM.2017.8402550

Kumar, R., & Malhotra, J. (2023). The interplay between Beamforming and Massive MIMO: A comprehensive analysis. *Journal of Advanced Wireless Systems, 5*(2), 120–134.

Lee, J. H., & Kim, D. W. (2022). An overview of 5G and beyond: Challenges and advancements. *IEEE Journal on Wireless Communication Systems, 39*(7), 2450–2463.

Lee, S., & Park, Y. (2021). Innovations in antenna technology: A pathway to robust wireless communication. *Advanced Communication Reviews, 14*(5), 486–499.

Li, P. K., Shao, Z. H., Wang, Q., & Cheng, Y. J. (2015). Frequency- and pattern reconfigurable antenna for multistandard wireless applications. *IEEE Antennas and Wireless Propagation Letters, 14*, 333–336. doi:10.1109/LAWP.2014.2359196

Li, W., Bao, L., Zhai, Z., Li, Y., & Li, S. (2015). An enhanced frequency and radiation pattern reconfigurable antenna for portable device applications. *Proceedings of the 31st International Review of Progress in Applied Computational Electromagnetics (ACES)*.

Liang, B., Sanz-Izquierdo, B., Parker, E. A., & Batchelor, J. C. (2015). A frequency and polarization reconfigurable circularly polarized antenna using active EBG structure for satellite navigation. *IEEE Transactions on Antennas and Propagation*, *63*(1), 33–40. doi:10.1109/TAP.2014.2367537

Lin, W., Chen, S. L., Ziolkowski, R. W., & Guo, Y. J. (2018). Reconfigurable, wideband, low-profile, circularly polarized antenna and array enabled by artificial magnetic conductor ground. *IEEE Transactions on Antennas and Propagation*, *66*(3), 1564–1569. doi:10.1109/TAP.2018.2790437

Lin, W., & Wong, H. (2015). Polarization reconfigurable wheel-shaped antenna with conical-beam radiation pattern. *IEEE Transactions on Antennas and Propagation*, *63*(2), 491–499. doi:10.1109/TAP.2014.2381263

Lin, W., Wong, H., & Ziolkowski, R. W. (2018). Circularly-polarized antenna with reconfigurable broadside and conical beams facilitated by a mode switchable feed network. *IEEE Transactions on Antennas and Propagation*, *66*(2), 996–1001. doi:10.1109/TAP.2017.2784452

Liu, J., Li, J., & Xu, R. (2018). Design of very simple frequency and polarization reconfigurable antenna with finite ground structure. *Electronics Letters*, *54*(3), 187–188. doi:10.1049/el.2017.4364

Lorenzo, P., & Di Marco, F. (2020). Harnessing the potential of Beamforming in urban communication networks. *Urban Communication Systems Journal*, *17*(8), 1123–1138.

Lu, Z.-L., Yang, X.-X., & Tan, G.-N. (2017). A multidirectional pattern reconfigurable patch antenna with CSRR on the ground. *IEEE Antennas and Wireless Propagation Letters*, *16*, 416–419. doi:10.1109/LAWP.2016.2581834

Majid, H. A., Rahim, M. K. A., Hamid, M. R., & Ismail, M. F. (2014). Frequency and pattern reconfigurable slot antenna. *IEEE Transactions on Antennas and Propagation*, *62*(10), 5339–5343. doi:10.1109/TAP.2014.2342237

Mak, K. M., Lai, H. W., Luk, K. M., & Ho, K. L. (2017). Polarization reconfigurable circular patch antenna with a C-shaped. *IEEE Transactions on Antennas and Propagation*, *65*(3), 1388–1392. doi:10.1109/TAP.2016.2640141

Mathews, A. S., & Rajan, P. (2021). Exploring Massive MIMO for next-generation networks: An empirical approach. *Wireless Networks*, *27*(4), 1291–1304.

Meng, L., Wang, W., Gao, J., & Liu, Y. (2017). Bandwidth reconfigurable antenna with three step-shaped slots. *Proceedings of the Sixth Asia-Pacific Conference on Antennas and Propagation (APCAP)*. 10.1109/APCAP.2017.8420865

Mistry, P., & Shah, D. (2021). From 5G to 6G: The role of advanced antenna systems. *Wireless Communication Reports*, *3*(3), 315–328.

Nachouane, H., Najid, A., Tribak, A., & Riouch, F. (2016). Reconfigurable and tunable filtenna for cognitive LTE femtocell base stations. *International Journal of Microwave Science and Technology*, *2016*, 1–10. doi:10.1155/2016/9460823

Nachouane, H., Najid, A., Tribak, A., & Riouch, F. (2016). Dual port antenna combining sensing and communication tasks for cognitive radio. *International Journal of Electronics and Telecommunications*, *62*(2), 121–127. doi:10.1515/eletel-2016-0016

Nadeem, Q.-U.-A., Kammoun, A., Debbah, M., & Alouini, M.-S. (2018). Design of 5G full dimension massive MIMO systems. *IEEE Transactions on Communications*, *66*(2), 726–740. doi:10.1109/TCOMM.2017.2762685

Nair, V., & Gupta, M. (2020). Pioneering antenna solutions for sustainable and efficient 5G systems. *International Journal of Mobile Communications*, *18*(5), 456–472.

Narbudowicz, A., Bao, X. L., & Ammann, M. J. (2014). Omnidirectional microstrip patch antenna with reconfigurable pattern and polarization. *IET Microwaves, Antennas & Propagation*, *8*(11), 872–877. doi:10.1049/iet-map.2013.0665

Nguyen-Trong, N., Hall, L., & Fumeaux, C. (2015). A frequency- and polarization-reconfigurable stub-loaded microstrip patch antenna. *IEEE Transactions on Antennas and Propagation*, *63*(12), 5235–5240. doi:10.1109/TAP.2015.2477846

Nguyen-Trong, N., Hall, L., & Fumeaux, C. (2016). A frequency- and pattern-reconfigurable center-shorted microstrip antenna. *IEEE Antennas and Wireless Propagation Letters*, *15*, 1955–1958. doi:10.1109/LAWP.2016.2544943

Nguyen-Trong, N., Piotrowski, A., & Fumeaux, C. (2017). A frequency reconfigurable dual-band low-profile monopolar antenna. *IEEE Transactions on Antennas and Propagation*, *65*(7), 3336–3343. doi:10.1109/TAP.2017.2702664

Nikolaou, S., Bairavasubramanian, R., Lugo, C., Carrasquillo, I., Thompson, D. C., Ponchak, G. E., Papapolymerou, J., & Tentzeris, M. M. (2006). Pattern and frequency reconfigurable annular slot antenna using PIN diodes. *IEEE Transactions on Antennas and Propagation*, *54*(2), 439–448. doi:10.1109/TAP.2005.863398

Niture, V., Govind, P. A., & Mahajan, S. P. (2016). Frequency and polarisation reconfigurable square ring antenna for wireless application. *Proceedings of the 2016 IEEE Region, 10 Conference (TENCON)*. 10.1109/TENCON.2016.7848223

Ojaroudi, N., Ghadimi, N., Ojaroudi, Y., & Ojaroudi, S. (2014). A novel design of microstrip antenna with reconfigurable band rejection for cognitive radio applications. *Microwave and Optical Technology Letters*, *56*(12), 2998–3003. doi:10.1002/mop.28754

Ojaroudi, N., Parchin, N. O., Ojaroudi, Y., & Ojaroudi, S. (2014). Frequency reconfigurable monopole antenna for multimode wireless communications. *Applied Computational Electromagnetics Society Journal*, *8*, 655–660.

Pan, J. Y., Ma, Y., Xiong, J., Hou, Z., & Zeng, Y. (2015). A compact reconfigurable microstrip antenna with frequency and radiation pattern selectivity. *Microwave and Optical Technology Letters*, *57*(12), 2848–2854. doi:10.1002/mop.29452

Pan, Y., Ma, Y., Xiong, J., Hou, Z., & Zeng, Y. (2017). A compact antenna with frequency and pattern reconfigurable characteristics. *Microwave and Optical Technology Letters*, *59*(10), 2467–2471.

Parchin, N. O., Al-Yasir, Y., Abdulkhaleq, A. M., Elfergani, I., Rayit, A., Noras, J. M., Rodriguez, J., & Abd-Alhameed, R. A. (2018). Frequency reconfigurable antenna array for mm-Wave 5G mobile handsets. *Proceedings of the 9th International Conference on Broadband Communications, Networks, and Systems*.

Pardhu, T., & Kumar, V. (2015). Reduction of clutter using TWI ultra wideband imaging. *International Journal of Ultra Wideband Communications and Systems*, *3*(2), 101–106. doi:10.1504/IJUWBCS.2015.077142

Pardhu, T., & Kumar, V. (2016). Implementation of TWI using UWB radar signals. *International Conference on Recent Trends in Engineering, Science and Technology*. 10.1049/cp.2016.1496

Pardhu, T., & Kumar, V. (2018). An investigation on human identification behind the wall. *Journal of Advanced Research in Dynamical and Control Systems*, *10*(5), 122–129.

Pardhu, T., & Kumar, V. (2021). Novel implementations of clutter and target discrimination using threshold skewness method. *TS. Traitement du Signal, 38*(4), 1079–1085. doi:10.18280/ts.380418

Pardhu, T., & Kumar, V. (2023). Human motion classification using Impulse Radio Ultra Wide Band through-wall RADAR model. *Multimedia Tools and Applications, 2023*(24), 36769–36791. Advance online publication. doi:10.1007/s11042-023-14496-w

Park, J., & Kim, H. (2022). A deep dive into Beamforming techniques: Past, present, and future. *Wireless Networks Review, 28*(6), 542–556.

Patel, N., & Desai, V. (2022). Adaptive Beamforming in modern wireless communication: Techniques and applications. *Journal of Communication Technology, 35*(2), 501–515.

Peroulis, D., Sarabandi, K., & Kateh, L. P. B. (2005). Design of reconfigurable slot antennas. *IEEE Transactions on Antennas and Propagation, 53*(2), 645–654. doi:10.1109/TAP.2004.841339

Purisima, M. C. L., Salvador, M., Augstin, S. G. P., & Cunanon, M. T. (2016). Frequency and pattern reconfigurable antennas for community cellular application. *Proceedings of the IEEE Conference TENCON*. 10.1109/TENCON.2016.7848765

Qin, P. Y., Wei, F., & Guo, Y. J. (2015). A wideband-to-narrowband tunable antenna using a reconfigurable filter. *IEEE Transactions on Antennas and Propagation, 63*(5), 2282–2285. doi:10.1109/TAP.2015.2402295

Rahman, M. A., Nishiyama, E., & Toyoda, I. (2017). A frequency diversity reconfigurable antenna with circular polarization switching capability. *Proceedings of the 2017 IEEE International Symposium on Antennas and Propagation & USNC/URSI National Radio Science Meeting*. 10.1109/APUSNCURSINRSM.2017.8072726

Rangan, S., & Ghosh, A. (2023). Harnessing the potential of Metamaterials in modern communication systems. *Journal of Radio Frequency Identification and Wireless, 21*(7), 789–802.

Rao, M., & Iyer, S. (2021). The revolution of antenna technology in the age of IoT and 5G. *Wireless Communication Reviews, 16*(5), 412–429.

Reddy, C. R., & Kumar, V. (2018). A Triple Band flexible Antenna using Multi-layer Graphene patch and polymer Substrate. *JSIR, CSIR, 77*(07), 394–398.

Rodrigo, D., Cetiner, B. A., & Jofre, L. (2014). Frequency, radiation pattern and polarization reconfigurable antenna using a parasitic pixel layer. *IEEE Transactions on Antennas and Propagation*, *62*(7), 3422–3427. doi:10.1109/TAP.2014.2314464

Row, J.-S., & Hou, M.-J. (2014). Design of polarization diversity patch antenna based on a compact reconfigurable feeding network. *IEEE Transactions on Antennas and Propagation*, *62*(10), 5349–5352. doi:10.1109/TAP.2014.2341271

Sahu, N. K., & Sharma, A. K. (2017). An investigation of pattern and frequency reconfigurable microstrip slot antenna using PIN diodes. *Proceedings of the 2017 Progress in Electromagnetics Research Symposium–Spring (PIERS)*. 10.1109/PIERS.2017.8261884

Sam, S., Kang, H., & Lim, S. (2014). Frequency reconfigurable and miniaturized substrate integrated waveguide interdigital capacitor (SIW-IDC) antenna. *IEEE Transactions on Antennas and Propagation*, *62*(2), 1039–1045.

Selvam, Y. P., Elumalai, L., Alsath, G., Kanagasabai, M., Kingsly, S., & Subbburaj, S. (2017). Novel frequency- and pattern-reconfigurable rhombic patch antenna with switchable polarization. *IEEE Antennas and Wireless Propagation Letters*, *16*, 1639–1642. doi:10.1109/LAWP.2017.2660069

Singh, L., Gupta, A., & Ahuja, B. (2020). Metamaterial antennas: A paradigm shift in wireless transmission. *Journal of Antenna and Propagation*, *48*(1), 64–78.

Tang, M.-C., Zhou, B. Y., Duan, Y., Chen, X., & Ziolkowski, R. W. (2018). Pattern-reconfigurable, flexible, wideband, directive, electrically small near-field resonant parasitic antenna. *IEEE Transactions on Antennas and Propagation*, *66*(4), 2271–2280. doi:10.1109/TAP.2018.2814220

Tawk, Y., Costantine, J., & Christodoulou, C. G. (2012). A varactor based reconfigurable filtenna. *IEEE Antennas and Wireless Propagation Letters*, *11*, 716–719. doi:10.1109/LAWP.2012.2204850

Thottempudi, P., Pardhu, T., & Kumar, V. (2021). Experimental study of through the wall imaging for the detection of vital life signs using SFWR. *Indonesian Journal of Electrical Engineering and Computer Science*, *24*(2), 825–830. doi:10.11591/ijeecs.v24.i2.pp825-830

Tong, K.-F., & Huang, J. (2008). New proximity coupled feeding method for reconfigurable circularly polarized microstrip ring antennas. *IEEE Transactions on Antennas and Propagation*, *56*(6), 1860–1866. doi:10.1109/TAP.2008.924736

Tran, H. H., Nguyen-Trong, N., Le, T. T., Abbosh, A. M., & Park, H. C. (2018). Low-profile wideband high-gain reconfigurable antenna with quad-polarization diversity. *IEEE Transactions on Antennas and Propagation*, *66*(7), 3741–3746. doi:10.1109/TAP.2018.2826657

Trong, N. N., Hall, L., & Fumeaux, C. (2017). A dual-band dual-pattern frequency-reconfigurable antenna. *Microwave and Optical Technology Letters*, *59*(11), 2710–2715. doi:10.1002/mop.30815

Trong, N. N., Mobashsher, A. T., & Abbosh, A. M. (2018). Reconfigurable shorted patch antenna with polarization and pattern diversity. *Proceedings of the 2018 Australian Microwave Symposium (AMS)*. 10.1109/AUSMS.2018.8346965

Vargas, L., & Sanchez, R. (2022). The transformative power of Massive MIMO in the 5G era. *Wireless Communication Trends*, *29*(6), 413–427.

Wang, F., & Zhang, H. (2023). The future of wireless communication: Insights from antenna technology trends. *International Journal of Communication Systems*, *36*(1), 15–29.

Wang, K. X., & Wong, H. (2017). A reconfigurable CP/LP antenna with cross probe feed. *IEEE Antennas and Wireless Propagation Letters*, *16*, 669–672. doi:10.1109/LAWP.2016.2598248

Wu, B., Okoniewski, M., & Hayden, C. (2014). A pneumatically controlled reconfigurable antenna with three states of polarization. *IEEE Transactions on Antennas and Propagation*, *62*(10), 5474–5484. doi:10.1109/TAP.2014.2357411

Wu, C., & Tang, X. (2022). Advancements in antenna technology for enhancing wireless communications. *Journal of Network Innovations*, *4*(3), 173–187.

Wu, F., & Luk, K. M. (2017). Widband tri-polarization reconfigurable magnetoelectric dipole antenna. *IEEE Transactions on Antennas and Propagation*, *65*(4), 1633–1641. doi:10.1109/TAP.2017.2670521

Yang, K., Loutridis, A., & Bao, X. (2016). Printed inverted-F antenna with reconfigurable pattern and polarization. *Proceedings of the 10th European Conference on Antennas and Propagation (EuCAP)*. 10.1109/EuCAP.2016.7481210

Yang, L., & Zhao, M. (2023). Insights into Metamaterial Antennas: From basic principles to practical applications. *Antenna Systems & Technology Magazine*, *45*(10), 25–38.

Ye, M., & Gao, P. (2015). Back-to-back F semicircular antenna with frequency and pattern reconfigurability. *Electronics Letters*, *51*(25), 2073–2074. doi:10.1049/el.2015.3029

Yi, X., Huitema, L., & Wong, H. (2018). Polarization and pattern reconfigurable cuboid quadrifilar helical antenna. *IEEE Transactions on Antennas and Propagation*, *66*(5), 2707–2715. doi:10.1109/TAP.2018.2816785

Zainarry, S. N. M., Nguyen-Trong, N., & Fumeaux, C. (2018). A frequency and pattern-reconfigurable two-element array antenna. *IEEE Antennas and Wireless Propagation Letters*, *17*(4), 617–620. doi:10.1109/LAWP.2018.2806355

Zhang, J., Zhang, S., Lin, X., Fan, Y., & Pedersen, G. F. (2018). 3D radiation pattern reconfigurable phased array for transmission angle sensing in 5G mobile communication. *Sensors (Basel)*, *18*(12), 4204. doi:10.3390/s18124204 PMID:30513658

Zhao, Y., & Liu, X. (2021). Design and analysis of metamaterial-based antennas for IoT applications. *IEEE Transactions on Antennas and Wireless Propagation*, *70*(9), 2045–2057.

Zhong, L., Hong, J.-S., & Zhou, H.-C. (2016). A novel pattern-reconfigurable cylindrical dielectric resonator antenna with enhanced gain. *IEEE Antennas and Wireless Propagation Letters*, *15*, 1253–1256. doi:10.1109/LAWP.2015.2504127

Zhong-Liang, L., Xue-Xia, Y., & Guan-Nan, T. (2014). A wideband printed tapered-slot antenna with pattern reconfigurability. *IEEE Antennas and Wireless Propagation Letters*, *13*, 1613–1616. doi:10.1109/LAWP.2014.2342737

Zhu, H. L., Cheung, S. W., Liu, X. H., & Yuk, T. I. (2014). Design of polarization reconfigurable antenna using metasurface. *IEEE Transactions on Antennas and Propagation*, *62*(6), 2891–2898. doi:10.1109/TAP.2014.2310209

Zhu, Z., Wang, P., You, S., & Gao, P. (2018). A flexible frequency and pattern reconfigurable antenna for wireless systems. *Progress in Electromagnetics Research Letters*, *76*, 63–70. doi:10.2528/PIERL18040401

Chapter 7
Design and Investigation of Line-Defected Photonic Crystal Antenna for Outstanding Data Transmission

Sathish Kumar Danasegaran
Sri Eshwar College of Engineering, India

Elizabeth Caroline Britto
IFET College of Engineering, India

S. Dhanasekaran
Sri Eshwar College of Engineering, India

G. Rajalakshmi
Sathyabama Institute of Science and Technology, India

S. Lalithakumari
Sathyabama Institute of Science and Technology, India

A. Sivasangari
Sathyabama Institute of Science and Technology, India

G. Sathish Kumar
Sri Eshwar College of Engineering, India

ABSTRACT

Wireless technology now employs 5G (generation) communication, while 6G is the subject of increased research. Various G's have evolved due to the need to achieve high data transfer for reliable communication. A tremendous data transfer rate expressed in terabits per second (Tbps) is facilitated by the terahertz (THz) spectrum. The photonic crystal (PhC) patch antenna has been proposed in this chapter, and the consequences of a line defect were examined. The defect has been embedded in the perpendicular and parallel directions of the antenna substrates. The antenna's

DOI: 10.4018/979-8-3693-0916-2.ch007

performance regarding return loss (RL), directivity, and voltage standing wave ratio (VSWR) is evaluated. The structure is simulated using CST, and the PhC structure results show excellent characteristics such as -48.02 dB RL, 1.007 VSWR, and 5.23 dB directivity at THz frequency.

I. INTRODUCTION

A PhC's (Danasegaran et al., 2024; Povinelli et al., 2021) dielectric constant varies periodically in any direction. When the PhCs dielectric constant varies in just a single direction, it is referred to as one-dimensional (1D) PhC. Two-dimensional (2D) PhC structures are those in which the material arrangement is periodic in two directions. The identical is exact for (3D) PhC, whose dielectric constant values contrast in 3D and the categorization of various PhC is shown in figure 1.

Figure 1. Types of PhC

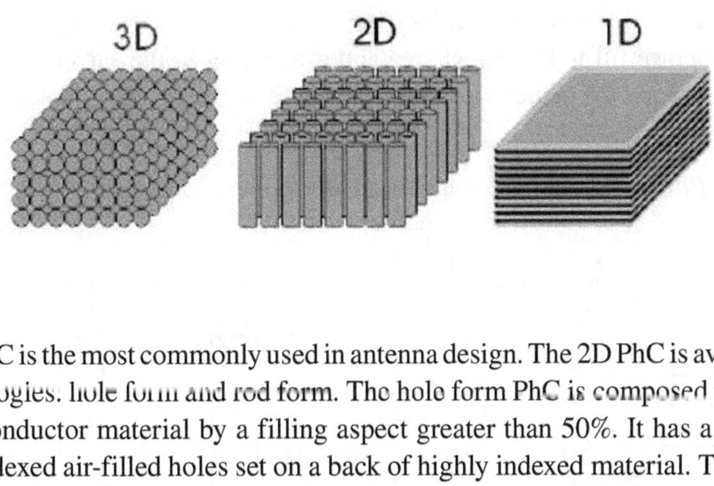

2D PhC is the most commonly used in antenna design. The 2D PhC is available in two topologies: hole form and rod form. The hole form PhC is composed primarily of semiconductor material by a filling aspect greater than 50%. It has a structure of less indexed air-filled holes set on a back of highly indexed material. The highly indexed rods in rod type PhC are enclosed by less index surface material, and the filling factor is < 50%. Figures 2 (a) and (b) show the rod and hole category PhC structures.

Figure 2. (a) Rod type PhC, (b) hole type PhC

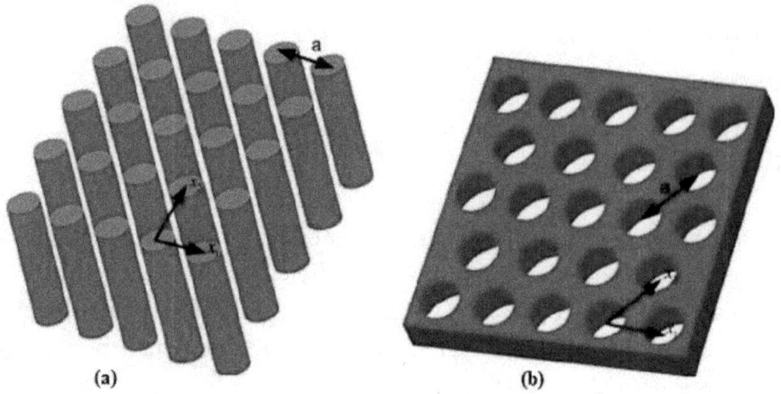

In optics, the defect is a change in the PhC structure's periodical dielectric material. The PhC defect comes in two different forms: point and line defect. The optical cavity in the point defect supports the capture of light. A line defect benefits to direct light from one to another location, providing a waveguiding effect. The EM waves are powerful in the defect area and evanescent in the surrounding regions. Figures 3 (a) and (b) depict a point and line defected structure sample.

Figure 3. PhC structure (a) point defect, (b) line defect

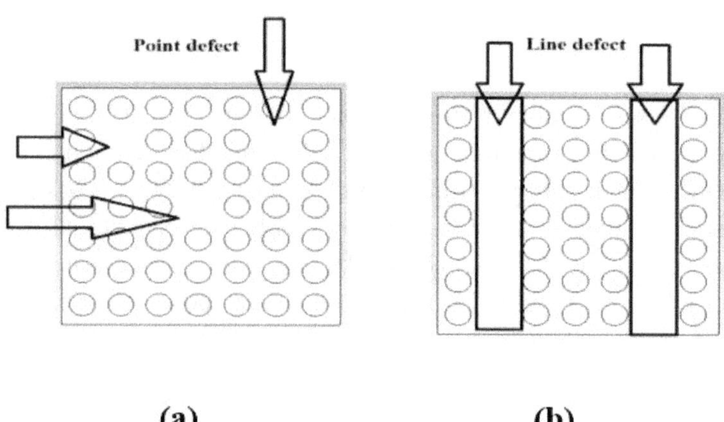

A "line defect" is a form of structural variability or alteration underneath a PhC topological structure. PhCs are materials with constant changes in refractive index on the wavelength scale. These PhC structures can be designed in various ways to regulate the light flow. In PhC, a line defect corresponds to a linear area where the periodicity of the structure is disrupted or altered. It can be achieved by creating a directed path for light inside the PhC by introducing a line defect. The idea is similar to the employing of waveguides in conventional optics, however, in this case, it takes benefit of the PhC lattice's unique properties.

The topology of the PhC defect is critical for controlling light flow and enabling various optical functionalities. Photonic crystals are periodic dielectric structures that, through their bandgap properties, may influence the propagation of electromagnetic waves including light. Defects in these structures provide localized states inside the bandgap, allowing optical devices to be created. The lattice of the PhC slab (Johnson et al., 1999) falls into two categories such as square and triangular as shown in figures 4 (a) and (b). In the PhC slab, the total internal reflections control the light localization caused by the high index contrast between the high and the low index surface. Conversely, the distributed Bragg reflector emerges from the 2D PhC structure and controls the confinement in a longitudinal direction.

Figure 4. PhC structure (a) square lattice, (b) triangular lattice

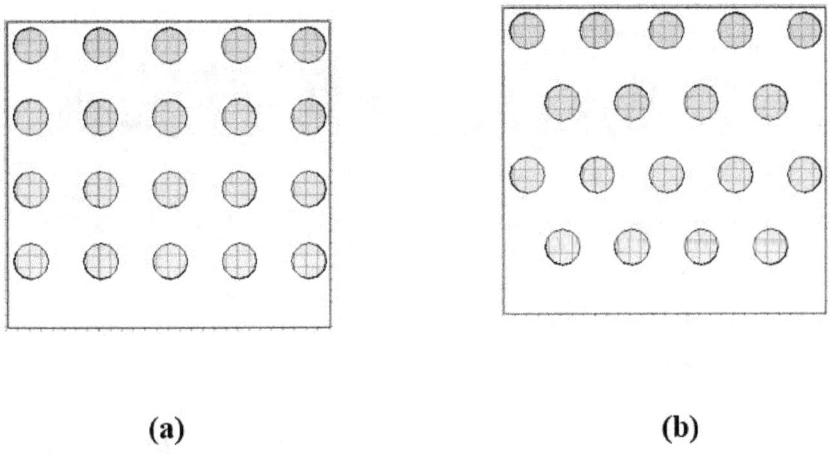

In the PhC slab, the structure is neither periodical nor unbounded in the 3D. The photons incident to the boundary amongst the semiconductor slab and air under angles lesser than the critical TIR angle, may leak from the structure and combine with the continuum radiation modus. In the study of the PhC slab, the radiation losses are considered by employing the light line. Leaky mode is referred to as the regime overhead the light line; TIR does not restrict. The modes under the light can be directed into the slab and referred to as guided mode. The PBG (Dyogtyev et al., 2010) exists between the X point at the lower portion of the II band and the M point at upper the band I, as illustrated in Figure 5. The electric fields of guided waves are often localized in the insulating region, called dielectric band modes. Likewise, the region above the guided waves is denoted as the airband mode (Urbonas et al., 2016) because the electric waves are primarily concentrated in PhC holes. The airband consumes more energy than the dielectric band in which the electric waves are localized in a low refractive index (Knight et al., 1998; Lalithakumari et al., 2023) regime. The PBG position and width are intensely dependent on PhC slab factors. Thus, it is possible to regulate PBG of PhC slabs using lithography by modifying the air-hole size and lattice constant values. When a defect is introduced into a periodic PhC structure, it leads to localized states inside the bandgap. It is termed defect modus, which can be designed to interact with specific light wavelengths.

Figure 5. Bandgap diagram

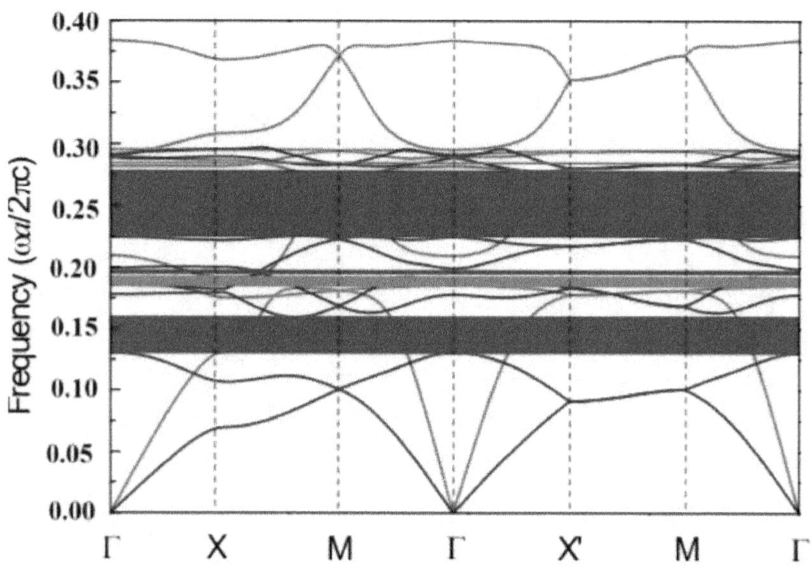

The defect is formed in a periodic PhC lattice by inducing random fluctuations. This disruption breaks the translation symmetry and leads to novel optical observations. These defects serve as scattering axes for the EM fields, leading to cross-coupling between different Bloch modes (Elsner et al., 2000; Wu et al., 2014). Multiple scattering encounters different loss paths, which reduces the propagation of light. However, depending on the nature of the defect (Liu & Wu, 2014), new states can be localized or guided with their frequencies in the PBG.

In the PhC lattice, the defects are classified as point and line defects. A point defect is made by amending the hole radius or slightly shifting its location. This induces a dielectric variation at the center of the defect, which ultimately results in the transfer of an air band into the bandgap. The defect attracts the majority of the field energy. The modus is not localized and leaks into the surrounding environment if a significant amount of reflective crystal centers do not exist around the point defect. Line defect is often recognized by removing a row/column after the regular PhC lattice. Removing rows or columns in PhC have the potential to firmly restrict light in small amounts, which has resulted in the development of several creative devices for commercial light entrapment applications (Mehta et al., 2023), such as wavelength filtering (Ohtera et al., 2007), tunable lasers high-quality cavities nonlinear optics and quantum information processors (Sergei & Pryde, 2019).

II. PROPOSED MODEL

In this chapter, a 1x2 array of trigonal patches is designed and placed on the rectangular Rogers substrate. The proposed design of conventional 1X2 TMPA is shown in Figure 6. In the organization of the patch array, the width and length of the ground and the substrate are placed with a value of 27.80X 25.40 μm^2. Copper is utilized for the ground, patch and feed because of its strong electrical properties and superior resistance to corrosion. The size of the antenna is determined using the resonant frequency (Benlakehal et al., 2022; Priyadharshini et al., 2023; Shi et al., 2021) (f_{res}). The formula of f_{res} is

$$f_{res} = \frac{2C_0}{\sqrt{\varepsilon r X_h}} \sqrt{i^2 + ij + j^2} \tag{1}$$

where 'c_0' is the speed of light in a vacuum, 'ε_r' is the relative permittivity of the substrate and 'x' is a side of the trigonal patch (let $x = L_3 = L_4$). The trigonal side 'a' can be obtained from the following equation. The equation for lowest order resonant frequency is

$$f_{res} = \frac{2C_0}{\sqrt{\varepsilon r} X_h} \tag{2}$$

Using the above standard equations, a 1x2 trigonal patch is designed with the help of copper material. The dimensions of the triangle sides are shown below. In this design, the rectangular substrate uses the Rogers RO4350 materials. It is very effective in achieving high directivity (Sathish Kumar et al., 2020) and the least RL at the THz (Sharma et al., 2009; Singhal, 2019; Temmar et al., 2020) frequency range than other materials. The dimension of the substrate is 25.4x27.8 μm^2

Figure 6. Proposed patch model

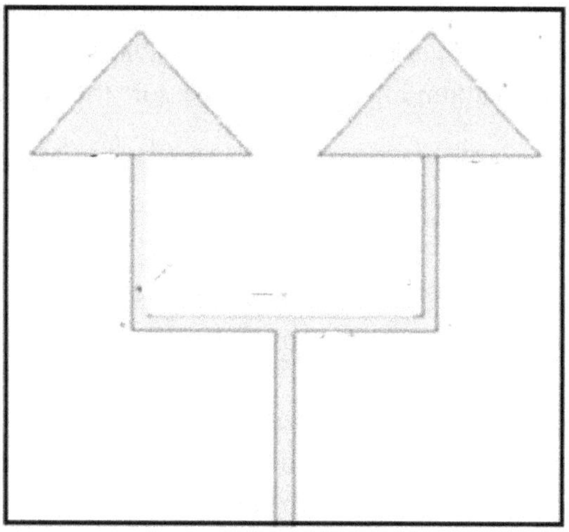

III. IMPACT OF T-SHAPED SINGLE-LINE AIR HOLE DEFECTS ON THE HORIZONTAL AXIS

The triangular patch is placed on the T-shaped air pores in the PhC and the line defects are introduced row-wise. The defected structures in T-shaped PhC structures are shown in Figure 7 (a) – (d). All the possible defective structures are designed by removing each row in the PhC structure. The simulated results of various structures are showcased in table 1. It is observed that the structure had defects on rows 1,2 and 3 resonated around the frequency of 9.6THz. Figure 8 illustrates the effect of the RL of defective structures with RL values of -45dB, -48dB, -24dB and -16dB.

Based on the antenna characteristics, the structure with a defect on row 2 seems ideal and produces reasonable output.

Figure 7. Single line defect in T shape air hole PhC structure (a), defect on Row 1, (b) defect on Row 2, (c) defect on Row 3, (d) defect on Row 4

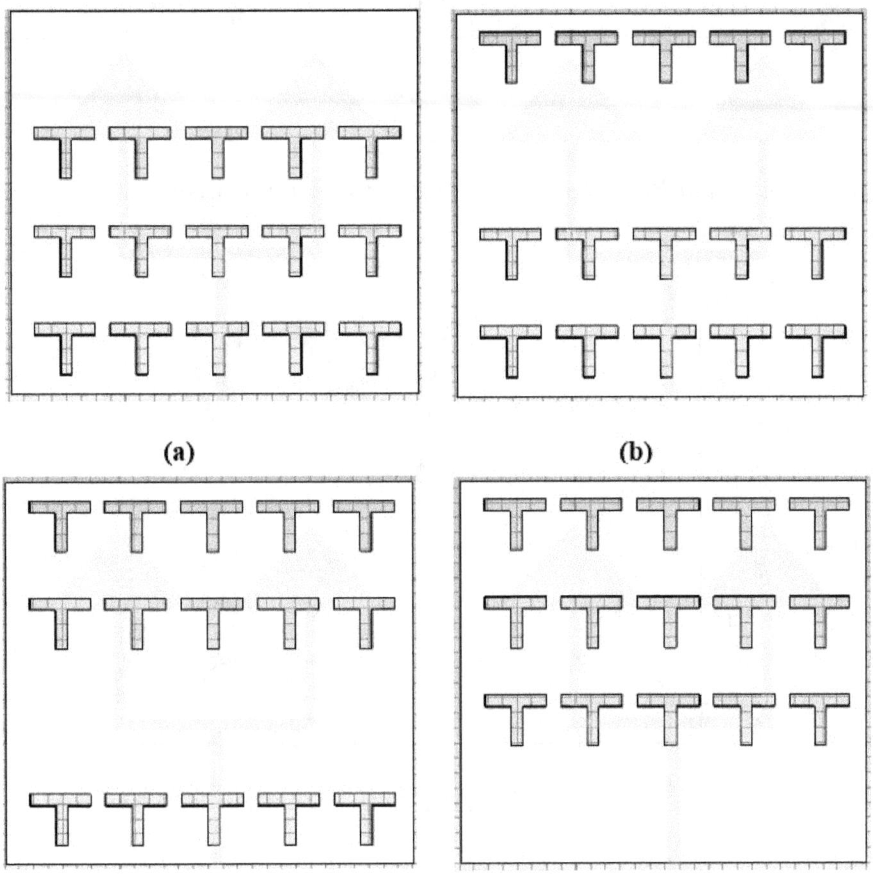

(a) (b) (c) (d)

Table 1. Impact of T-shaped single-line air hole defects on the horizontal axis

Defect on T Air Holes Rows	Antenna Parameters			
	f_r (THz)	RL (dB)	VSWR	Directivity (dBi)
1	9.664	-45.67	1.010	2.31
2	9.645	-48.02	1.007	5.23
3	9.679	-24.46	1.127	2.21
4	18.48	-16.25	1.363	3.18

Figure 8. Impact of RL of defected T shape air hole in row wise

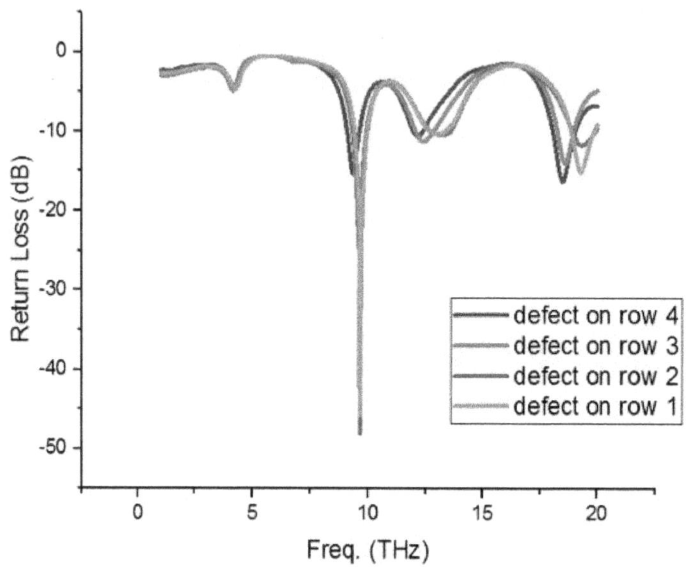

IV. IMPACT OF T-SHAPED SINGLE-LINE AIR HOLE DEFECT ON THE VERTICAL AXIS

The square lattice PhC structure consists of T-shaped air holes combining 4 rows and 5 columns. Here the line defects are introduced in the column manner; hence each column is removed one by one for each structure. The defective sample structures are shown in figure 9 (a) - (d). The proposed defected PhC antennas are simulated and the executed results are shown in Table 2. The table displays that all defective structures are resonated around 9 THz. The poor RL (-18dB) for the third column defected structure is attained. The value of -30dB RL attained for the second and fourth column defected structures. The excellent value of -41 dB is achieved for the first column defected structure, illustrated in figure 10. Also, this structure offers a superior value of VSWR of 1.01.

Figure 9. Single line defect in h shape air hole PhC structure (a) defect on Column 1, (b) defect on Column 2, (c) defect on Column 3, (d) defect on Column 4, (e) defect on Column 5

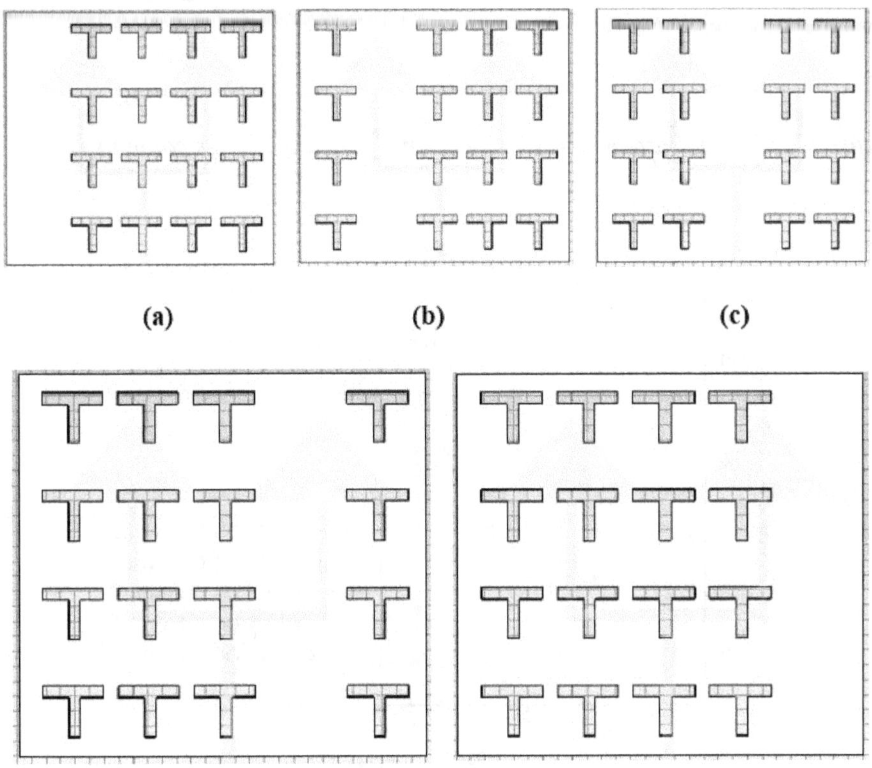

Table 2. Impact of single line column defects in T Shape air hole PhC structure

Defect on T Air Holes Columns	Antenna Parameters			
	f_r (THz)	RL (dB)	VSWR	Directivity (dBi)
1	9.64	-41.66	1.016	2.2
2	9.58	-30.54	1.061	2.1
3	9.30	-18.61	1.265	1.9
4	9.58	-30.95	1.058	2.1
5	9.62	-40.41	1.019	2.2

Figure 10. RL plot for defected structures in T shape air holes columns

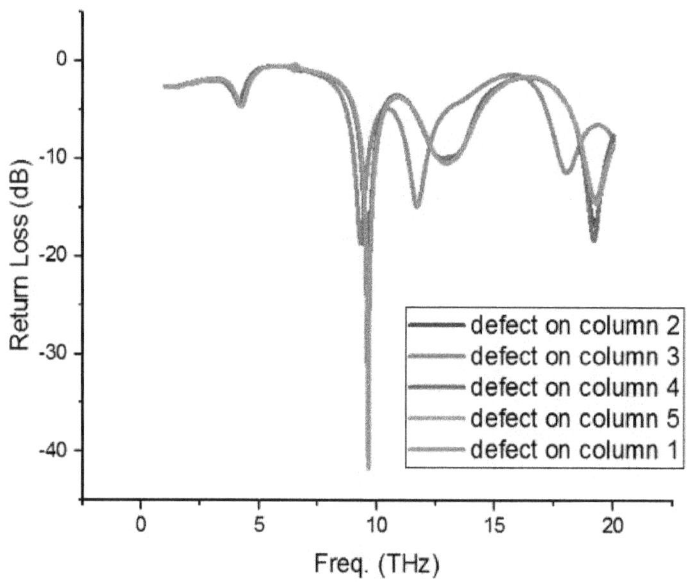

V. IMPACT OF T-SHAPED DOUBLE LINE AIR HOLE DEFECT ON THE HORIZONTAL AXIS

The triangular patch is placed on the T-shaped air pore in the PhC layer and the line defects are introduced row-wise. The defected structures in T shape PhC structures are shown in figures 11 (a) – (d). All the possible defective structures are designed by removing each row in the PhC structure. The simulated results of various structures are showcased in table 3. It is observed that the structure had defects on rows 1,2 and 3 resonated around the frequency of 9.6THz. Figure 12 illustrates the impact of RL on defective structures with RL values of -45dB, -48dB, -24dB and -16dB. Based on the antenna characteristics, the structure with a defect on row 2 seems ideal one and produces good output.

Figure 11. Double line defect in T shape air hole PhC structure (a) defect on Row 1, (b) defect on Row 3

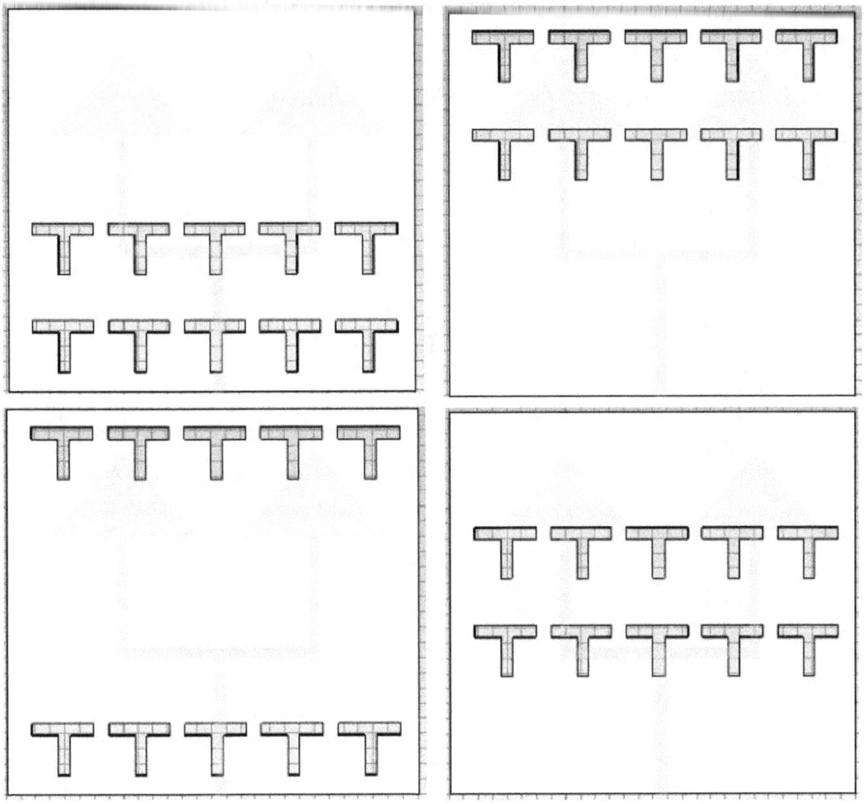

Table 3. Impact of double line row defects in T shape air hole PhC structure

Defect on T Air Holes Rows	Antenna Parameters			
	f_r (THz)	RL (dB)	VSWR	Directivity (dBi)
1, 2	9.64	-42.22	1.015	5.23
2, 3	9.65	-21.45	1.184	4.99
3, 4	9.679	-24.70	1.123	2.21
4,1	9.47	-17.80	1.29	4.49

Figure 12. Impact of RL of defected T shape air hole in row wise

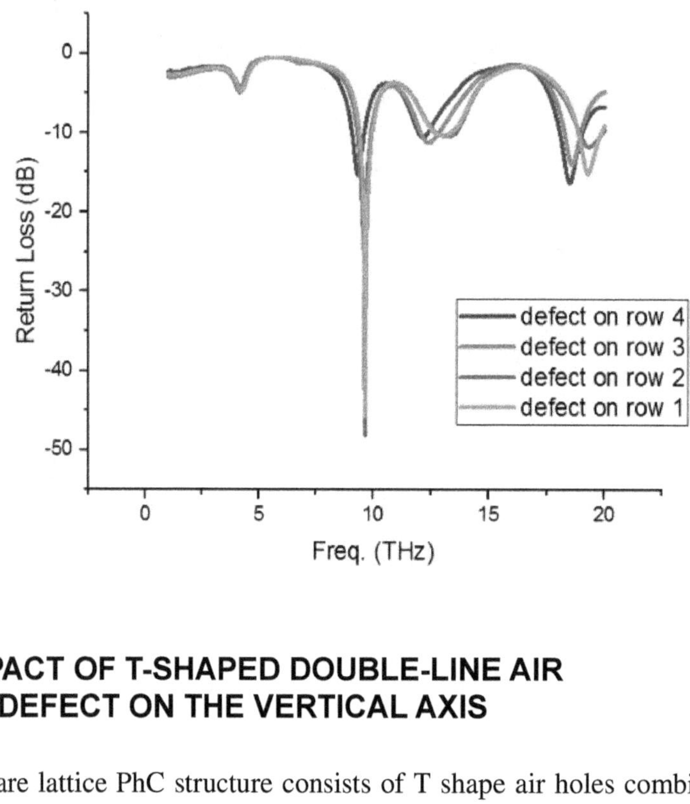

VI. IMPACT OF T-SHAPED DOUBLE-LINE AIR HOLE DEFECT ON THE VERTICAL AXIS

The square lattice PhC structure consists of T shape air holes combining 4 rows and 5 columns. Here the line defects are introduced in the column manner; hence each column is removed one by one for each structure. The defective structures are shown in figures 13 (a) - (d). Table 4 shows that the simulated result of defective structures resonated around 9 THz. The poor RL (-18dB) for the third column defected structure is found. The value of -30dB RL was attained for the second and fourth column defective structures. The excellent value of -41 dB is achieved for the first column defected structure, illustrated in figure 14. Also, this structure offers a superior value of VSWR of 1.01.

Figure 13. Double line defect in H shape air hole PhC structure (a) defect on Column 1, (b) defect on Column 2, (c) defect on Column 3, (d) defect on Column 4, (e) defect on Column 5

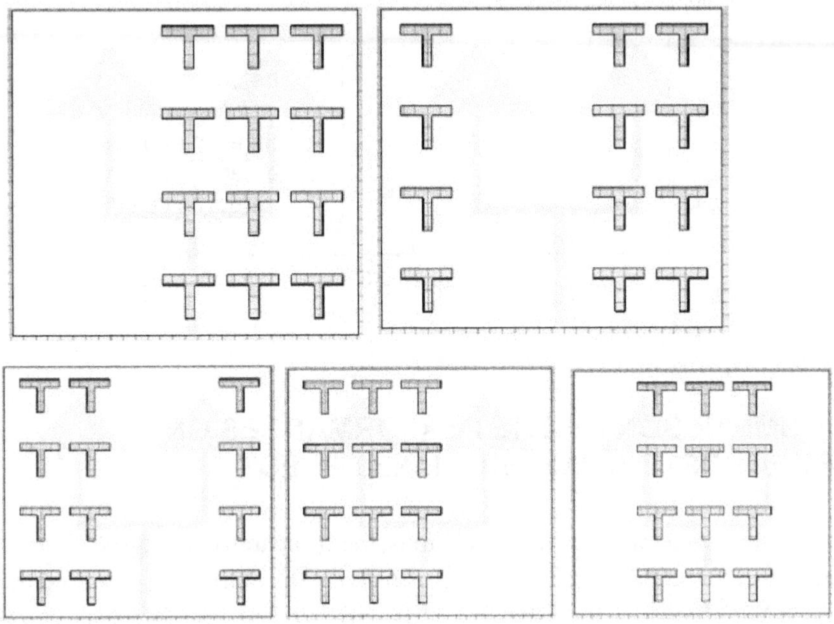

Table 4. Impact of double line column defects in T shape air hole PhC structure

Defect on T Air Holes Columns	Antenna Parameters			
	f_r (THz)	RL (dB)	VSWR	Directivity (dBi)
1, 2	9.56	-32.75	1.047	2.2
2, 3	9.22	-16.98	1.329	1.8
3, 4	9.22	-16.94	1.331	1.8
4, 5	9.56	-32.12	1.050	2.2
5, 1	9.60	-39.63	1.021	2.3

Figure 14. RL plot for defected structures in T shape air holes columns

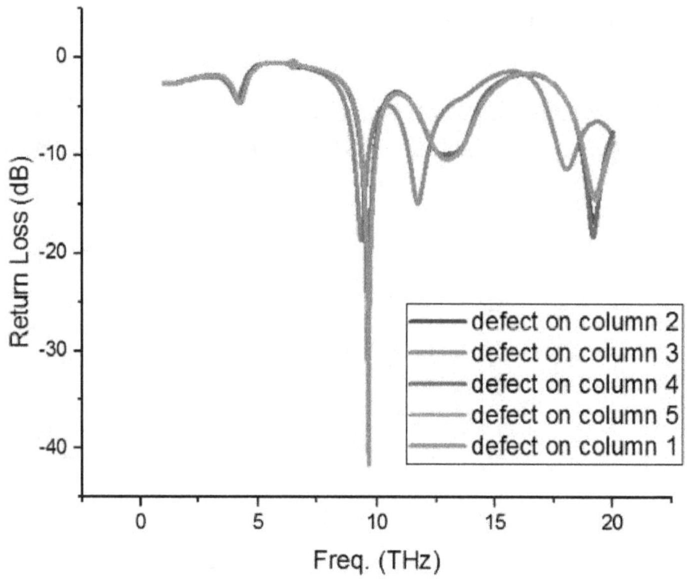

VII. COMPARISON OF THE PERFORMANCES OF THE PROPOSED MPAS WITH LINE DEFECT

From the discussion of the previous sections, the comparative analysis of the impact of line defect for the proposed structures is listed in the table 5. It is inferred that the proposed MPA with defects in the lattice rows offers the best performance compared to the defected column structures.

Table 5. Comparison of proposed MPAs with PhC defected structure

Antenna Designs	Defects on the Lattice	Antenna Parameters			
		f_{res} (THz)	Return Loss (dB)	VSWR	Directivity (dB)
Single line defect	Row 2	9.64	-48.02	1.007	5.23
	Column 1	9.64	-41.66	1.016	2.20
Double line defect	Row 1 and 2	9.64	-42.22	1.015	4.90
	Column 5 and 1	9.60	-39.63	1.021	2.30

When the line defects are introduced in the parallel direction of the PhC, the periodicity of the crystal is disordered and the propagation of EM waves is disturbed. Hence the different PBG bands define the occurrence of EM modes at the discrete frequencies. The discrete guided band is localized because it ensues within the bandgap where no extended modes can be integrated. Such states transit EM energy laterally the defective line path with a group velocity. Hence in line defect, lights are confined in the hose with impeccably reflecting walls. Due to this consequence, the defected PhC antenna radiated well. The line defect improved the antenna performances, such as enhanced RL, VSWR, etc. Also, the defect in horizontal directions produced better antenna performances than the vertical axis line defect PhC structure.

VIII. CONCLUSION

This chapter analyzes the triangle patch antenna structures with defected PhC substrates based on their performances. This chapter proposed the unique T-shaped air holes on the PhC square lattice structure and embedded with the line defect in both parallel and perpendicular. The line defect enhanced the antenna performances, such as improved RL, VSWR, etc. It is observed that single-line defects produce better output than double-line defects, especially when the single-line defects introduced in the second row of the PhC lattice offer the best antenna characteristics. Wireless communication with rapid connectivity data rates up to Tbps can be achieved with the PhC-defected patch antenna.

REFERENCES

Benlakehal, M. E., Hocini, A., Khedrouche, D., Temmar, M. N., & Denidni, T. A. (2022). Design and analysis of a 1 × 2 microstrip patch antenna array based on photonic crystals with a graphene load in THz. *Journal of Optics*. Advance online publication. doi:10.1007/s12596-022-01006-8

Britto, E. C., Danasegaran, S. K., & Sagadevan, K. (2023). Performance Analysis of SSRR in High-Speed Terahertz Antenna for Biomedical Applications. In S. Mehta & A. Abougreen (Eds.), *Metamaterial Technology and Intelligent Metasurfaces for Wireless Communication Systems* (pp. 180–199). IGI Global. doi:10.4018/978-1-6684-8287-2.ch008

Danasegaran, S. K., Britto, E. C., Sagadevan, K., Paranthaman, M., Poonguzhali, S., & Krishnakumar, M. (2024). Design and Investigation of Photonic Crystal Antenna Performance for Expeditious Data Rate in Wireless Communication. *Brazilian Journal of Physics*, *54*(1), 31. doi:10.1007/s13538-023-01408-4

Dyogtyev, A.V., Sukhoivanov, I.A., & De La Rue, R.M. (2010). Photonic band-gap maps for different two dimensionally periodic photonic crystal structures. *Journal of Applied Physics, 107*(1), 1-7.

Elsner, J., Blumenau, A., Frauenheim, T., Jones, R., & Heggie, M. (2000). Structural and Electronic Properties of Line Defects in GaN. *Materials Research Society Internet Journal of Nitride Semiconductor Research*, *5*(1), 481–492. doi:10.1557/S1092578300004671

Johnson, S. G., Fan, S., Villeneuve, P. R., Joannopoulos, J. D., & Kolodziejski, L. A. (1999). Guided modes in photonic crystal slabs. *Physical Review B: Condensed Matter*, *60*(8), 5751–5758. doi:10.1103/PhysRevB.60.5751

Knight, J. C., Birks, T. A., Russell, P. J., & Sandro, J. P. (1998). Properties of photonic crystal fiber and the effective index model. *Journal of the Optical Society of America. A, Optics, Image Science, and Vision*, *15*(3), 748–752. doi:10.1364/JOSAA.15.000748

Lalithakumari, S., Danasegaran, S. K., Rajalakshmi, G., Pandian, R., & Britto, E. C. (2023). Analysis of Wave Propagation in Hybrid Metamaterial Structure for Terahertz Applications. *Brazilian Journal of Physics*, *53*(5), 140. doi:10.1007/s13538-023-01351-4

Liu, C. C., & Wu, C. C. (2014). Analysis of defect mode in a dielectric photonic crystal containing ITO defect. *Optik (Stuttgart)*, *125*(24), 7140–7142. doi:10.1016/j.ijleo.2014.07.120

Ohtera, Y., Onuki, T., Inoue, Y., & Kawakami, S. (2007). Multichannel Photonic Crystal Wavelength Filter Array for Near-Infrared Wavelengths. *Journal of Lightwave Technology*, *25*(2), 499–503. doi:10.1109/JLT.2006.889094

Povinelli, M. L., Johnson, S. G., Fan, S., & Joannopoulos, J. D. (2021). Emulation of two-dimensional photonic crystal defect modes in a photonic crystal with a three-dimensional photonic band gap. *Physical Review. B*, *64*(7), 1–8.

Priyadharshini, A. S., Arvind, C., & Karthikeyan, M. (2023). Novel ENG metamaterial for gain enhancement of an off-set fed CPW concentric circle shaped patch antenna. *Wireless Personal Communications*, *130*(4), 2515–2530. Advance online publication. doi:10.1007/s11277-023-10390-8

Sathish Kumar, D., Elizabeth Caroline, B., Sagadevan, K., Sakthiganesh, G., & Saravanan, R. (2020). Investigation Of High Directional Gain Pentagonal Shaped Patch Antenna. *7th IEEE International Conference on Smart Structures and Systems*. 10.1109/ICSSS49621.2020.9202223

Sergei, S., & Pryde, J. P. (2019). Photonic quantum information processing: A concise review. *Applied Physics Reviews*, *6*(4), 041303. doi:10.1063/1.5115814

Sharma, A., Dwivedi, V. K., & Singh, G. (2009). THz Rectangular MPA on multilayered substrate for advance wireless communication systems. *Progress in Electromagnetics Research Symposium Proceedings*, 627-631.

Shi, Y., Zhang, X., Qiu, Q., Gao, Y., & Huang, Z. (2021). Design of Terahertz Detection Antenna With Fractal Butterfly Structure. *IEEE Access : Practical Innovations, Open Solutions*, *9*(1), 113823–113831. doi:10.1109/ACCESS.2021.3103205

Singhal, S. (2019). Ultrawideband elliptical microstrip antenna for terahertz applications. *Microwave and Optical Technology Letters*, *61*(10), 2366–2373. doi:10.1002/mop.31910

Temmar, M. N. E., Hocini, A., Khedrouche, D., & Denidni, T. A. (2020). Enhanced Flexible Terahertz Microstrip Antenna Based on Modified Silicon-Air Photonic Crystal. *Optik (Stuttgart)*, *217*(1), 164897. doi:10.1016/j.ijleo.2020.164897

Urbonas, D., Balcytis, A., Vaskevicius, K., Gabalis, M., & Petruskevicius, R. (2016). Air and dielectric bands photonic crystal microringresonator for refractive index sensing. *Optics Letters*, *41*(15), 3655–3658. doi:10.1364/OL.41.003655 PMID:27472642

Wu, M. R., Wu, C. J., & Chang, S. J. (2014). Investigation of defect modes in a defective photonic crystal with a semiconductor metamaterial defect. *Physica E, Low-Dimensional Systems and Nanostructures*, *64*(1), 146–151. doi:10.1016/j.physe.2014.07.023

Chapter 8
Design Considerations of EBG Structures for Wearable Applications

Mehaboob Mujawar
https://orcid.org/0000-0002-8260-7062
Bearys Institute of Technology, India

D. Vijaya Saradhi
Malineni Lakshmaiah Women's Engineering College, India

ABSTRACT

This chapter shows and evaluates the design of a compact dual-band microstrip patch antenna working at 2.5 and 5.2 GHz for a number of important parameters. The antenna is designed to operate at these frequencies. By using not one but two distinct EBG ground planes, the proposed dual-band conventional ground plane antenna achieves a higher level of overall performance in comparison to its predecessor (in terms of directivity, gain, efficiency, and other features). It has been discovered that antennas modeled after EBGs perform much better than antennas constructed using conventional ground planes. In addition, the authors examined the performance of the recommended dual-band conventional ground-plane antenna under a variety of bending conditions.

DOI: 10.4018/979-8-3693-0916-2.ch008

INTRODUCTION

Research on antennas was essential to the quick advancement of civilization and resulted in the development of a number of innovative software packages as well as antenna technology on the cutting edge. The invention of a low profile antenna design allowed for the creation of very small electrical devices that may be used for a variety of wireless applications. One example of this kind of study is the creation of EBG structures, which provided scientists, researchers, and engineers with an intriguing core. The study of microstrip (Mujawar, 2021) antennas has opened up a wide variety of new paths for the development of the design as well as the enhancement of the antenna's efficiency, bandwidth, return loss, and other characteristics (Li et al., 2018).

Since an antenna is the primary component of every wireless device, including mobile phones, laptops, satellite receivers, Televisions, navigators, and so on, the development of the microstrip antenna presented a challenging problem. Because of computational electromagnetics, the development of one-of-a-kind antenna designs has accelerated in recent years. As a result of these improvements, professionals in antenna design were able to construct and perfect new antenna configurations. The design of low power antennas makes use of a wide variety of numerical approaches; nevertheless, the methods of moments (MOM), the finite element method (FEM), and the finite difference time domain (FDTD) are the most often used numerical techniques. In addition to the more advanced numerical methods, there are also commercially available software tools that are available for use in the design process of antennas. These software tools allow researchers to make use of the process of developing complex EM materials. The task entails coming up with the most effective antenna structures for a variety of applications. While the design of a microstrip antenna offers numerous advantages, there are also some disadvantages to using this kind of antenna. The development of the microstrip antenna was significantly aided by the incorporation of EBG (Yang & Rahmat-Samii, 2003) structures.

The key categories that are described up above are some of the primary areas in which it was challenging to build an MSP antenna. The development of electromagnetic band gap structures for microstrip antennas offered a reliable solution to the issues described above. This innovation gave a way to address these issues. Nevertheless, due to the more complex nature of EBG construction, it is not viable to describe them using traditional analytical methods.

For the EBG analysis, more sophisticated numerical approaches are used; the interaction between the antenna and the EBG is meticulously investigated, and the most significant results are chosen for a variety of antenna design parameters.

In the realm of electromagnetic research, the materials that are utilized in the construction of EBG structures are making rapid strides. These materials have the ability to address issues involving EM waves that have never been solved before. Researchers that are focusing on microwaves, antennas, or photonics have access to a large array (Xia et al., 2015) of different design options. The unique characteristics of Electromagnetic Band Gap (EBG) structures offer tremendous potential and can be applied for diverse purposes within the field of electromagnetics.

EBG Structure Definition

EBG Structures are very well-liked among scientific researchers. When brought into touch with electromagnetic radiation, the EBG formations, which are periodic in nature, display awe-inspiring occurrences. It was also discovered that there were band gaps and characteristics such as frequency stopband, pass bands. The findings of the literature review shed light on the many terminologies that are put to use in a variety of fields and applications. This umbrella term encompasses various uses and serves as a collective reference for these diverse applications.

EBG structures serve as integral components in a wide range of microwave devices, contributing to overall performance enhancement and specifically optimizing radiation patterns. This helps to minimize noise in the network lines as well. EBGs are recognized as high impedance surfaces owing to their capacity to dampen surface waves at a certain conditional frequency. Their ability to dampen surface (Zhu et al., 2019) waves earned them this name. The antenna is the most important part of EBG structures, which are employed in a wide range of different applications. EBG structures are founded on the principle of total internal reflection, and many periodic EBG structures further take into consideration the phenomena associated with optical band gaps.

In 1987, Eli Yablonovitch and Sajeev John presented the first EBG (Tan et al., 2019) structures to the scientific community. They hypothesized that if the structures of the refractive index were changed on a periodic basis, it would be possible to remove spontaneous photon emission. Piercing periodic circles on a metallic material resulted in the original design for the structure. This design allowed microwave impulses to go in the desired direction but stopped them from moving in the other way. This led researchers to pursue the creation of a periodic sequence of Electromagnetic Band Gap (EBG) structures, aiming to impede the undesired flow of electromagnetic waves in different directions. The primary focus of this investigation revolves around devising methods to obstruct microwave broadcasts originating from undesirable directions, often represented as microwave cavities and waveguides (Zhao et al., 2014). These structures exclusively permit microwave propagation along their axes or below a specified frequency threshold. This pioneering work has inspired numerous scientists and engineers to develop EBG structures. Research indicates that these photonic crystals possess the capacity to govern electromagnetic waves both within and outside the microwave domain. Incorporating various dielectric media to mimic wave paths and cavities, these photonic crystals exhibit the capability to perform diverse frequency division operations. Since these structures provide a substantial potential for electromagnetic spectrum management, the advent of EBG marked the commencement of a new era characterized by groundbreaking advancements in the realms of material science and technology. So, we are able to reach the conclusion that gaps are wider when there are more dielectric dissimilarities present between the medium (Mujawar, Gunasekaran, & Rashid, 2021).

Types of EBG Structures

Within EBG, the 3D structures are primarily exemplified by the wood pile configuration, constituting the first type. These structures exhibit periodicity in all dimensions. A notable advantage of this type lies in its comprehensive band gaps, effectively prohibiting propagation states in any direction.

Figure 1. EBG structure with three-dimensional design (wood pile configuration)

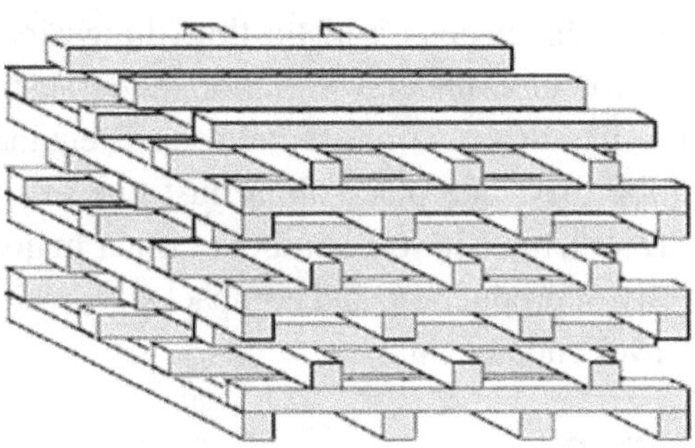

Despite the 3D structure's ability to obstruct unwanted waves from all directions, its fabrication poses challenges, and integration remains a significant hurdle. Constructing these complex structures is intricate, and connecting them with microstrip antennas proves challenging across various applications. A literature survey indicates that 2D EBG structures are more advantageous in terms of fabrication compared to their 3D counterparts. Furthermore, studies highlight that 2D structures exert similar control over wave propagation as observed in 3D structures.

Figure 2. EBG structures with two-dimensional design

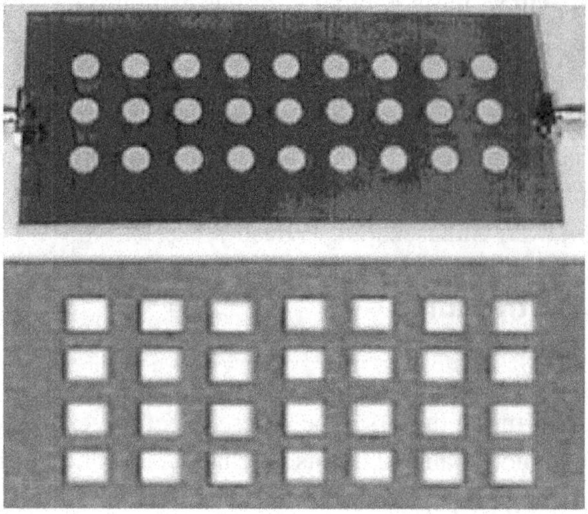

EBG Structures with Two-Dimensional Design: The periodicity of 2D structures is limited to two dimensions, with the third dimension remaining uniform. In simpler terms, these structures undergo variations and transformations in the first two dimensions, while all transformations remain consistent in the third dimension. As a result, propagation may take place along just one of the crystal's axes. In addition to this, the 2D structures have a number of other major benefits, including their stability, simplicity of production, and compact size, which makes them more appealing for microwaves and antennas

1D-Structures in EBG

The microstrip technology makes use of the vast majority of EBG's one-dimensional structures. The periodicity of the one-dimensional structure may be seen on two distinct kinds of media that have just one orientation.

Figure 3. 1D EBG structures

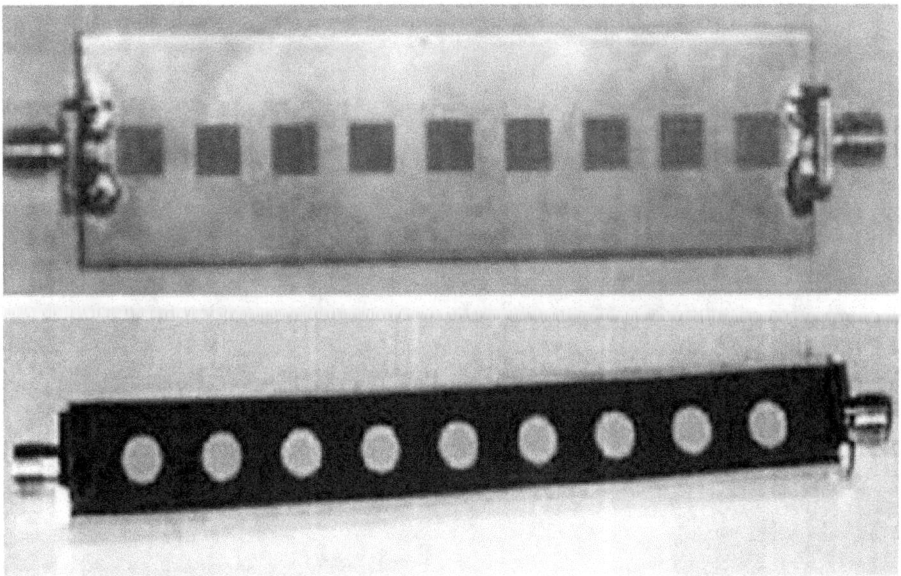

Both metallic and dielectric slabs may be used in the construction of the fundamental structures described before. Once referred to as Photonic Band Gap (PBG) structures, electron band gap (EBG) structures are now more often known as photonic crystals. EBG structures trace their roots back to solid-state physics and optical theory. The fundamental characteristics of these structures are consistently integrated (Sun et al., 2011) into the designs of novel electromagnetic structures, now extensively employed as EBG structures in microwave and radio frequency applications. Experiments conducted in the 1990s employing periodic dielectric structures led to the experimental discovery of these structures. These structures were then included into wood pile constructions using dielectric components. Figure 4a depicts the wood pile dielectric slab, while Figure 4b depicts the multi-layered steel tripod structure. Figure 5 illustrates that the display also includes the two-dimensional structures.

Figure 4. Three-dimensional EBG structures include (a) dielectric structure in a wood pile configuration and (b) metallic tripod with multiple layers

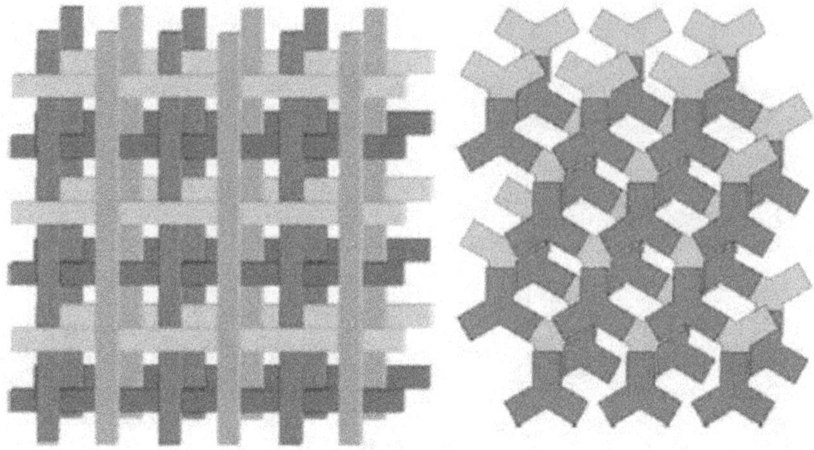

Figure 5. Two-dimensional EBG structures comprise (a) surface with mushroom-like features and (b) uni-planar surface

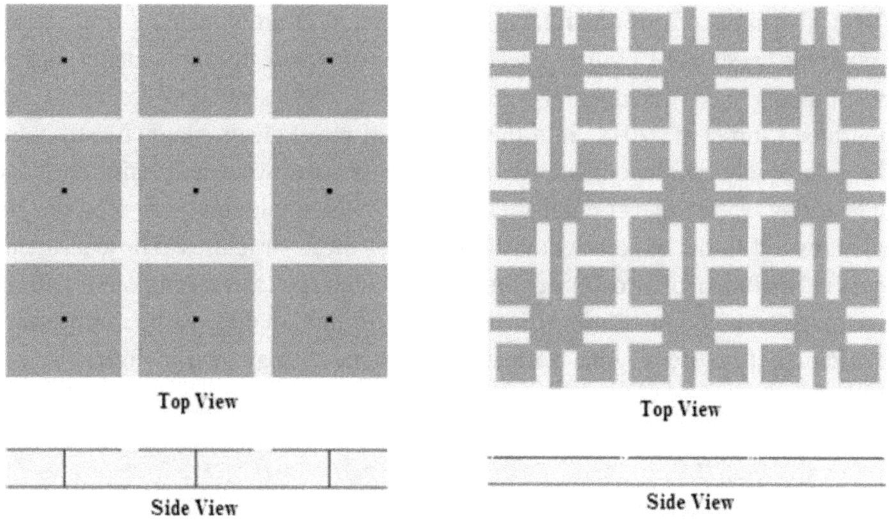

They are determined by the application as well as the specifications of the output that is wanted. Every one of these EBG structure types comes with its own unique set of advantages and analyses. First and foremost, researchers typically employ

EBG structures in mushroom forms because of the high surface impedance that these structures have. Piper, Sieven. D. was the first person to build an EBG structure that resembled a mushroom, and ever since then, every design that has come after it has been inspired by it. Every EBG design will have a band gap frequency that is one of a kind and will vary. There is a choice between a capacitive and an inductive design for the option.

Fundamental Concepts of EBG and Considerations in Design

The mushroom-like Electromagnetic Band Gap (EBG) structures, first introduced by Sievenpiper, demonstrate improved surface impedance and facilitate a more space-efficient design in antenna applications. At their resonant frequency, these EBG structures act as Perfect Magnetic Conductors (PMCs), reflecting the entire electric field. The inherent feature of EBG structures makes them well-suited for augmenting forward gain in antennas by reducing back radiations. The mushroom-like structures distinguish themselves through their attributes of low loss, low profile, and adaptability for integration into diverse antenna systems. Arranged in a sequence, as depicted in Fig 6, they possess a unique configuration. According to the definition, periodic structures referred to as EBG structures have the capability to either enhance or completely inhibit electromagnetic wave propagation within a specific frequency band, regardless of angle or polarization. These structures typically exhibit a pure periodic design and are constructed from metallic or dielectric conductors. Moreover, when a particular design is applied to conductive surfaces, it holds the potential to alter the RF surface properties. If the period of the surface design is shorter than the wavelength, the effective medium and its properties can efficiently encapsulate the quality of the structure.

Sievenpiper's proposed mushroom-like EBG structures showcase elevated surface impedance, offering the capability to facilitate compact designs in antenna applications. At its resonance frequency, the EBG structure mimics the attributes of a Perfect Magnetic Conductor (PMC), effectively reflecting the entire electric field. These structures, resembling mushrooms, excel in enhancing forward gain by minimizing back radiation in antennas, leveraging the advantages mentioned earlier for EBG. Mushroom-like structures stand out as superior due to their versatility in attachment to various antennas and their minimal loss. As illustrated in Figure 6, metal structures connected to a ground are commonly employed to arrange components into a continuous two-dimensional structure. The figure demonstrates the utilization of connected metal structures to achieve a seamlessly continuous two-dimensional configuration.

Figure 6. Cross-sectional view of mushroom-like EBG structure

Plate constructed using what are known as through vertical pillars. Since these projections mimic the structure of a mushroom, the EBG that looks like a mushroom was given that name. When attempting to visually represent these EBG topologies, grouped circuit components such as inductors and capacitors are used. At lower frequencies, transverse magnetic (TM) waves may flow through the structure, but transverse electric (TE) waves may travel through the structure at higher frequencies (TE). Within the range of frequencies that have been established, the structures prohibit transverse electromagnetic (TEM) and transverse magnetic (TM) waves from passing through them.

It is impossible for electricity to flow across the EBG structure since it is shown by a parallel resonant LC circuit. This circuit is used to show the equivalent circuit of the EBG structure. The element that is physically near to its adjacent element is responsible for the capacitive action, whereas the element that is made up of a single stretch of metal across its whole is responsible for the inductive action. Fig. 7 depicts an analogous circuit for an EBG that looks like a mushroom. The position of the band gap will move closer together as the inductance or capacitance rises.

Figure 7. Equivalent circuit of mushroom-like EBG

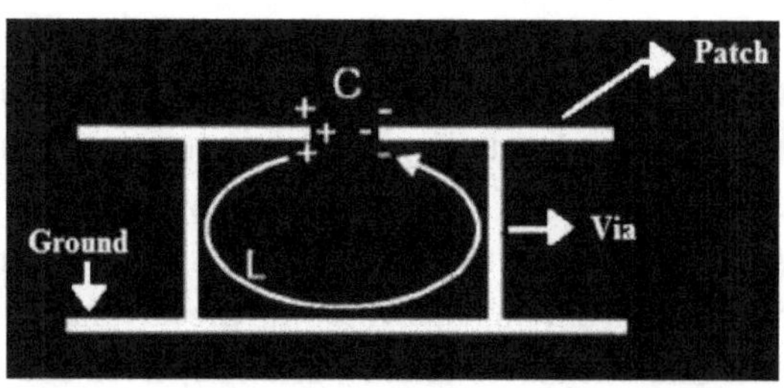

EBG Structures and Their Arrangements

Several of the EBG structures that are utilized for microstrip antennas have the best directivity, lobe pattern, and return loss at broadside results, despite the fact that microstrip antennas employ more than one EBG structure. This elucidates why the input frequency band remains within the surface wave band gap, even with the utilization of a thin grounded dielectric slab. Consequently, the grounded slab to which the microstrip antenna is affixed becomes ineffective. The well-known PEC and PMC have reflection phases of 180^0 and 0^0, respectively, when the plane wave is incident on them in normal incidence. EBG structures made it possible to satisfy the PMC-like requirement for specific frequency bands by opening up new possibilities.

The reflection phase of the EBG structures changes with respect to frequency from $+180^0$ to -180^0, which distinguishes them from PMC surfaces in that they are not only the same as PMC surfaces but also distinct from them. This characteristic, distinctive to EBG for microstrip antennas, establishes a conductive surface essential for mobile and wireless communication devices and gadgets. These devices and gadgets are used in mobile and wireless communication (Mujawar, Saradhi, Desai et al, 2021).

The blueprint for the EBG structures is yet another essential concept. These structures are conceived of as inductive and capacitive devices with a range of geometries; each of these geometries has a frequency bandgap that is distinct from the others. The following list contains a number of significant EBG structures:

Figure 8. Diverse EBG structures featuring the following arrangements: A) mushroom-like, B) uni-planar, C) fork-like, and D) bent strip line

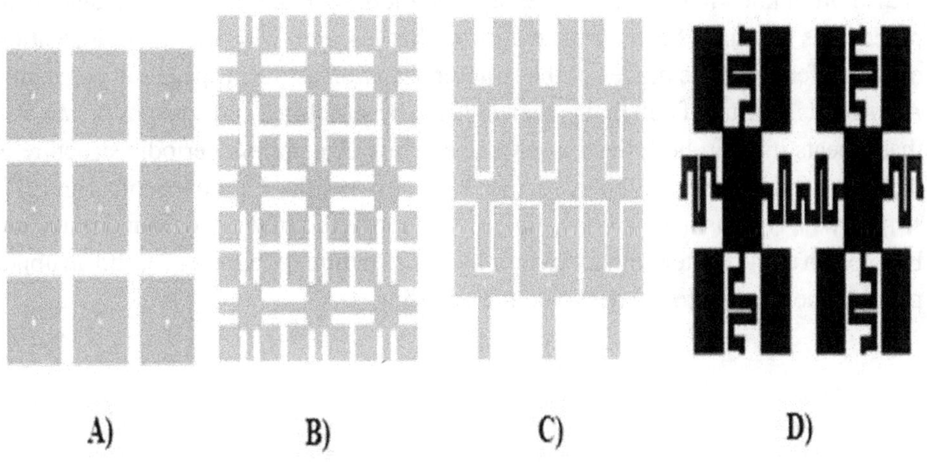

Microstrip antennas are used in a wide variety of applications, and each of these uses calls for a one-of-a-kind EBG structure. The EBG, which has the appearance of a mushroom, is one of the most essential structures among them. Sieven, Piper,

Yang, and others first put out the idea for this approach in the year 1999. Up until that moment, the only types of structures that had been employed were those that resembled rod or wood piles.

The Mushroom-shaped EBG structure and the Fork-shaped EBG structure individually cover 10% and 7% of the available unoccupied regions, respectively. The uni-planar design requires substrates with high stop band frequencies and dielectric constants. This is necessary since the configuration drastically reduces the amount of open space available. Though the bent strip line EBG structure shares a lattice structure period similarity with the uni-planar EBG structure, it necessitates the use of expensive substrates, particularly for low-frequency antenna applications. With the aid of this research, the EBG structure need to be reduced in size, and for low frequency antenna applications, substrates that are available at more reasonable prices have to be used. Comprehensive mathematical examinations of the structures are looked at in order to make a decision about which EBG structures should be used for antenna applications.

Mushroom-Like EBG Analysis

Transmission line circuits are used in microwave and radio frequency engineering applications in order to produce EBG structures. The analysis and design of the Mushroom-shaped EBG structure require the application of transmission line theory and Bloch's theorem. Floquet's It is possible to get the dispersion equation, which is also often known as the characteristic equation, by making use of the analytical procedures that have been outlined above. This research will offer the information on the different excite modes of the structures as well as the propagation constants. A unit cell is a kind of cell that is seen in periodic structures and consists of an array that repeats itself in both the x and y dimensions. An endless periodic structure is the term used to describe this particular kind of structure. Figure 9 depicts an EBG structured as a mushroom, consisting of three metallic planes: a ground plane, an EBG patch at the center with a mushroom-like structure (conductor 2), and an upper metal plane (conductor 1) designed for shielding.

Figure 9. Representation of a unit cell in a two-dimensional mushroom-like EBG structure

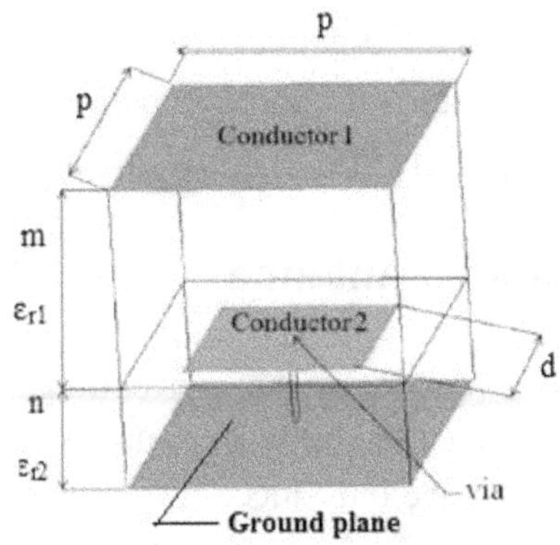

Each structure in the form of a mushroom contains a cylindrical component that's been referred to as a through, and it links the center to the ground plane (Mujawar, 2020). This is an example of a periodic 2D unit cell structure that looks like an EBG shaped like a mushroom. The conductor's length and width are denoted by the symbol "p," which appears on both the square EBG structure and the dimensions of the mushroom-shaped 2D unit cell EBG. There is a distance of one conductor between conductor1 and conductor "m". The dielectric constants pertaining to the substrate between Conductor 1 and Conductor 2, as well as the ground plane, are denoted as r1 and r2, referring to the aforementioned sentence. These dimensions describe the space between Conductor 2 and the ground plane. Multi-Conductor Transmission Line (MTL) is employed, functioning as a parallel circuit. Figure 10 depicts the equivalent circuit for reference.

Figure 10. Formulating the equivalent circuit for the mushroom-like EBG structure using (MTL)

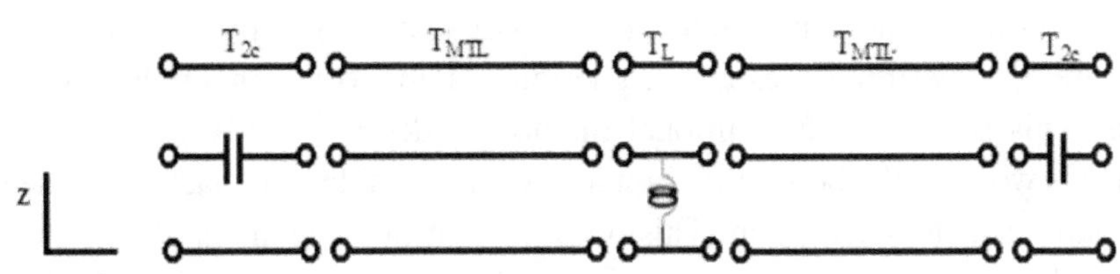

Passive lumped components are employed in constructing the equivalent circuit, and the transfer matrix of the unit cell is illustrated through transmission line analysis.

$$T_{unitcell} = T_{2c} T_{MTL(1)} T_L T_{MTL(1)} T_{2c}$$

The symbol T_{2c} signifies the series gap capacitance. T_L is the inductance resulting from the ground plane and conductor 2 through a via. $T_{MTL(1)}$ indicates the Multi-Conductor Transmission Line (MTL) with a length of $l = p / 2$.

Dual Band EBG Wearable Antenna

Wearable systems are an extension of the conventional concept of clothes that may be worn, which is to protect the body from the effects of its surroundings by covering it with fabric. They have introduced new services in order to completely incorporate these technologies into the wearable (Mujawar et al., 2022) garment without sacrificing the ease of use or comfort of the garment itself. At the core of a wearable system lies the fundamental component of a wearable antenna. This antenna plays a crucial role in establishing a vital link between the electronic circuitry situated on the body and the external electronic equipment. Utilized across various applications such as security, search and rescue, athletics, and entertainment, these antenna-based systems serve diverse purposes. As a consequence of this, it is anticipated that they will be very small and able to withstand the harsh climate. The ground plane reduction technique is often employed in the process of antenna miniaturization. The reduced size of the ground plane increases the susceptibility of spurious currents, as these currents circulate within the ground plane. Various downsizing techniques, such as incorporating lumped components, introducing slots and making the antenna flexible, are employed to minimize the antenna's dimensions. Other techniques include these. Patch antennas, on the other hand, are often ideal for wearable (Mujawar & Pramono, 2023) applications because to the peculiar characteristics they possess.

Patch antennas are an optimal choice for various applications due to their low Specific Absorption Rate (SAR) and a radiation pattern perpendicular to the antenna plane. The full ground plane, an integral part of a patch antenna, effectively shields the wearer's body from undesirable radiation, leading to a further reduction in SAR values. This serves as an additional method to decrease SAR values. When an antenna is worn on the body, radiation lobes from its sides and back enter the body, potentially causing absorption. This occurs when the antenna is connected to the wearer's body, leading to tissue damage and hotspot formation. The calculation of Specific Absorption Rate (SAR) involves electromagnetic waves absorbed by human tissue. The value of E, representing the tissue electrical conductivity (measured in S/m), dictates the level of exposure to electromagnetic fields that can be tolerated.

The increasing prevalence of wireless body area networks (WBAN) applications since the 1990s has led to heightened attention on the specific absorption rate (SAR) values associated with antenna interactions with the human body. It is crucial to ensure that the SAR value remains within predefined limits and does not exceed recommended exposure thresholds. To address this concern, researchers have explored various strategies to effectively reduce the SAR value to acceptable levels. One such approach involves leveraging the electromagnetic band-gap (EBG). This alternative, discussed in numerous papers, presents a viable option for lowering SAR values without relying on EBG.

The antenna's performance is commonly compromised when pressed against or bent around the wearer's body. This degradation is attributable to the heightened conductivity and permittivity of human tissue. Beyond the impact of tissue properties, bending the antenna can alter its effective electrical length, leading to an impedance mismatch. As a result, this may induce variations along the return path, leading to either an increase or decrease. As a consequence of this, it is essential to analyze the different characteristics of a wearable antenna when it is attached to a person and subjected to a variety of bending conditions. In order to slow down or stop this performance decline, EBG structures or high impedance surfaces are used. Antennas often use an artificial material type known as an EBG structure to serve as its ground plane. The in-phase reflection and surface wave suppression that are provided by these components contribute to an improvement in an antenna's overall performance. A further use for EBG is in array antennas. When there are a large number of antenna systems or antenna arrays, mutual coupling becomes a significant problem. EBG is a technique that might be used to avoid the problem of mutual coupling that occurs between several antenna components. By using substandard ground constructions, it is possible to achieve the same result without employing the usage of EBG (DGS). Yet, in other situations, such as in, making use of DGS may reduce gain that the array antenna provides. EBG is a more advantageous choice than DGS in this situation.

The existing body of research on antenna performance and Specific Absorption Rate (SAR) measurements has predominantly focused on flat antenna configurations. However, the inherent flexibility of wearable (Mujawar, 2023) antennas poses challenges as they may not always maintain a flat orientation when worn on the user's body. A larger bending radius is associated with a lower SAR value. There is also a suggestion that crumpling and bending may have detrimental effects on antenna performance. Consequently, our paper provides a comprehensive investigation into SAR, considering the implications of various scenarios. Additionally, we take into account the conditions associated with antenna bending based on the diverse radii diameters of wearable antennas.

To enhance the working of the conventional antenna in terms of various antenna parameters, two different Electromagnetic Band Gap (EBG) were designed and fabricated. These ground planes were attached to the antenna using epoxy. Polyethylene foam, a kind of cloth that has been used in the construction of the antenna, has a relative permittivity of 1.05, It has a low profile and can fit to different shapes. When the recommended antenna's functionality is studied in a range of bent configurations, it is bent in different radii around human arms and legs. For on-body simulations, two different phantom bodies are employed to replicate real bodies. The initial phantom represents a three-layer flat body model utilized for the chest, whereas the second phantom constitutes a four-layer cylindrical body model applied to the arms and legs. Both phantoms serve the same purpose. SAR analysis of the proposed antenna is conducted after placing it on the flat body phantom, bending it around arms (with varying radii), and around legs. The CST MWS program was used for all of the simulations and analysis that are presented in this research.

Design of the Structure

All designs and simulations for this study were conducted using the CST. The proposed model is illustrated in Figure 11. The substrate was constructed using polyethylene foam, with a thickness of 3 millimeters, compact dimensions of $68mm \times 73mm$. The radiating patch's overall dimensions were $44.75mm \times 47mm$.

Figure 11. The proposed model

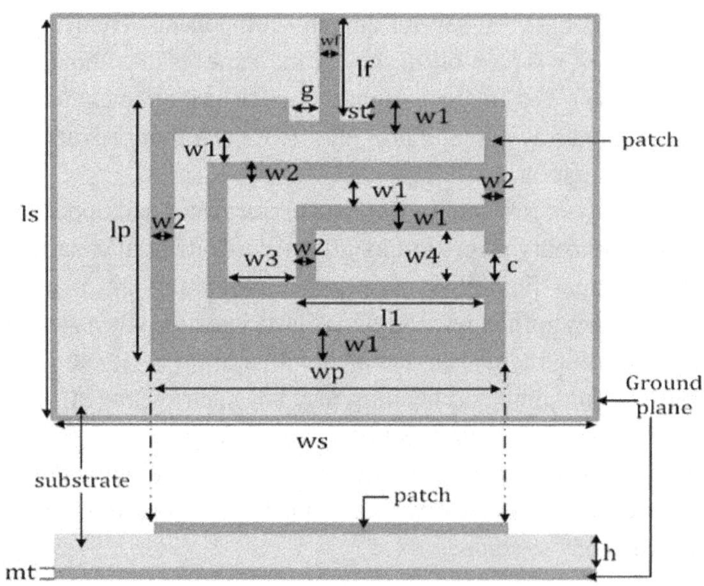

Table 1. Catalog of dimensions and their description

Variable	Explanation	Measurement (in Millimetres)	Variable	Explanation	Measurement (in Millimetres)
ls	Dimension of substrate or ground plane	68	g	Spacing at the sides of the feedline	4
ws	Width of the substrate or ground plane	73	st	Depth of the gap adjacent to the feedline	3
lp	Length of the patch	44.75	c	Length of the cut on the right side	8
wp	Width of the patch	47	$w1$	Width of the upper, centered, or bottom strip	4
h	Thickness of the substrate	2	$w2$	Width of the right-side or left-side strip	3
mt	Thickness of the patch or ground plane	0.04	$w3$	Width of the "P"-shaped cut	8
lf	Length of the microstrip feedline	14.1	$w4$	Width of the "T"-shaped cut	11
wf	Width of the microstrip feedline	4	lI	Dimension of the "T"-shaped cut	25

Unit Cells for Dual-Band Electromagnetic Band Gap (EBG) Structures

This section offers a synopsis of the development derived to accommodate specified frequency. Both EBG structures, employ polyethylene as the substrate material. The resonant frequency, or natural frequency, of each unit cell is determined by (C) and (L), where C and L are influenced by free space. Figure 12 depicts the design used. The first unit cell follows a conventional mushroom-shaped EBG structure, measuring 32.5 millimeters by 32.5 millimeters with a square shape and a through radius of 1.5 millimeters. The other design added after modification, resembles a mushroom but offers enhanced capacitance, resulting in a more compact design. This unit cell measures 26.5 millimeters by 26.5 millimeters, making it approximately 18.46% smaller than the conventional. The second EBG (Yang & Rahmat-Samii, 2001) has dimensions of 3 millimeters in width and 13 millimeters in length. Figure 13 illustrates the in-phase reflection as previously mentioned.

Figure 12. Unit cells for electromagnetic band gap (EBG) structures: (a) mushroom-like EBG unit cell, (b) unit cell with plus-shaped slots

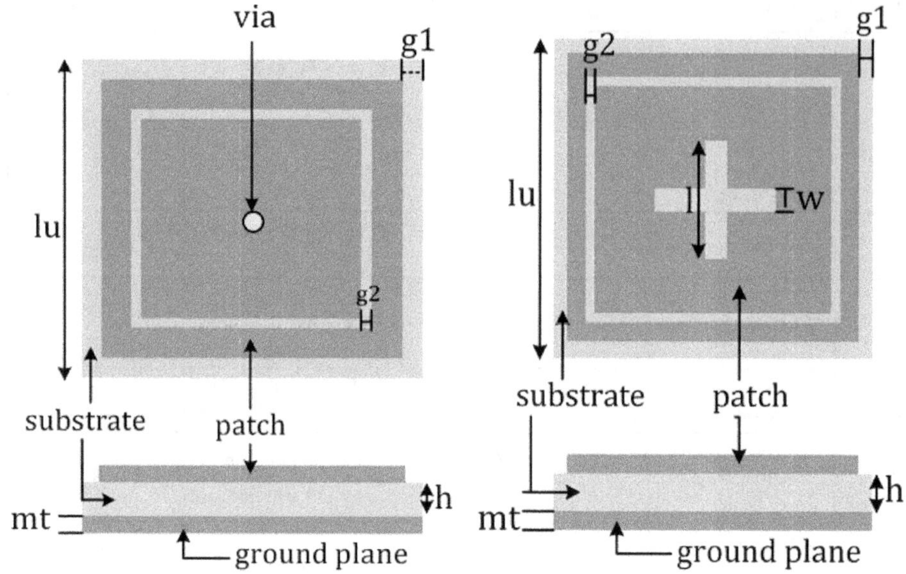

Figure 13. In-phase reflection characteristics of EBG unit cells: (a) mushroom-like EBG unit cell, (b) unit cell with plus-shaped slots

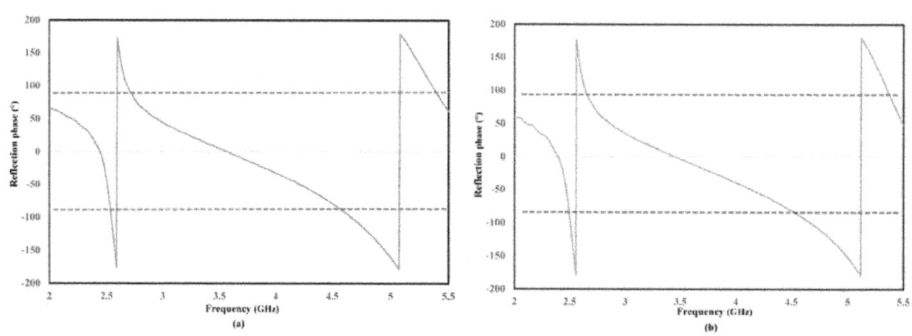

Table 2. Overview of the dimensions of the EBG unit cells

Category	Length (in Millimetres)	g1(mm)	g2(mm)	h(mm)	r(mm)	l(mm)	w(mm)	Repetition
Unit cell with a mushroom-like EBG structure	32.5	2	1.5	3	1.5			32.5
Unit cell featuring a plus-shaped slot in the EBG structure	26.5	0.5	1.1	3		13	3	26.5

Array of Electromagnetic Band Gap (EBG) Structures

One of the critical functions is preventing the propagation of waves. Suppressing surface waves may minimize the size of an antenna's side and back lobes while simultaneously boosting its gain, directivity, and other attributes. The utilization of two-port transmission line techniques enables the determination of scattering characteristics for the two Electromagnetic Band Gap (EBG) surfaces (Zhang et al., 2023). It is presumed that one of the two ports is the source of the excitation, and the other port, which is located on the right, is thought to be a matched load. The second EBG(Bao et al., 2007) array, featuring plus-shaped slots, has a size of $136mm^2$, while the overall size of the mushroom-shaped EBG array is $176.5mm^2$.

The calculation of transmission and reflection coefficients for the two distinct EBG structures demonstrates that S_{21} remains below 15 dB at both frequencies, for each EBG. Consequently, both EBG exhibit the capability to attenuate within the required range. As compared to the second EBG structure, which has slots in the form of a plus, the conventional mushroom-shaped EBG array is about 22.94% more compact. As can be seen in table 3, it was possible to obtain better antenna performance results with EBG structures as compared with the conventional antenna design. Furthermore, more improvement in the antenna design parameters could be observed with the plus shaped slots on the EBG (Farahani et al., 2010) structures.

Figure 14. Antenna configurations situated over electromagnetic bandgap (EBG) ground planes, illustrated as (a) mushroom-like EBG and (b) EBG featuring plus-shaped slots

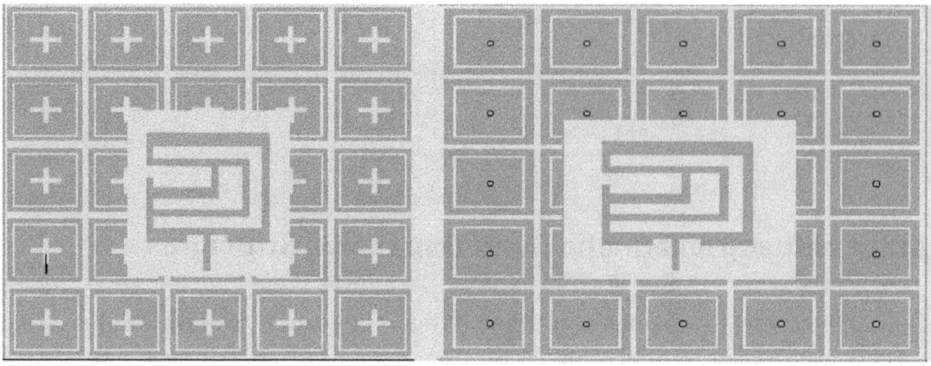

Figure 15. Return loss comparisons in the absence of electromagnetic bandgap (EBG) and with EBG ground planes

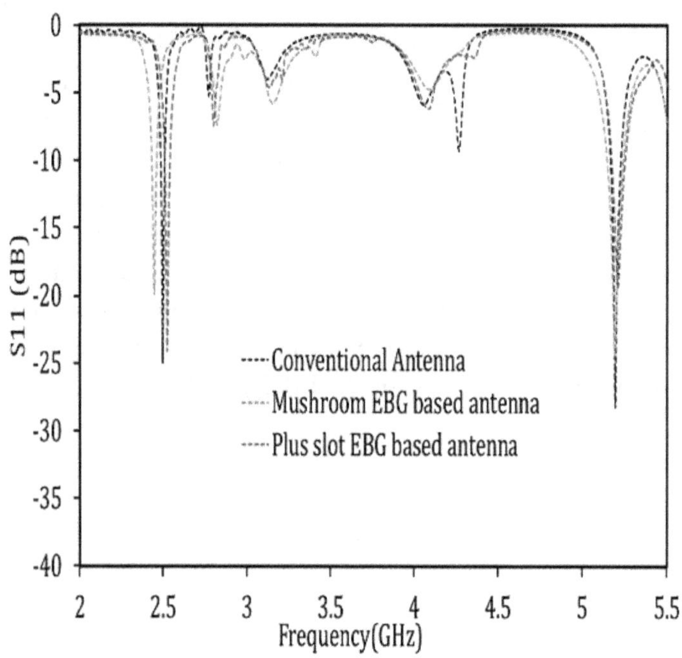

Table 3. Comparative assessment of the performance, considering the presence and absence of electromagnetic bandgap (EBG) ground planes

Factor	Traditional Dual Band Antenna		Antenna Utilizing Mushroom-Like EBG STRUCTURE		Antenna Featuring EBG With Plus-Shaped Slots	
F (in Gigahertz)	**2.5**	**5.16**	**2.2**	**5.2**	**2.8**	**5.8**
B (in megahertz)	23.8	60.4	29	94	31.7	77.1
R_L (in decibels)	-24.5	-28.4	-19.2	-24.1	-24.6	-19.5
D (in decibels)	8.39	9.01	8.55	11.6	9.6	9.3
G (dBi)	8.08	8.74	8.4	11.4	9.43	9.1
R_E (%)	93.06	93.4	97.3	96.7	94.9	95.1
Total efficiency (%)	92.76	93.8	96.41	95.2	94.2	94.8
Angular width (Deg)	83.4	76	80.2	99.4	62.8	70.2

Figure 16. SAR values of a conventional antenna on a flat body phantom at (a) 2.5 GHz and (b) 5.2 GHz

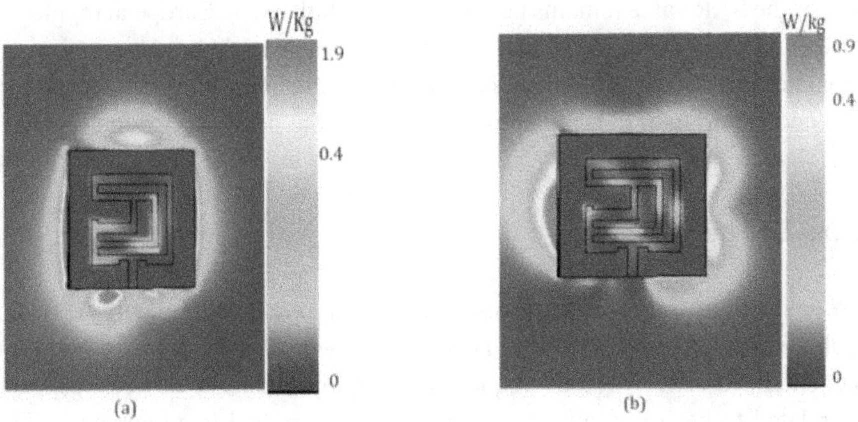

Figure 17. SAR assessment for a conventional antenna on an infant arm

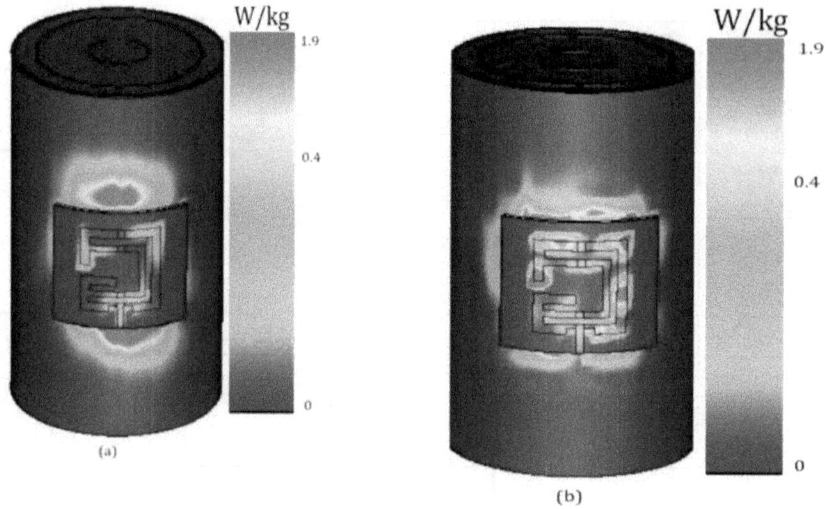

CONCLUSION

The results of our research have led us to the conclusion that the intended antenna will, despite changes in bending radius, continue to keep its tuning. Furthermore, we conducted an extensive study on Specific Absorption Rate (SAR) and examined how the suggested antenna design interacts with diverse human body components, including both flat and rounded areas. The relevant findings indicate that, in both scenarios the SAR value remains below 2 W/kg mandated by European regulations. The creation of a wearable medical device that could monitor the antenna performance in a manner that is true to life would make for an interesting avenue of study in the future, and would help to evaluate whether or not the performance measurements are accurate.

Future Scope

Future developments in wearable antennas may focus on enhancing energy efficiency to prolong device battery life. This is crucial for wearable devices, as users often expect extended usage without frequent recharging. Wearable devices, especially those related to health monitoring, may benefit from dual-band antennas. These antennas can be optimized for specific frequency bands used in medical applications, enabling better connectivity and data transmission for health-related information. Ongoing research in antenna design, metamaterials (Mujawar & Pramono, 2023), and electromagnetic structures may lead to breakthroughs in the development of wearable antennas. Innovation in these areas could open up new possibilities for dual-band EBG antennas in terms of performance, size, and integration.

REFERENCES

Bao, X. L., Ruvio, G., & Ammann, M. J. (2007). Low-profile dual-frequency GPS patch antenna enhanced with dual-band EBG structure. *Microwave and Optical Technology Letters*, *49*(11), 2630–2634. doi:10.1002/mop.22815

Farahani, H. S., Veysi, M., Kamyab, M., & Tadjalli, A. (2010). Mutual coupling reduction in patch antenna arrays using a UC-EBG superstrate. *IEEE Antennas and Wireless Propagation Letters*, *9*, 57–59. doi:10.1109/LAWP.2010.2042565

Li, M., Li, Q. L., Wang, B., Zhou, C. F., & Cheung, S. W. (2018). A low-profile dual-polarized dipole antenna using wideband AMC reflector. *IEEE Transactions on Antennas and Propagation*, *66*(5), 2610–2615. doi:10.1109/TAP.2018.2806424

Mujawar, M. (2020). Design and Analysis of Log-Periodic Dipole Antenna as a Proximity Fuse Antenna. In *2020 International Conference on Industry 4.0 Technology (I4Tech)* (pp. 182-185). 10.1109/I4Tech48345.2020.9102636

Mujawar, M. (2021). Compact microstrip patch antenna design with three I-, Two L-, One E-and One F-Shaped patch for wireless applications. In *Microstrip Antenna Design for Wireless Applications* (pp. 57–68). CRC Press. doi:10.1201/9781003093558-7

Mujawar, M. (2023). Reconfigurable Wearable Antennas and IoT. In Digital Transformation–Modernization and Optimization of Wireless Networks (p. 97). Academic Press.

Mujawar, M., Gunasekaran, T., & Rashid, A. (2021). Design and Analysis of X Band Pyramidal Horn Antenna using FEKO. In *2021 International Conference on Advances in Electrical, Computing, Communication and Sustainable Technologies (ICAECT)* (pp. 1-4). 10.1109/ICAECT49130.2021.9392523

Mujawar, M., & Pramono, S. (2023). Millimeter Wave Antennas for Wearable Applications. In Innovative Smart Materials Used in Wireless Communication Technology (pp. 44-54). IGI Global. doi:10.4018/978-1-6684-7000-8.ch003

Mujawar, M., & Pramono, S. (2023). Metamaterial-Based Absorber. In Metamaterial Technology and Intelligent Metasurfaces for Wireless Communication Systems (pp. 110-120). IGI Global. doi:10.4018/978-1-6684-8287-2.ch005

Mujawar, M., Saradhi, D. V., Desai, S. L., & Venkateswararao, M. (2021). *Performance Analysis of Dipole and Bow-Tie Antenna for Underwater Communication Using FEKO. In 2021 Emerging Trends in Industry 4.0 (ETI 4.0)*. doi:10.1109/ETI4.051663.2021.9619355

Mujawar, M., Saradhi, D. V., Naz, M. A., & Muduli, A. (2022). Design and performance comparison of arrays of circular, square and hexagonal meta-material structures for wearable applications. *Journal of Magnetism and Magnetic Materials*, *553*, 169235. doi:10.1016/j.jmmm.2022.169235

Sun, G. Y., Chen, Y. B., Dong, X. Y., Jin, O. Y., & Xie, Y. (2011). *Three-Dimensional Integrated Circuits: Design EDA and Architecture*. Now Foundations and Trends. doi:10.1561/9781601984630

Tan, X., Wang, W., Wu, Y., Liu, Y., & Kishk, A. A. (2019). Enhancing isolation in dual-band meander-line multiple antenna by employing split EBG structure. *IEEE Transactions on Antennas and Propagation*, *67*(4), 2769–2774. doi:10.1109/TAP.2019.2897489

Xia, R. L., Qu, S. W., Li, P. F., Jiang, Q., & Nie, Z. P. (2015). An efficient decoupling feeding network for microstrip antenna array. *IEEE Antennas and Wireless Propagation Letters*, *14*, 871–874. doi:10.1109/LAWP.2014.2380786

Yang, E., & Rahmat-Samii, Y. (2001). A low profile circularly polarized curl antenna over electromagnetic band-gap (EBG) surface. *Microwave and Optical Technology Letters*, *31*(4), 264–261. doi:10.1002/mop.10006

Yang, F., & Rahmat-Samii, Y. (2003). Microstrip antennas integrated with electromagnetic band-gap (EBG) structures: A low mutual coupling design for array applications. *IEEE Transactions on Antennas and Propagation*, *51*(10), 2936–2946. doi:10.1109/TAP.2003.817983

Zhang, H. H., Chao, J. B., Wang, Y. W., Liu, Y., Xu, Y. X., Yao, H., Jiang, L., & Li, X. H. (2023). Electromagnetic–Thermal Co-Design of Base Station Antennas With All-Metal EBG Structures. *IEEE Antennas and Wireless Propagation Letters*, *22*(12), 3008–3012. doi:10.1109/LAWP.2023.3308585

Zhao, L., Qian, K. W., & Wu, K. L. (2014). A cascaded coupled resonator decoupling network for mitigating interference between two radios in adjacent frequency bands. *IEEE Transactions on Microwave Theory and Techniques*, *62*(11), 2680–2688. doi:10.1109/TMTT.2014.2358202

Zhu, Y., Chen, Y., & Yang, S. (2019). Decoupling and low-profile design of dual-band dual-polarized base station antennas using frequency-selective surface. *IEEE Transactions on Antennas and Propagation*, *67*(8), 5272–5281. doi:10.1109/TAP.2019.2916730

Chapter 9
Design and Analysis of Dual Band Frequency Selective Surface

Devendra Kumar Somwanshi
https://orcid.org/0000-0003-4331-0917
Poornima College of Engineering, Jaipur, India

Payal Bansal
Poornima Institute of Engineering and Technology, India

ABSTRACT

Recent progress in microstrip antennas has improved their size and efficiency, particularly in mobile, satellite, and Wi-Max technologies. These enhancements are crucial for advancing wireless communication systems, benefiting applications like telemedicine and GPS technology. The chapter introduces a dual-band unit cell of frequency selective surface (FSS) resonating at 4 GHz and 5.5 GHz, using a combination of I-shaped and modified I-shaped metal strips for dual-band filtering. The research focuses on individual shape analyses and aims to contribute to the development of compact and efficient dual-band FSS designs. The research introduces a dual-band FSS unit cell resonating at 4 GHz and 5.5 GHz, combining I-shaped, H shaped, and modified I-shaped and H shaped metal strips for dual-band filtering. The proposed FSS design, with a 10x10 mm^2 unit cell dimension and FR-4 dielectric, exhibits broad frequency band characteristics, aligning well with simulated and measured results.

DOI: 10.4018/979-8-3693-0916-2.ch009

1. INTRODUCTION

In the 19th century, the significant discovery of electromagnetic waves, predicted by James Clark Maxwell, paved the way for the development of radio communication. Heinrich Hertz demonstrated radio communication in 1888, leading to the success of commercial radio in the 20th century (Dadgarpour et al., 2014). Antennas have since become an indispensable technology in various aspects of human life. The IEEE standard defines an antenna as a crucial component in wireless communication systems, serving as a link between guiding media and free space. Researchers have introduced various antenna types to accommodate the evolving wireless communication system, with the microstrip patch antenna being widely recognized. Known for its cost-effectiveness, lightweight, compact size, and planar structure, it has gained popularity due to advancements in VLSI and semiconductor technology (Dhouibi et al., 2012).

Antennas play a crucial role in various applications, including satellite systems and telecommunications, by transmitting and receiving radio waves. Instead of relying on a single large antenna, arrays of smaller antennas are often organized to enhance directivity and gain (Dhouibi et al., 2012).

The study of frequency selective surfaces dates back to the early 1960s, with similar structures patented by Marconi in 1919. The term "frequency selective surface" was later patented by Yee in 1993, with clear definitions and discussions provided by Munk (Foroozesh & Shafai, 2006). In satellite platforms like Voyager, Galileo, and Cassini, the integration of FSS sub-reflectors with dual-reflector antennas has proven effective, enabling the allocation of the main reflector among various frequency bands (Sabban, 2014). The growing demand for multifunctional antennas in communication systems has led to the development of FSS with multiband features.

In specific applications, signal filtering is critical for extracting necessary signals from space. FSS plays a vital role in this context by reducing interference between signals and acting as a filter (Foroozesh & Shafai, 2006).

1.1 Frequency Selective Surface

Filters are pivotal in nearly all electronic or RF circuits, serving to control frequency signals by diminishing noise and undesirable interference once integrated into a design. They are classified as—low-pass, band-pass, and high-pass filters—based on their function. For instance, a low-pass filter permits lower frequencies to pass while obstructing higher frequencies. Complementary to filters in transmission lines are frequency selective surface (FSS) or dichroic configurations for space waves. Acting as a spatial filter when exposed to electromagnetic radiation, an FSS transmits specific frequency bands while reflecting others (Foroozesh & Shafai, 2006).

To grasp the idea of spatial filtering more comprehensively, envision a wave impinging on a metal surface. This concept falls into two main categories: absorbing energy or redirecting the energy. From a dynamic perspective, these represent the two possible outcomes for power in those frequencies – either absorption or redirection.

Figure 1. FSS mechanism (Foroozesh & Shafai, 2006)

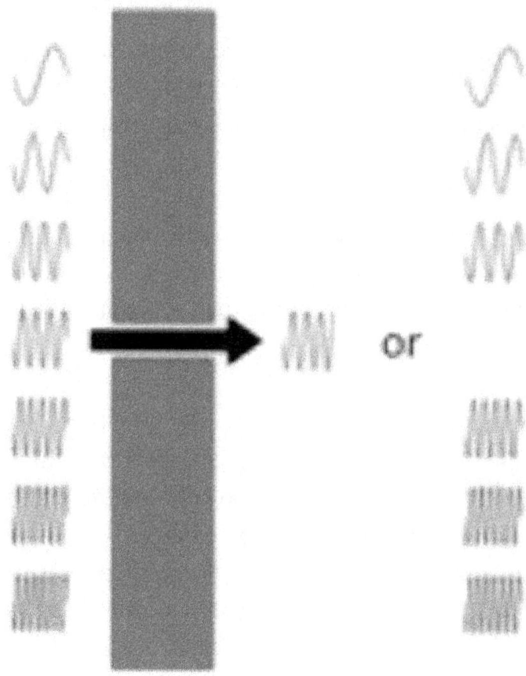

An FSS can serve as a concealment for communication facilities, marking one of its initial applications. These structures, known as radomes, operate as band-pass filters to reduce Radar Cross Section (RCS) (Foroozesh & Shafai, 2006).

1.2 Frequency Selective Surface: Brief History

Frequency Selective Surfaces (FSS) serve as specialized filters that can either allow or block specific frequencies. Designed with simplicity, spatial filters utilize periodic metallic patterns on dielectric materials. In the 18th century, physicist David Rittenhouse found that viewing a street lamp through a silk handkerchief suppressed specific colors in the visible spectrum, showcasing the selective transmission behavior of non-continuous surfaces for various frequencies. This discovery gave rise to the term "frequency selective surfaces" (FSS) (Jouvaud et al., 2013).

The concept of FSS expanded into radio frequencies, and in 1919, G. Marconi and B. Franklin used a reflector made from horizontal conductors to create a non-continuous parabolic surface for a specific frequency. Over the last 40 years, FSS has garnered extensive attention. The Voyager 77 satellite utilized an FSS for a dual-frequency reflector, demonstrating its application in dual-frequency bands. In 1996, the Cassini satellite extended this principle to four frequency bands, showcasing the demoralization of selective surfaces, even in military applications (Ourir et al., 2007).

Frequency Selective Surfaces (FSS), originating in the early 1900s with Guglielmo Marconi, passively filter electromagnetic waves through periodic arrays of materials or metallic elements. This interference induces a frequency-selective response, allowing specific frequencies to transmit while blocking others. The Lockheed F-117 Nighthawk during the Gulf War demonstrated FSS application in stealth technology, utilizing diffraction, RF absorbent materials, and FSSs. Despite their effectiveness in filtering microwaves, metallic FSSs face limitations in handling power generated by modern space transportations and pulsed power systems, leading to destructive phenomena like flashovers and arcs. Recent efforts focus on developing miniature dielectric FSSs with high breakdown voltage to enhance electric field handling. Additionally, FSS applications in wireless security have gained prominence with advancements in wireless networks. Table 1 summarizes various FSS designs found in the literature.

Table 1. Comparison of various literatures

Ref. No.	Year	Frequency Range	Unit Cell Size	Substrate Used	No. of Array	State of Art
(Kataja & Nikoskinen, 2011)	1993	7.5 to 16 GHz	7.0*3.2mm	RT Duroid with Ɛr = 2.3	NA	A square frequency selective surface and pin diodes used to control the switching.
(Li, 2013)	2004	9 to 13 GHz	14.5*10mm	FR4 with Ɛr=4.4	15*12	Single layer active microwave absorber and pin diodes used to control the switching.
(Lee & Itoh, 2006)	2011	2.35 to 2.7 GHz	32.8*32.4mm	Polyester with Ɛr= 3.0	5*5	Square loop slots used here to design a reconfigurable frequency selective surface. PIN diodes pin diodes used to control the switching with separate biasing circuit.
(Manh, 2014)	2014	10 to 12 GHz	9.44*9.44mm	FR4 with Ɛr=3.5	25*25	A novel microwave absorber and pin diodes used to control the switching.
(Markley & Eleftheriades, 2011)	2015	2.45 GHz	57*25mm	Flexible polyester	NA	An cylindrical slot frequency selective surface incorporating with an antenna for beam swiping application at WLAN frequency.
(Labadi et al., 2011)	2014	60 GHz	2.3*5.2mm	ROGERS RT/ duroid with Ɛr=2.2	NA	Millimeter-wave (MMW), frequency selective surface with Dielectric Resonator Antenna. Reconfigurable radiation pattern using only one PIN diode.
(Li, 2008)	2018	0.54 to 2.50 GHz	12*12mm	FR4 with Ɛr=3.5	10*10	Circular loop used here to design a reconfigurable shielding frequency selective surface. PIN diodes pin diodes used to control the switching for wideband.
(Saeed Majedi, 2012)	2016	2.45 and 5.2 GHz	36.6*65mm and 18.8*30mm	RT/duroid 5880 with Ɛr=2.2	NA	Two independent shape cylindrical frequency selective surfaces incorporating with a dual band antenna for beam swiping application. Switching can be controlled by reverse bias current at PIN diode.
(Abdalla et al., 2012)	2015	3.3 GHz	11.4*11.4mm	Rogers4350B with Ɛr=3.66	17*17	An innovative method by using induced voltage as biasing and switching control. Active Frequency selective surface at 3.3 GHz.
(Malathong, 2011)	2011	2.45 GHz	38.16*38.16 mm	Polyester with Ɛr= 3	2*2	Ring element used here to design a reconfigurable frequency selective surface. PIN diodes used to control the two state switching either reflective or transmissive at 2.45 GHz.
(Oliaei & Abrishamian, 2014)	2014	3 GHz	12.4*12.4mm	Rogers4350B with Ɛr=3.66	17*17	Square AFSS consist of slots etched on upper side substrate while the biasing circuits on another side.The inductive voltages created thru incident wave come to be bias control signal.
(Nakano, 2014)	2016	3.71 GHz	28*28mm	FR4 with Ɛr=4.4	5*5	An active square shape unit cell constructed for switchable absorber or reflector frequency selective surface. FSS performed either or broadband applications.
(Zani et al., 2010)	2017	2.3.68,4.02 and 6.37	26*26mm	FR4 with Ɛr=4.4	6*5	Annular ring and cross dipole shape used. Independent switching used here for dual band and individual band response. PIN diodes used to control the switching.
(Abdalla & Ibrahim, 2013)	2017	0.54 to 2.50 GHz	12*12mm	FR4 with Ɛr=4.4	10*10	A low-profile FSS used for wideband shielding. The FSS comprises of periodic arrangements of circular loops connected amongst themselves thru varactor diodes for tunable operation
(Li, 2000)	2017	5.2 GHz	12.5*12.5mm	Rogers RO3003 substrate (Ɛr = 3) and RO2929 (Ɛr = 2:95).	61*46	Low profile multilayer FSS, periodic scatters in the shape of loaded dipoles (dogbones) to produce a single stopband.Multiple stopband can be created by cascaded such layers

Continued on following page

Table 1. Continued

Ref. No.	Year	Frequency Range	Unit Cell Size	Substrate Used	No. of Array	State of Art
(Mishra & Patnaik, 1998)	2004	2.5 GHz	5*17.5mm	dipole CCMEBG Ɛr=2.2	NA	Closely-coupled-metallodielectric electromagnetic band-gap (CCMEBG) arrangements is presented and explored using tripole and dipole arrays.
(Mercier, 2005)	2008	5 GHz	28*28mm	FR4 with Ɛr=4.4	6*6	Double layer FSS to improve the bandwidth of an EBG antenna.
(Majid & Rahim, 2007)	2011	4 to 12 GHz	15*15mm	FR4 with Ɛr=4.4	6*6	A multilayer FSS and its potential used in UWB antennas. The multilayer FSS offers an suitable reflection phase to doing as a reflector and is capable to improve its gain
(Mittra, 2008)	2011	7 to 14 GHz	17*17mm	FR4 with Ɛr=4.4	7*7	Multi octave double sided FSS used to enhance gain of UWB antenna
(Park & Lee, n.d.)	2014	3 Ghz	24*24mm	Rogers RT/ duroid 5880 with Ɛr=2.2	37*50	Miniaturized-element-frequency-selective-surfaces(MEFSSs), Multiple capacitive and inductive patch. Harmonics suppressed at higher passbands.
(Patnaik, 2005)	2014	8.6 to 11.4 GHz	9.3*9.3mm	Rogers RT/ duroid 5880 with Ɛr=2.2	9*9	Square elements are used in EBG structure as patch and slot. Two FSS combined to improve the gain up to 13.8 dBi.
(Paul, 2013)	2014	8.4 GHz	17.86*17.86mm	Roger and Taconic	NA	Square elements, cross dipole, slot dipole are used as multilayer FSS for transmitarry FSS design.
(Bhagat, 2012)	2015	44 GHz	4.4*4.4mm	FR4 with Ɛr=4.4	NA	Ring elements used on upper layer and wheel shape in lower layer double layer for multiple dual band response at the aim of bandpass and bandstop
(Pandeeswari, 2012)	2015	X- and Ka-band	3*2.85mm	F4B-2 with Ɛr=2.65	100*100	Square loop elements used on multilayer FSS. Dual pass band for X and Ka band
(Venkatrami Reddy et al., 2014)	2004	10.7 to 11.9 GHz	25*37.5mm	polyfoam Ɛr=1.01	NA	Ferrite Ba(MnTi)$Fe_{10}O_{19}$ used to develop broadband microwave absorber
(Subramanian, 2012)	2005	4.5 to 5.5 GHz	29.7* 29.7mm	FR4 with Ɛr=4.4	NA	A circular fractal element used to develop broadband microwave absorber. Reflective response measured using the decrement method.
(Luo et al., 2014)	2017	3.28, 4.2, and 5.4 GHz	6.2*6.2mm	F4B-2 with Ɛr=2.65	50*50	Center connected sprial triangular loop elements used on top and metallic cross dipole in bottom to produce triple band.
(Li, 2000)	2017	6 to 16 Ghz	9.8*9.8mm	F4B with Ɛr=2.2	50*50	Loaded dipole elements arranged in cascade to develop multi stop band response.
(Pan, 2012)	2017	8 to 12 Ghz	7.7*7.7mm	Arlon Di 880 Ɛr=2.2	21*14	Incurved arm cross dipole element for X band. Excellent angular stability
(Lee et al., 2011)	2014	2.4 and 5.0 GHz.		F4B-2 with Ɛr=2.65	30*52	Anchor-shaped looped element with dual-band stop response at WLAN application
(Jiang, 2014)	2015	2.54 and 3.54 GHz	10.4*10.4mm	FR4 with Ɛr=4.4	28*28	Two individual crooked cross elements joined to provide dual band response. Good angular stable performance.
(Zhang et al., 2008)	2014	1.0 to 13.5 GHz	80*60 and 17*17mm	FR4 with Ɛr=4.4	11*11	Teragon fractal element embedded with monopole UWB antenna.
(Werner et al., 1999)	2015	3 to 11 GHz	50*50mm	Ɛr=3.48	NA	Square loop elements used for unit cells. Improvement in Gain of UWB antenna by the help of FSS
(Delaune et al., 2008)	2018	900 and 1790 MHz	24*23mm	Pure PTFE with Ɛr = 2.2	8*8	Double square loop used here in 2.5D to realize band-stop performance. The main target area of designing is to protect from GSM signal

Continued on following page

Table 1. Continued

Ref. No.	Year	Frequency Range	Unit Cell Size	Substrate Used	No. of Array	State of Art
(Choi et al., 2005)	2017	5.198 to 5.882 GHz	6*6mm	FR4 with Ɛr=4.4	66*66	A miniaturized absorptive FSS permitting power transmission in 5.5 GHz band. The via-based structure is adopted to make unit-cell size only 0.109λ0 × 0.109λ0, where λ0 is the wavelength in free space at operating frequency.
(Gopalakrishnan & Gunasekaran, 2005)	2015	2.40 and 5.30 GHz	45.19*46	FR4 with Ɛr=4.4	6*6	A double ring element used here to develop angular stable FSS thru functioning in the WLAN range.
(Huang et al., 2015)	2014	1.647 to 2.65 GHz	9.8*9.8mm	FR4 with Ɛr=4.4	NA	Modified square loop aperture. Equivalent circuit method used for analysis, Miniaturization and band pass performance
(Griguer, 2009)	2015	3.25 to 3.38 GHz	6*6mm	Rogers RT/ duroid 5880 with Ɛr=2.2	30*30	A new sort of FSS with the help of $0.067\lambda_0$ miniaturized resonator element, designed at WiMAX band
(Hassan Mirzaei & Eleftheriades, 2011)	2015	8.84 to 10.74 GHz	23.25*23.25mm	MNG with Ɛr = 1.15	NA	Electromagnetic techniques and design strategies for FSS structure applications.
(Kim, 2008)	2011	2.45 GHz	38.16* 38.16mm	Polyester substrate with Ɛr=3	NA	Ring element in kind of all polarizations unaffected and worthy angular stability.
(Joshi et al., 2011)	2017	9.76 to 10.57 GHz	8.2*8.2mm	FR4 with Ɛr=4.4	17*17	Polarization rotator design is demonstrated theoretically as well as experimentally based on frequency selective surface (FSS).
(Abraham et al., 2003)	2014	2.4 and 5.2 GHz	30*30mm	FR4 with Ɛr=4.4	7*7	Square shaped patch loaded with circular and square slot, Double slot FSS aimed at 2.4 GHz and 5.2GHz. Design recommended for ISM band
(Turpin, 2011)	2016	3.4 to 3.7 GHz	18*14mm	R5800 with Ɛr = 2.2	NA	A pattern-reconfigurable antenna through enriched bandwidth and fine beam width is offered with switchable FSS. Alteration of beam using AFSS
(Turpin, 2012)	2014	20 dB	35*28mm	Bekinox® BK yarn	NA	Three models of FSS appreciated from conducting yarns and bare copper wire arranged on textiles. The FSS arrays formed using the procedures of weaving, hand looming, and embroidery.
(Zhu & Eleftheriades, 2009)	2015	4.7 THz	4.5*4.5mm	Graphene Ɛr = 1	NA	graphene patch FSS has demonstrated by an similar tonsorial surface conductivity. The rotation angle and total transmission amplitude are then predicted for normal incidence with different values of the biasing magnetic field.
(Joshi et al., 2009)	2015	8.29 GHz	5*5mm	Arlon Diclad 522 Ɛr = 2.5	15*15	A fast analysis process depends on Contour-FFT, Macro basis function methodology is suggested to compute the reflected field from printed FSS.
(Pintos et al., 2010)	2013	5.34 and 5.94 GHz	7.5*7.5mm	FR4 with Ɛr=4.25	4*4	The combination of ELC resonator and cave ELC the structure behave like dual band absorber work on C band by means of single dielectric layer
(Wang et al., 2013)	2015	4, 6.54 and 11.15 GHz	13.6*13.6	FR4 with Ɛr=4.4	17*17	A tetra-arrow ring shaped element aimed for thin microwave absorber Three different absorption peaks at 11.15 GHz, 4 GHz, 6.54 GHz through absorptivities of 97.96%, 95.27% and 98.35%
(Ranjan & Singh, 2013)	2015	2 to 12 GHz	16*16mm	RT/Duroid 6202 with Ɛr=4.4	NA	Dissimilar Sierpinski Fractal unit cell for simple, compact, and multiband FSS

1.2 Band Stop FSS

The Band Stop FSS stands out as one of the most commonly employed types. Its configurations include circular, square, or hexagonal loops, cross dipoles, tripoles, and more. These diverse shapes are systematically arranged in periodic arrays of conductive elements. Figure 2 illustrates an example of a band-stop filter featuring a cross dipole structure.

Figure 2. Band stop FSS

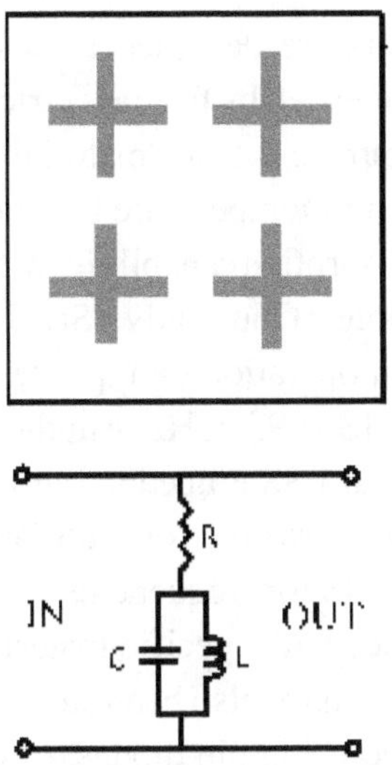

Several innovative Frequency Selective Surface (FSS) designs have been proposed for wireless LAN (WLAN) applications, targeting dual-band reject responses. One approach, presented in Gopalakrishnan and Gunasekaran (2005), employs conducting patches and slots on the unit cell to achieve cascaded dual-band rejection in the ISM and UNII bands. This FSS design demonstrates improved attenuation levels, ranging from 10 to 25 dB in the ISM band and 10 to 20 dB in the UNII band.

Another design, detailed in Lee et al. (2011), utilizes an anchor-formed loop unit cell arrangement for dual-band reject performance in both the 5.0 GHz and 2.4 GHz WLAN bands. This FSS also exhibits excellent miniaturization with 0.065λ*0.076λ unit cells and provides stable performance for both TE and TM polarizations under incident angles of 0°-60°.

In Pan (2012), a crooked cross arrangement geometry is explored for an FSS with angular-independent functioning. The symmetrical shape dual-band FSS achieves rejection at 2.54 and 3.54 GHz, serving as a shielding component for S-band frequencies. The design ensures equal stop-band response for TE and TM modes.

In Lu et al. (2013), a suggestion is presented for shielding against WLAN bands (2.4-2.5 GHz and 4.98-5.825 GHz). This approach can be implemented with paired window glasses, providing defense against WLAN signal intrusion in indoor settings and buildings.

1.3 Frequency Selective Surface as Absorber

Electromagnetic wave absorbers are designed to absorb incident or transmitted waves on periodic surfaces, influenced by the dielectric properties of their elements. Typically, these absorbers comprise a filler material within a matrix, chosen for specific physical attributes such as temperature resistance and weatherability.

Several studies explore the absorption capabilities of Frequency Selective Surfaces (FSS) in different configurations. One study (Srivastava & Somwanshi, 2015) presents an FSS designed as an operation to a quarter-wave absorber, achieving a shielding behavior of about 30 dB at 9.7 GHz. Another (Pyo et al., 2009) discusses microwave absorbers utilizing an FSS embedded in an impedance-graded carbon fiber compound, with evaluations performed on multilayer absorbers. In Peng (2010), a quarter-wave absorber using FSS for frequencies 9 GHz and 15 GHz is reported, demonstrating good performance with specific material selections.

Tunable microwave absorbers have also been studied, with Li (2013) presenting a single-layer absorber switched using pin diodes, offering tunability from 9 GHz to 13 GHz. Circular fractal-shaped FSS (Subramanian, 2012) and FSS layered with ferrite (Venkatrami Reddy et al., 2014) are explored for broadband microwave absorption, while Wang (2005) discusses two types of FSS-based planar microwave absorbers, including a high impedance surface (HIS) and a Salisbury absorber with resistively loaded HIS.

1.4 Reconfigurable FSS

FSS can be categorized based on their switching characteristics, defining how they dynamically reconfigure filtering properties. Two main types are identified:

- Passive FSS:

These surfaces have fixed properties that cannot be altered.

Suited for specific frequencies but require different configurations for various applications.

Drawback: Lack of reconfigurability.

- Active FSS:

Comprised of periodic structures with active devices

Properties can be changed by external power supply, making them reconfigurable.

Drawbacks: Higher manufacturing costs, bulkier structures.

literature explores reconfigurable FSS designs:

- A PIN diode-based FSS for broadband applications (Nakano, 2014), offering polarization independence and a shift-able absorber/reflector.
- Switchable FSS for millimeter-wave applications (Odabasi & Fernando, 2013), integrated into a dielectric resonator antenna for pattern diversity and beam steering.
- A ring-shaped FSS (Kim, 2008) offering reflective or absorptive states, applicable in the 2.45 GHz WLAN band.
- FSS in a pattern reconfigurable antenna (Turpin, 2011) using PIN diodes for beam control, demonstrating notable performance in bandwidth, gain, and radiation pattern.
- Tunable FSS using graphene (Zhu & Eleftheriades, 2009) for adjustable transmission and reflection by altering chemical potential and biasing magnetic field.
- Dual-band alterable FSS with PIN diode-loaded active doubly periodic flat strip (Sarmah & Sarma, n.d.) serving as a dynamic screen.
- FSS-based switchable microwave absorbers (Li, 2013; Manh, 2014), employing PIN diodes for reflectivity control and behavior modulation by changing pin diode current.
- Dual-band agile absorber (Zani et al., 2010) with switchable frequencies through embedded switchable pin diodes.
- FSS using varactor diodes (Li, 2008) for wideband shielding with tunable operation.
- A novel design using induced voltage (Abdalla et al., 2012) for active FSS with controllable transmission properties.
- Switchable slot-type FSS (Lee & Itoh, 2006; Markley & Eleftheriades, 2011) utilizing PIN diodes for frequency characteristics, introducing novel biasing tracks and incorporating Fabry-Perot approach for improved selectivity.

These advancements highlight diverse applications and innovative approaches to enhance the functionality of FSS in various scenarios.

1.5 Multilayer Frequency Selective Surface

Frequency Selective Surfaces (FSS) are often designed as planar, periodic arrays in 2-D space, commonly known as single-layer FSS. However, to achieve multi-band applications, researchers cascade two or more surfaces, forming multilayer frequency selective surfaces. Several studies explore the advancements and applications of multilayer FSS:

- Dual/Triple Band Response:

Cascaded Jerusalem cross structure and double square element FSSs for dual/triple band response (Mishra & Patnaik, 2000).

Dual layer dipole and tripole arrays (Mishra & Patnaik, 1998).

Dual layer FSS combined with EBG resonator for bandwidth improvement (Mercier, 2005).

- Wide Bandwidth and Gain Enhancement:

Multilayer FSS with ultra-wideband antenna for better reflection phase and 140% gain enhancement (Majid & Rahim, 2007).

UWB antenna gain enhancement using a multioctave double layer FSS (Mittra, 2008).

- Optimized Designs:

Hexagonal unit cell-based FSS for wideband applications with optimized coupling (Mussetta, 2010).

Equivalent circuit analysis of alternate double-layered FSS (Fhafhiem et al., 2012).

- Bandpass FSS with Harmonic Suppression:

Miniaturized-element FSS (MEFSS) with bandpass response and harmonic suppression (Park & Lee, n.d.).

- Directivity Enhancement:

Multilevel FSS for directivity enhancement, utilizing a planar lens (Patnaik, 1997).

- Fabry-Perot Resonator Antenna:

FP resonator antenna with complementary patch-type and slot-type FSSs, achieving a 3 dB gain and 2.8 GHz bandwidth (Patnaik, 2005).

- Applications in Transmitarray Antennas:

Limiting the transmission phase of transmitarray antennas using multilayer FSS (Paul, 2013).

- Angular Stability Improvement:

2nd order multilayer FSS for improved angular stability, particularly in oblique incidence (Pelosi, 2004).

- Metasurface and Dual Band Response:

Multilayer FSS in the application of metasurface, changing the reflection way of incident waves (Mookiah & Dandekar, 2009).

Multilayer FSS with dual-band pass and dual-band stop response acting as a dichroic reflector (Bhagat, 2012).

- Miniaturized Dual Band FSS:

Dual-band FSS with 2nd order band-pass behavior, designed for wide-ranging band response and stability (Pandeeswari, 2012).

These studies highlight the versatility and applications of multilayer FSS in achieving various functionalities such as bandwidth improvement, gain enhancement, angular stability, and selective reflection.

2. DESIGN METHODOLOGY

The optimization of the proposed structure in this research work was achieved through a systematic theoretical and experimental design approach. The strategy employed for streamlining the design and development procedures can be outlined in the following steps:

- **Substrate Selection:** Choose the appropriate substrate for the FSS.
- **Geometry Selection:** Determine the candidate radiator geometry.

- **Initial Parameter Calculation:** Calculate the initial geometrical parameters using empirical formulas predicting the lower edge frequency.
- **Simulation and Analysis:** Examine FSS behavior over an extended bandwidth and analyze results comprehensively.
- **Parameter Adjustment:** Modify parameters based on intuition and iteration until generally satisfactory results are achieved, especially for the lower band. Consider changing the candidate geometry if needed.
- **Bandwidth Enhancement:** Implement effective bandwidth-enhancement techniques for wider bandwidth.
- **Miniaturization Attempt:** Reduce geometry size by eliminating non-contributing parts after assessing current distribution.
- **Simulation Validation:** After successful simulation, obtain results such as impedance, radiation patterns, and gain.
- **Fabrication:** Fabricate the FSS and measure reflectivity and transmissivity to confirm simulated results.

3. FSS DESIGN PROCESS

The performance of a frequency selective surface is influenced by various factors. Let's explore each of these factors individually.

- Shape of elements (apertures/patches):

Oblique incidence and parallel polarization in a dipole array lead to double resonance issues. The unloaded tripole shape, with added end capacitance, serves as an anchor element, influencing the frequency response, particularly the lowest second resonance frequency and null.

- Size of the elements:

For loop-type elements, it is crucial to maintain a distance between two opposite ends within the range of $\lambda/3$ to $\lambda/4$. Additionally, ensuring that the total circumference or perimeter corresponds to one complete wavelength is essential. Failure to adhere to these size conditions can lead to null interactions in the frequency response.

- Inter-spacing between elements:

FSS sensitivity to oblique angles of incidence leads to resonance frequency and bandwidth variations. Maintaining inter-element spacing below 0.4λ minimizes changes in resonance frequency.

- Thickness of dielectric/metallic screen:

To preserve a stable phase difference, the dielectric or metallic screen should have a thickness equal to or less than λ/2, allowing reflections from both ends of the FSS to cancel each other.

- Oblique angle of incidence and polarization:

To stabilize the bandwidth of Frequency Selective Surfaces (FSS) under oblique incidence and polarization, one or more FSSs are strategically placed within a layered dielectric medium. This entails the integration of extra dielectric slabs on both sides of the Frequency Selective Surface (FSS), with the surfaces being notably thicker, slightly surpassing λ/4, compared to the thin substrate adjacent to the FSS. The thicker slabs aim to improve bandwidth stability, while the thinner ones offer mechanical protection and contribute to dielectric loading.

The fabrication of the FSS aligns with budgetary constraints and material availability. Project implementation involves both software and hardware design. The comprehensive design and development process are encapsulated in Figure 3, providing a step-by-step overview of the actions taken to successfully conclude the entire design cycle.

Figure 3. FSS design flow chart

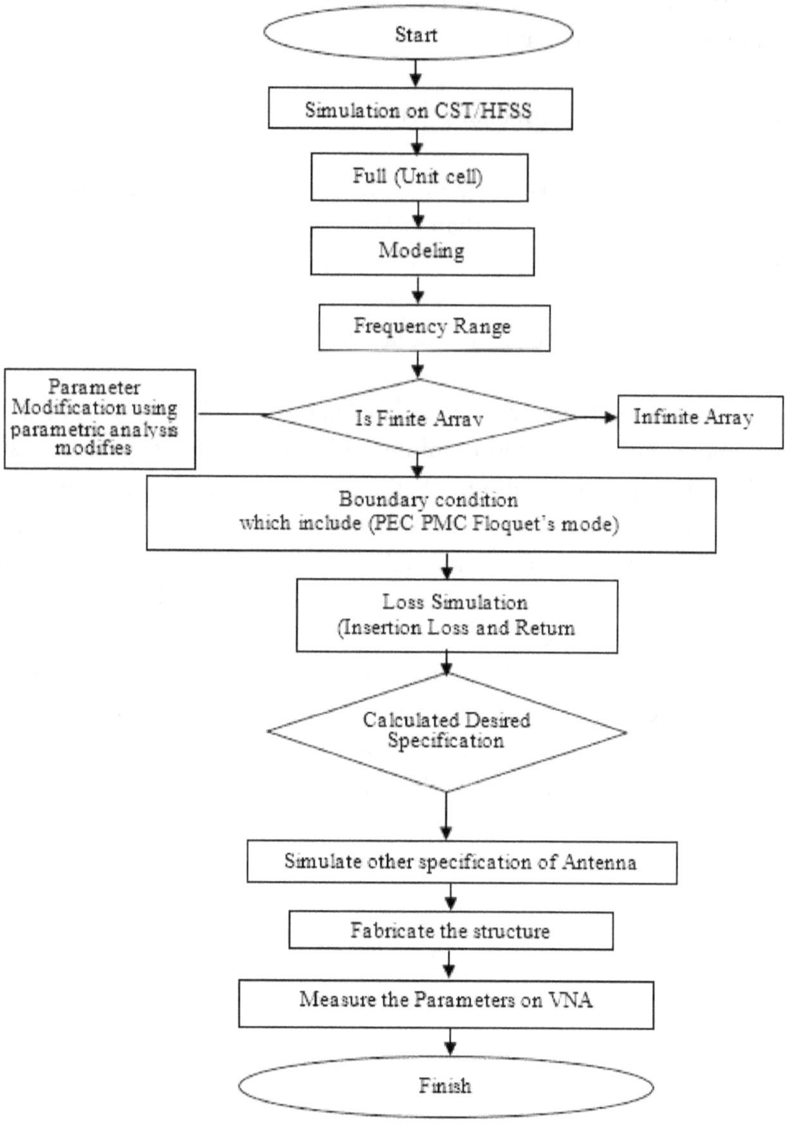

4. RESULT AND DISCUSSIONS

In the exploration of frequency selective surface (FSS) structures, resonant unit cells have been extensively studied for diverse applications in communication, including band pass, band stop, spatial filters, absorbers, and electromagnetic band gap materials. In this research a compact dual-band unit cell for FSS is introduces, resonating at the C-band and Wireless Local Area Network (WLAN) frequencies. The proposed design combines I shape and modified I shape metal strips, demonstrating dual-band filtering characteristics with good agreement between simulated and measured results. The concept of miniaturized-element FSSs is also discussed as a novel approach, utilizing small unit cells for enhanced performance in specific applications.

Upon comparing the data presented in Table 2, it is evident that the recommended Dual-Band Frequency Selective Surface (DBFSS) exhibits uniformity throughout its entire array structure, such as the I shape and modified I shape. The analysis considered both TE and TM characteristics. For TE polarization, the resonant frequency remains nearly constant, with a decrease in bandwidth observed at 45° and 60°. Similarly, for TM waves, the resonant frequency remains almost the same, with a reduction in bandwidth at 45° and 60°. The analysis indicates a shift in resonant frequency from 4.1 GHz to 3.8 GHz and an increase in the upper band from 5.6 GHz to 6.5 GHz. The detailed comparison of resonant frequency and bandwidth for TE and TM is summarized in Table 2.

Table 2. Parameter calculations

Angle	Resonance Frequency (GHz)		Frequency Band (GHz)	
	TE	TM	TE	TM
0°	4.080 & 5.60	3.82& 6.58	3.67-4.30 & 5.32-5.86	3.54-3.96 & 5.43-7.72
15°	4.069 & 5.61	3.82& 6.58	3.67-4.28 & 5.33-5.86	3.56-3.96 & 5.46-7.67
30°	4.078 & 5.60	3.83& 6.55	3.72-4.27 & 5.36-5.83	3.61-3.95 & 5.54-7.55
45°	4.078 & 5.60	3.84 & 6.51	3.80-4.24 & 5.40-5.79	3.67-3.94 & 5.68-7.36
60°	4.078 & 5.60	3.84 & 6.48	3.90-4.21 & 5.45-5.70	3.73-3.92 & 5.88-7.12

4.1 Fabrication and Measurement

To verify the proposed DBFSS design, a prototype was fabricated on an FR-4 substrate (1.6 mm thickness) with dimensions of 15 cm x 10 cm, featuring an array of 20 x 20 elements, as shown in Fig.4.

Figure 4. A fabricated prototype of FSS array

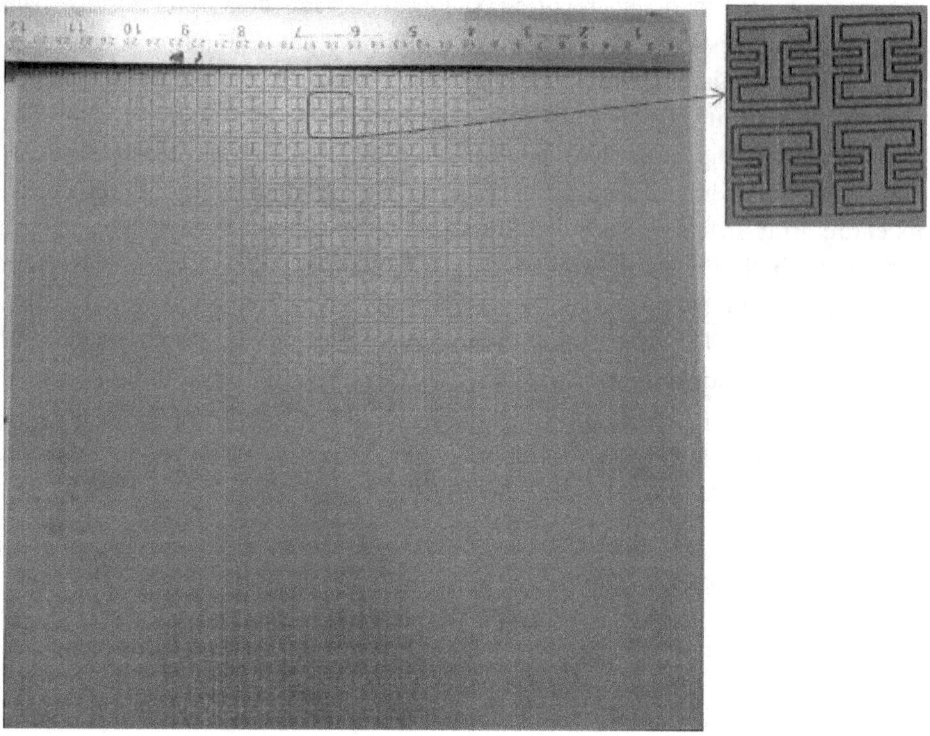

The measurement of S_{11} and S_{21} properties of I shape FSS is made in the anechoic chamber. Two horn antennas with having same properties have been used to reflection coefficient measurement even though the transmission coefficient. The DBFSS model placed in between the horns to calculate S_{21} and S_{11} characteristics measured by placing both horn in same direction. The experimental setup to measure characteristics of the prototype has been illustrated in Fig.5.

Figure 5. (a) Measurement setup of proposed antenna S_{21} measurement, (b) measurement setup of proposed antenna S_{11} measurement

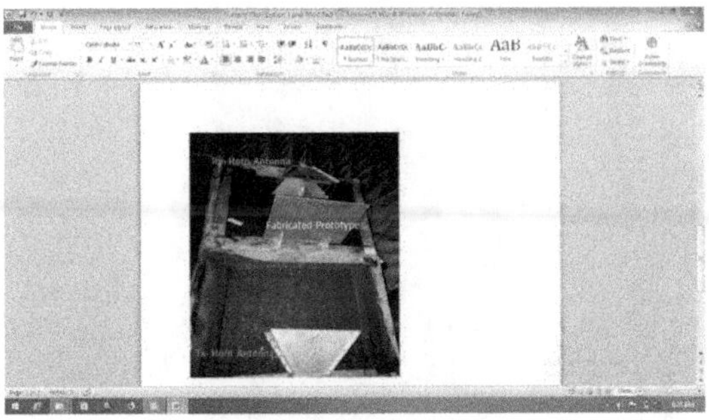

Figures 6 and 7 depict the comparison between measured and simulated results of the proposed Dual-Band Frequency Selective Surface (DBFSS) for S11 (in dB) and S21 (in dB). Discrepancies are attributed to boundary condition errors in simulation and calibration inaccuracies in cable and connectors during measurement.

Figure 6. Analysis of S_{21} (dB) vs. frequency

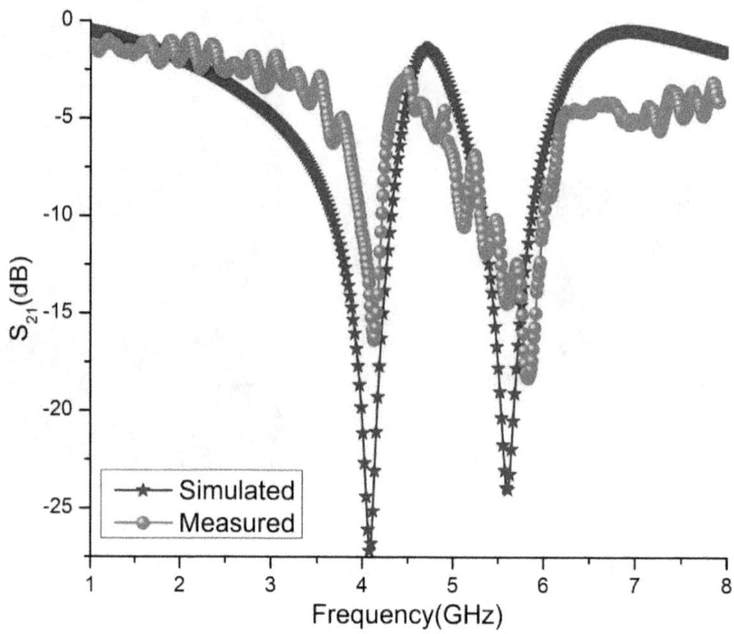

Figure 7. Analysis of S_{11} (dB) vs. frequency

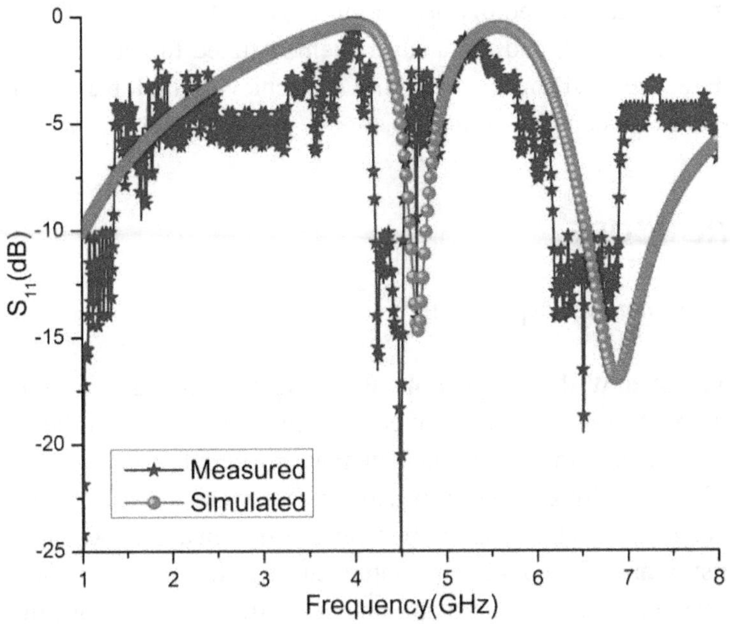

5. CONCLUSION

Based on the study of the different of different-different unit-cell with dual band response, it is found that the outer ring shows an essential part in incident signal excitation. Moreover, the transmission coefficient can be controlled by the optimization of geometry. Loop structures are found suitable for microwave band stop filter designing. The main objective of our research was to design dual band frequency selective surfaces with excellent performance. The above-mentioned objectives were satisfied by CST simulation. Also, we have investigated the reflection, transmission and polarization insensitive capability of frequency selective surface. The entire surface dimension has calculated by basic formulas.

Using the methodology two band stop frequency selective surface with band stop performance are investigated in this research. Initially work has been carried out by designing of IFSS structure. It was observed that creating of loop in IFSS gives miniaturization. Further this IFSS modified to produce MIFSS structure. Finally both structure embedded on same plane to create DBFSS structure. The investigation into the proposed Dual-Band Frequency Selective Surface (DBFSS) revealed its effective operation in dual frequencies—C band (4.0 GHz) and WLAN (5.5 GHz). The FSS design is characterized by a simple structure and a miniaturized unit cell of 10x10 mm2, with detailed discussion on individual unit cells. The fabricated prototype was measured, and the analysis demonstrated the effectiveness of the suggested FSS, emphasizing its suitability for achieving multiband performance without the need for complex structures.

6. FUTURE SCOPE

The future work involves exploring three key areas:

- **Conformal and Reconfigurable FSS:** Developing FSS for antennas on moving platforms like aircraft or cars to minimize air drag while addressing design challenges for low-profile antennas on curved surfaces.
- **Focal Length:** Utilizing the proposed surfaces to create lenses with very short focal lengths, expanding possibilities in lens design and optical applications.
- **RF Insulators:** Creating encapsulating surfaces to insulate wireless devices from the human body, improving antenna efficiency and reducing Specific Absorption Rate (SAR) by designing miniaturized surfaces as pure reactance shields, a solution not achievable with metallic surfaces.

REFERENCES

Abdalla, M., Abdelnaby, U., & Mitkees, A. A. (2012). Compact and Triple Band Meta-material Antenna for All WiMAX Applications. *Proceedings of ISAP2012.*

Abdalla & Ibrahim. (2013). Compact and Closely Spaced Metamaterial MIMO Antenna With High Isolation for Wireless Applications. IEEE Antennas and Wireless Propagation Letters, 12.

Abraham, J. K., Vinoy, K. J., Tellakula, R. A., & Varadan, V. K. (2003). Experimental Results of Planar Meta Materials. *IEEE-APS International Symposium.*

Bhagat, P. P. (2012). Analysis and Synthesis of Microstrip Patch Antenna using Artificial Neural Networks. *IEEE Asia-Pacific Conference on Antennas and Propagation.* 10.1109/APCAP.2012.6333177

Choi, D., Shin, C., Kim, N., & Shin, H. (2005). Design and SAR Analysis of Broadband PIFA with Triple Band. *Progress in Electromagnetics Research Symposium, 1*(3), 290–293. doi:10.2529/PIERS041212211650

Dadgarpour, A., Zarghooni, B., Virdee, B., & Denidni, T. A. (2014). Beam Tilting Antenna Using Integrated Metamaterial Loading. *IEEE Transactions on Antennas and Propagation, 62*(May), 2874–2879. doi:10.1109/TAP.2014.2308516

Delaune, D., Guan, N., & Ito, K. (2008). *Simple Multiband Antenna for Mobile Phone Application Based on a Dual-arm Monopole Structure*. PIERS Online, 4(1).

Dhouibi, A., Burokur, S. N., Lustrac, A., & Priou, A. (2012). Compact Metamaterial-Based Substrate-Integrated Luneburg Lens Antenna. *IEEE Transactions On Antennas And Propagation, VOL., 11*, 1504–1507.

Fhafhiem, N., Krachodnok, P., & Wongsan, R. (2012). The Circularly Polarized Resonator Antenna using Double Polarizing Metallic EBG. *IEEE Asia-Pacific Conference on Antennas and Propagation*. 10.1109/APCAP.2012.6333133

Foroozesh, A., & Shafai, L. (2006). Size reduction of a microstrip antenna with dielectric superstrate using meta-materials: Artificial magnetic conductors versus magneto- dielectrics. *Proc. IEEE Antennas Propag. Soc. Int. Symp.*, 11–14. 10.1109/APS.2006.1710439

Gopalakrishnan, R., & Gunasekaran, N. (2005). Design of equilateral triangular microstrip anten using artificial neural networks. *IEEE International Workshop on Antenna Technology: Small Antennas and Novel Metamaterials*, 246 - 249.

Griguer, H. (2009). Patch Antenna Bandwidth Enhancement through the Use of Metamaterials. *International Conference on Telecommunications*, 323-327.

Hassan Mirzaei & Eleftheriades. (2011). Frequency Reconfigurable Metamaterial Inspired Antenna. IEEE Antennas and Wireless Propagation Letters, 10

Huang, Liu, & Gong. (2015). Multiband Metamaterial-Loaded Monopole Antenna for WLAN/WiMAX Applications. *IEEE Antennas and Wireless Propagation Letters, 14*, 662-665.

Jiang. (2014). A Low-Profile High-Gain Substrate-Integrated Waveguide Slot Antenna Enabled by an Ultrathin Anisotropic Zero-Index Metamaterial Coating. *IEEE Transactions On Antennas And Propagation, 62*(March), 1173–1184.

Joshi, J. G., Pattnaik, S. S., & Devi, S. (2011). *Metamaterial loaded dual band microstrip patch antenna. Antenna Week.*

Joshi, J. G., Pattnaik, S. S., Devi, S., Lohokare, M. R., & Vidyasagar, C. (2009). Offset fed diamond shaped split ring (DSSR) planar metamaterial antenna. *IEEE Conference on Applied Electromagnetics*, 1-4. 10.1109/AEMC.2009.5430686

Jouvaud, C., de Rosny, J., & Ourir, A. (2013). Adaptive metamaterial antenna using coupled tunable split-ring resonators. *Electronics Letters*, *49*(8), 518–519. doi:10.1049/el.2013.0398

Kataja, J., & Nikoskinen, K. (2011). The Parametric Optimization of Wire Dipole Antenna. *IEEE Transactions on Antennas and Propagation*, *59*(2), 350–356. doi:10.1109/TAP.2010.2096181

Kim. (2008). The Design of a Small Monopole Type Antenna using Negative Permittivity Metamaterial Structure. *Antenna Technology: Small Antennas and Novel Metamaterials, iWAT 2008. International Workshop on*, 275-278.

Labadi, M., Jamel, B. T., & Choubani, F. (2011). New design of antenna array using Left Handed Meta-Material "LHM" based on circular Split Ring Resonator "SRR". *Mediterranean Microwave Symposium (MMS),* 52-56.

Lee, C. H. L., & Itoh, T. (2006). Compact dual-band antenna using an anisotropic metamaterial. *Proceedings of the 36th European Microwave Conference*, 1044-1047. 10.1109/EUMC.2006.281112

Lee, Y., Ga, D., & Choi, J. (2011). Design of a MIMO Antenna with Improved Isolation Using MNG Metamaterial. *International workshop on Antenna Technology (iWAT),* 231 – 234.

Li. (2013). Design of Electrically Small Metamaterial Antenna With ELC and EBG Loading. *IEEE Antennas and Wireless Propagation Letters, 12*, 678-681.

Li, L.-W. (2008). Design of a Novel Rectangular Patch Antenna with Planar Metamaterial Patterned Substrat. *International Workshop on Antenna Technology: Small Antennas and Novel Metamaterials,* 124-126.

Li, Z. (2000). PBG, PMC and PEC Ground Planes: A Case Study of Dipole Antennas. *IEEE Antennas and Propagation Society International Symposium (APSURSI),* 2, 674 - 677.

Manh. (2014). Antenna optimization based on Artificial Neural Network. *The 8th European Conference on Antennas and Propagation (EuCAP 2014),* 3172-3175.

Lu, Z. Q., Zhao, M., Yang, Z. Y., Wu, L., & Zhang, P. (2013, August 15). Helical Metamaterial Absorbers: Broadband and Polarization-Independent in Optical Region. *Journal of Lightwave Technology*, *31*(16), 2762–2768. doi:10.1109/JLT.2013.2272632

Luo, Y., Han, Z., & Toshiyoshi, H. (2014). A Novel J-Shape Active MEMS Metamaterial Antenna. *IEEE Antennas and Propagation Society International Symposium (APSURSI), 569-570.* 10.1109/APS.2014.6904615

Malathong. (2011). Chiral Metamaterial Based Circularly Polarized Microstrip Antennas. *IEEE-APS Topical Conference on Antennas and Propagation in Wireless Communications*, 898 - 901.

Majid, H.A. & Rahim, M.K.A. (2007). Investigation of LH-meta material in microstrip antenna. *Asia-Pacific Conference on Applied Electromagnetics*, 1-4.

Markley, L., & Eleftheriades, G. V. (2011, March). Meta-screens and near-field antenna-arrays: A new perspective on subwavelength focusing and imaging. *Metamaterials (Amsterdam)*, *5*(2-3), 97–106. doi:10.1016/j.metmat.2011.03.004

Mercier, L. (2005). Steerable and tunable "EBG resonator antennas" using smart metamaterials. *IEEE Antennas and Propagation Society International Symposium, 4B*, 406-409.

Mishra, R. K., & Patnaik, A. (1998). Neural network-based CAD model for the design of square patch antennas. *IEEE Transactions on Antennas and Propagation, 46*(12), 1890–1891. doi:10.1109/8.743842

Mishra, R. K., & Patnaik, A. (2000). Neurospectral Analysis of Coaxial Fed Rectangular Patch Antenna. *IEEE Antennas and Propagation Society International Symposium*, 2, 1062 – 1065. 10.1109/APS.2000.875405

Mittra, R. (2008). To Use or Not to Use the Effective Medium Approach for Designing Performance-Enhanced Small Antennas-That is the Question. *IEEE International Workshop on Antenna Technology: Small Antennas and Novel Metamaterials*, 55-58. 10.1109/IWAT.2008.4511289

Mookiah, P., & Dandekar, K. R. (2009). Metamaterial-Substrate Antenna Array for MIMO Communication System. *IEEE Transactions on Antennas and Propagation, 27*(October), 3283–3292. doi:10.1109/TAP.2009.2028638

Mussetta, M. (2010). Characterization of microstrip reflect array square ring elements by means of an Artificial Neural Network. *Proceedings of the Fourth European Conference on Antennas and Propagation (EuCAP)*, 1-4.

Nakano, H. (2014). Natural and Metamaterial-based Spiral and Helical Antennas. *Loughborough Antennas and Propagation Conference (LAPC)*. 10.1109/LAPC.2014.6996405

Odabasi, H., & Fernando, L. (2013). Complementary Electric-Field-Coupled (CELC) Based Resonator Antennas. *Antennas and Propagation Society International Symposium (APSURSI)*, 772-773.

Oliaei, M., & Abrishamian, M. (2014). Corrugated SIW K band horn antenna. *AEÜ. International Journal of Electronics and Communications*, *68*(12), 1199–1204. doi:10.1016/j.aeue.2014.06.009

Ourir, A., Burokur, S. N., & de Lustrac, A. (2007). Phase-Varying Metamaterial for Compact Steerable Directive Antennas. *Electronics Letters*, *43*(9), 493–494. doi:10.1049/el:20070298

Pan, Z. Y. (2012). Terahertz Metamaterials Fabrication from Narrowband to Broadband. *IEEE Asia-Pacific Conference on Antennas & Propagation*, 68 - 69. 10.1109/APCAP.2012.6333151

Pandeeswari, R. (2012). Artificial Neural Network Model for MNG-Metamaterial Spiral Resonator Progress. *Electromagnetics Research Symposium Proceedings*, 29-33.

Park & Lee. (n.d.). Omnidirectional circularly polarized antenna utilizing zeroth-order resonance of epsilon negative transmission line. *IEEE Trans. on Antennas and Propag.*, *59*, 2717-2720.

Patnaik, A. (1997). An Artificial Neural Network Model for Effective Dielectric Constant of Microstrip Line. IEEE Trans. Antennas Propag., 45.

Patnaik, A. (2005). A Frequency Reconfigurable Antenna Design Using Neural Networks. *IEEE Antennas and Propagation Society International Symposium*, 409-412. 10.1109/APS.2005.1551830

Paul P.M. (2013). Miniaturization of Square Patch Antenna using Complementary Split Ring Resonators. *Advances in Computing and Communications (ICACC)*, 122 – 125.

Pelosi, G. (2004). Inter-Element Coupling In Large Arrays by an ANN Approach. *IEEE Antennas and Propagation Society International Symposium*, 1, 523 – 526. 10.1109/APS.2004.1329711

Peng, Y. (2010). Ultra thin low profile U band folded meta-material wideband dipole antenna for multi-Gb/s data transmission using 65 nm CMOS technology. *Antennas and Propag Society Inter Symposium (APSURSI)*, 1-4.

Pintos, J. F., Louzir, A., Minard, P., Perraudeau, J., Mattei, J. L., Souriou, D., & Queffelec, P. (2010). *Ultra Miniature UHF Antenna Using Magneto-Dielectric Material*. ANTEM. doi:10.1109/ANTEM.2010.5552480

Pyo, Baik, Cho, & Kim. (2009). Metamaterial-based antenna with triangular slotted ground for efficiency improvement. *IEEE Electronics Letters, 45*(3).

Ranjan, K., & Singh, G. (2013). Effect of low dielectric permittivity on microstrip antenna at terahertz frequency. *Optik (Stuttgart), 124*(22), 5777–5780. doi:10.1016/j.ijleo.2013.04.061

Subramanian. (2012). Design of Dual Log-Spiral Metamaterial Resonator for X-band Applications. *International Conference on Computing, Communication and Applications (ICCCA)*, 1-6.

Sabban, A. (2014). New microstrip metamaterial antennas. In *Antennas and Propagation Society International Symposium (APSURSI)*. IEEE.

Saeed Majedi. (2012). *Compact MNG zeroth order resonator antenna*. Elsevier B.V.

Sarmah, K., & Sarma, K. K. (n.d.). Design Optimization of Loop Antenna Using Competitive Leakning ANN. *National Conference on Emerging Trends and Applications in Computer.*

Srivastava, S., & Somwanshi, D. (2015). Design and analysis of rectangular microstrip patch antenna for zigbee applications. *2015 IEEE International Symposium on Nanoelectronic and Information Systems*, 257-261. 10.1109/iNIS.2015.25

Turpin, J. P. (2011). Anisotropic Metamaterial Realization of a Flat Gain-enhancing Lens for Antenna Applications. *Symposium on Antennas and Propagation (APSURSI), IEEE International*, 2886-2889. 10.1109/APS.2011.5997130

Turpin. (2012). Low Cost and Broadband Dual-Polarization Metamaterial Lens for Directivity Enhancement. *IEEE Transactions On Antennas And Propagation, 12*.

Venkatrami Reddy, S., Saurav, K., Sarkar, D., Singh, A., & Srivastava, K. V. (2014). Design of a Novel Dual Band Patch Antenna Loaded with CELC Resonator for WLAN/WiMAX Applications. *Twentieth National Conference on Communications (NCC)*, 1-6.

Wang. (2005). Low Profile Highly Directive Antennas using EBG superstrates and Metamaterial Ground Planes. *IEEE Antennas and Propagation Society International Symposium, 4B*, 335 – 338.

Wang, J., Zhu, Z., Sun, Y., Shen, T., & Gong, L. (2013). Study on left-handed effect of composite helices and its application in square frame patch antennas. *Optik (Stuttgart), 124*(21), 5189–5192. doi:10.1016/j.ijleo.2013.03.052

Werner, D. H., Haupt, R. L., & Werner, P. L. (1999, October). Fractal Antenna Engineering "The Theory and Design of Fractal Antenna Arrays". *IEEE Antennas & Propagation Magazine, 41*(5), 37–59. doi:10.1109/74.801513

Zani, M. Z. M., Jusoh, M. H., Sulaiman, A. A., Baba, N. H., Awang, R. A., & Ain, M. F. (2010). Circular Patch Antenna On Metamaterial. *IEEE International Conference on Electronic Devices, Systems and Applications (ICEDSA),* 313-316. 10.1109/ICEDSA.2010.5503053

Zhang, Z., Xie, H., & Zheng, L. (2008). V-Shape wired structure H.F bowtie antenna. *International Symposium on Antennas, Propagation and EM Theory*, 207–210.

Zhu, J., & Eleftheriades, G. V. (2009). A Compact Transmission-Line Metamaterial Antenna With Extended Bandwidth. *IEEE Antennas and Wireless Propagation Letters, 8*, 295–298. doi:10.1109/LAWP.2008.2010722

Chapter 10
Phenomenon of Excess of Artificial Intelligence:
Quantifying the Native AI, Its Leverages in 5G/6G and Beyond

Snehasis Dey
https://orcid.org/0000-0002-0490-8628
College of Engineering, BPUT University, India

ABSTRACT

There is a challenging era for wireless communication especially after commercialization of AI/ML tool. Specifically, the phenomenon of mobile communication has evolved completely from being used as a mere communicating device to computing device or small messaging medium to manifest unparalleled comfort like proposing and composing podcasts, presentations, and scripting social posts with the help of AI-enabled ChatGPT. This chapter explores AI-enabled 5G, its standardization, and leveraging AI/ML in 6G and beyond.

INTRODUCTION

Starting from traditional 5G networks that have different processing units designed by humans to advanced 5G networks that have encapsulated AI and ML technology to optimize different functions, the roadmap of advancement is loud and clear. The 5G networks have graced the human race with a strong focus on digital transformation to support and accommodate the network with new age capabilities like enhanced mobile broadband (eMBB), ultra-reliable low latency (uRLLC), massive MIMO

DOI: 10.4018/979-8-3693-0916-2.ch010

(mMIMO), etc. (Shehzad et al., 2022). That progression of technology in 5G, which supports AI and ML, is basically AI-infused in nature. But the 6G network that demands a hybrid network like a terrestrial-underwater-aerial-to-space network will definitely propose complete automation from system design to network architecture, from signal processing to data optimization. Thus AI/ML must be used excessively, efficiently and effectively. AI, ML, and DL are extensively used to provide the ultimate solution to optimize large data sets and enhance network resource applications for better productivity. This educates the researchers to design and propose a network for 6G in such a way that it creates a place for AI/ML technology to be everywhere it can ultimately be. That network is thus AI native.AI/ML is all over then. In this chapter, we vividly review the AI/ML leverages in 5G/6G networks, and the excess uses of AI can be called excess AI or x-AI. 6G implements a neural transmitter and a neural receiver as the architecture for x-AI networking. Before going into 6G system architecture design, networking, and data governance, we should get the complete literature survey on it.

6G networks will replicate and enhance the 5G applications, along with the full utilization of AI and ML technology (S. Dang et al., 2020).

Key features of 6G that have to be maintained and developed in order to satisfy hybrids networking, including terrestrial and space, are as follows:

- It is expected to achieve a higher data rate per user, approximately 10–100 times more than 5G.
- Implements intelligent reflective surface technology (IRS) as a solution to energy and spectral efficiency issues.
- Advocates low-latency communications.
- Effectively reduces energy consumption.
- Support m-IOT and distribute m-MIMO.
- Bulk amount of data generated through the IOE (internet of everything).
- High reliability and connectivity are guaranteed.

So, these features require and demand the implementation of AI/ML techniques at the grass-roots level (Rekkas, Vasileios et al., 2021). Hence, this chapter dives into a detailed study regarding all. In addition, 6G supports applications like virtual reality (VR), smart healthcare solutions, augmented reality (AR) and extended reality (XR). This chapter explores the advantages and limitations of various AI/ML techniques.

1. Key Issues in 5G/6G and Respective Solutions to Overcome

Issue 1: Huge Traffic Accommodation

Traffic control and monitoring is a huge task in 5G and beyond networks, as it was around 7 exabytes per month in 2010 and is expected to be around 5k exabytes per month in 2030. To accommodate huge traffic congestion in 5G networks, which is around a thousand times more than in the above statistics, is quite a challenge.

Solution: AI/ML-Based Intelligent Reflecting Surface in 6G Networks

IRS, or intelligent reflecting surface, in 6G networks is a practical solution to the huge traffic congestion in recent networks. When a base station sends a signal to an IRS, the signal is reflected with phase changes, which are controlled by a controller. Consequently, the receiver gets the boosted or attenuated version of the signal. The IRS technologies are optimized by the AI/ML principle to consume less energy, which is required to accommodate huge traffic congestion.

Issue 2: Network Service for Different Cases and Different Verticals

The quality of service to the huge traffic is again a challenge to overcome in next-generation wireless communication. Network stability and services for different cases and different sectors are most important now, as and after the realization of 5G, to meet the hybrid network services from terrestrial to underwater to space.

Solution

Network slicing, excessive use of AI or x-AI, customized apps, and the AI-native principle will help in 6G n/w to achieve the required network stability and services. They are exclusively useful in computer vision, signal processing, speech analysis, image analysis, etc. The AI native principle is responsible for complete automation and transformation in 6G and beyond networks, which have big data computing and handling capabilities.

Issue 3: Security, Secrecy, and Stability (3S)

Security, secrecy, and privacy are the key demands that must be maintained in 6G networks (Dang et al. 2020). As 5G has reached its globalization, the next question is to work on its drawbacks to reach a conclusive solution for 6G, which will provide a constant 3S network with security, secrecy, and stability for maintaining the privacy of personalized data. That's why this chapter urges 3S solutions to 6G, or stability and maintaining general privacy. To get these 3S in 6G, we should study how it will be achievable in a frequent time interval in 6G.

Solution

Quantum computing and quantum cryptography in 6G will give ultimate security and secrecy to the network that is required. Deploying technology like block chain and federated AI increases security and privacy in 6G, which is essential in healthcare and business sectors. The integration of excess AI, or x-AI, into the physical layer in 6G ensures the highest level of security and privacy in a constant manner. Also, heavy computation in 6G due to the multidimensional network topology ensures stability and security in 6G networks. With the leverage of AI and ML in 6G, the self-healing of networks is done in cases of network breakdown. That empowers connectivity and stability issues in networks.

Issue 4: Creating a 6G Ecosystem for a Hybrid Network

Designing a 6G ecosystem for a complete hybrid network with every possible way to communicate for all is another task in hand after the realization. of 5G.The ecosystem must contain terrestrial, underwater, and space networking. This is possible when the 6G ecosystem brings communication from the satellite level to the cellular level to the radio level.

Solution

6G ecosystems can be effectively designed by taking various use cases, data types, and network operations into account. The inclusion of VR, AR, and XR applications and an excess of AI will give 6G a real boost for getting quick and user-friendly decision-making capabilities. AI/ML algorithms like KNN and random forest are basically used for path loss optimization, which is required to be monitored for developing robust ecosystems for 6G. The Internet of Everything (IOE) will play a vital and crucial role in designing ecosystems that caters to the large use cases of 6G in both vertical and horizontal fields. It further ensures the unmatched capabilities of 6G, like remote sensing, positioning, and remote communications. The terrestrial network generally includes M2M,D2D,IOT, mmWave/THz, etc. whereas the aerial network includes satellite-based communication like LEO,MEO,GEO, drones, etc.; and the underwater network is basically the AUVs (autonomous underwater vehicles) conducting sensing and data collection, etc. This is like a huge ecosystem built up for 6G networking with the full support and inclusion of AI and ML techniques.

Issue 5: Self-Healing Networking and Towards Multi-Directional Optimization

Industry 4.0 and beyond has excelled at the challenges of 5G and propelled the research to concentrate on quantifying the networks in every possible way so that self-healing of the network can be done when ever required and desired. This is a huge challenge in front of 6G to face and work on multi-directional optimization to accommodate the ecosystem it has built up.

Solution

This can only be done by the different algorithms of AI and ML, like supervised, unsupervised, reinforced learning, and deep neural networks. Generative adversarial networks (GAN) have been applied to data augmentation and security cases. Data classification and prediction are done using support vector machines (SVM), K-nearest neighbor (KNN), deep neural networks (DNN), convolutional neural networks (CNN), etc. Different multi-directional network optimization techniques, such as learning algorithms and softwarization for continuity of the network, load balancing, multipath routing, automatic network healing, etc. Consequently, AI and ML leverage 6G to create a self-healing network concept that makes for a healthy 6G ecosystem.

LITERATURE SURVEY

As we know, AI/ML has played a pivotal role in 5G/6G and beyond networks for all possible ways to achieve complete automation and optimal optimization. Different AI/ML learning perspectives, like supervised, unsupervised, and reinforced learning, provide the desired solution to achieving the same (Dey, S. et al., 2022; Ji, B. et al., 2021). In this literature survey, we propose to study vividly the algorithms of these AI/ML techniques, their basic applications, and their following limitations so that there is enough scope to design and demonstrate the AI native and x-AI principles and their subsequent inclusion in 6G and beyond networks. Let us discuss some machine learning algorithms like supervised, unsupervised, and reinforced learning and their respective inclusions in current-generation networks. Global mobile traffic is growing rapidly. By 2030, it is estimated to reach 5016 exabytes per month, which in turn will not be sufficient for 5G to address the requirements of such a huge traffic. The current issue is to address the limitations of 5G and introduce 6G networks that will completely induce AI and satisfy global traffic requirements. 6G networks support intelligent reflecting surfaces (IRS), which helps achieve massive MIMO 2.0 in 6G networks. 6G will enhance 5G applications and must satisfy a higher data rate 100 times more than 5G, provide global coverage, reduce energy consumption, and support safer and more reliable connectivity. This can only be possible by incorporating different learning algorithms and AI techniques. The following literature survey gives a proper roadmap for designing a complete AI/ML algorithm-based 6G network.

Supervised Leaning

Supervised learning algorithms have a predetermined set of input and output attributes that can be trained using labeled data sets. Supervised leaning algorithms are best suited for classification and regression problems, and in 5G/6G networks, the algorithms that are extensively used are artificial neural networks (ANN), k-nearest neighbor (kNN), random forest (RF), decision trees, and convolutional neural networks (CNN) (Kaur, Jasneet et al., 2021).

Yang, G. et al. (2019) described the basics of supervised learning and the different algorithms associated with supervised learning for path loss and delay spread prediction in mm waves. The K-nearest neighbor algorithm is a regression and classification algorithm that is extensively used in this study of optimization problems. The prediction of KNN is realized and applied by evaluating the predicted sample and historical samples. The predicted values of path loss and delay spread are calculated by averaging the best k-samples. This KNN is popularly termed lazy learning because there is no specific learning process.

Sliwa, B. et al. 2020 and Sliwa, B. et al. 2021 employed supervised machine learning-based data rate prediction using support vector machines (SVM), random forests (RF), and different neural networks like artificial neural networks (ANN) algorithms for 5G networks. This satisfies the basic quality of service requirements of next-generation wireless communication.

Mahbooba, B. et al. (2021) gave a comprehensive idea for a decision tree-based supervised learning algorithm for enabling sufficient security management in wireless networks.

Unsupervised Learning

Unsupervised learning uses a set of unlabeled data that predicts the output correctly, and it is generally used for clustering, aggregation, and regression problems. Algorithms like k-means, self-organizing maps (SOM), hidden markov models

(HMM), principal component analysis (PCA), and long-short-term memory (LSTM) are used in 5G/6G beyond networks.

Hassan, N. et al. (2021) used unsupervised learning techniques for data rate management and traffic congestion in wireless networks like RF and THz frequencies. They incorporated UL-based different association algorithms for the correct prediction and management of traffic congestion.

Liu, L. et al. (2020) and Farsad, N. et al. (2017) proposed LSTM-based unsupervised learning algorithms for channel estimation that were designed with inter-symbol interference for molecular communication cases. It gave a suitable channel estimation technique for 6G communication.

Cui, J. et al. (2018) and Ren, J. et al. (2019) studied unsupervised machine learning techniques for user clustering in mm waves. It proposed a k-means-based online user clustering algorithm to minimize the computational complexity and enhance the performance of the desired system.

Reinforced Learning

Reinforced learning is developed and constructed on the principles of behaviorist psychology, and it learns effectively to perform a new task. It is a trial-and-error method where an agent interacts with the surrounding environment to get feedback on what is good and bad with a reward and penalty system. RL algorithms like Q-learning and deep reinforced learning are generally used in 5G/6G and beyond networks (Rekkas, Vasileios et al., 2021).

Kang, J.M. et al. 2020, and Rekkas, V.P. et al. 2021 proposed Q-learning, k-means, and PCA algorithms in reinforced learning resource allocation, channel estimation, data traffic, and security, respectively, in their studies. They quantified their respective studies to boost the next generation of wireless networks.

Kong, J. et al. (2019) used a Q-learning algorithm-based reinforced learning approach to evaluate power allocation in RF systems, ensuring QoS satisfaction at the best optimum level.

In the below table 1, the detailed literature study and key applications of these algorithms in 5G/6G networks are discussed.

Table 1. Differenr AI/ML schemes and models for 5G/6G and beyond network

Research paper	AI Schemes	ML algorithms	DL algorithms	Key Application/solution
Yang, G. et al. 2019	Supervised learning	RF,KNN	-	Path loss and delay spread prediction and optimization in mm-Wave system
Sliwa, B. et al. 2020	Supervised learning	RF,SVM	ANN	Data rate prediction with lowest possible error
Sliwa, B. et al. 2021	Supervised learning	RF,SVM	-	Data rate in 6G network
Mahbooba, B. et al.2021	Supervised learning	Decision Tree		XAI or explainable AI for security management
L.-H.Shen et al. 2023	Supervised, Unsupervised and Reinforced learning	SVM,KNN, DT,EM, K-means, PCA, Q-leaning etc	DNN,CNN,RNN, LSTM,GAN, DQN etc	Optimum beam forming, channel prediction, big data processing, network optimization, quality of service, security etc
Hassan, N. et al. 2020	Unsupervised learning	UE association algorithm	-	Data rate and traffic congestion in RF and THz frequency .
Chen, Y. et al.2019	Unsupervised learning	Trained DRL	-	Security issues and possible attacks in 6G
Su, Y. et al. 2019	Reinforced learning	Deep RL	DQN	Energy consumption and reduction with lower convergence time
Kang, J.M. et al. 2020	Reinforced learning	Q-learning	-	Resource allocation by WPC method.
Rekkas, V.P. et al. 2021	Supervised, Unsupervised and Reinforced learning	SVM,KNN, DT,EM, K-means, PCA, Q-leaning etc	DNN,CNN, RNN, ANN,LSTM, GAN,DQN etc	Review of all algorithms in channel estimation, data traffic, security, energy consumption etc

Problem Statement

- There is enough study in AI/ML-based wireless networking, but the real use cases in 5G and 6G have enough voids to fill.
- The concepts of AI native and excess AI have not yet been addressed properly in any of the previous studies. So must be addressed in this chapter.

- By investigating the complete and systematic literature survey regarding the service of artificial intelligence, ML, and DL in 5G/6G and beyond and finding out the potential growth in architectures, networking, and applications enrooting towards the state-of-the-art of space air ground integrated networks (SAGINs),
- A defining comparison in technologies must be made between 5G and 6G from an AI/ML perspective.
- Learning-driven optimization and solutions for real-time issues must be addressed.
- Leverages of AI/ML in 5G/6G and beyond, with AI native as an important concept, should be prioritized.

Objective of the Research

The objective of the chapter is to study the systematic technological enhancement in 5G/6G and beyond the core AI/ML concept. This gives the opportunity to revisit the study from 5G to 6G. The leverages of AI and ML in 5G/6G and beyond, with a focus on AI native and excess AI, The real-time solution and different algorithms in different stages of networking must be well studied.

Novelty of the Research

The novelty of the proposed research lies in studying the inclusion of native AI principles in beyond 5G communication, which, if possible, can provide the SAGIN scenario for complete globalization of networking. In order to manifest the SAGIN scenario, different AI schemes must be taken into consideration, and along with that, the quality of service must be maintained. The different technical issues are taken into consideration, highlighting the native AI principles in next-generation wireless communication. An AI-native workflow model is designed to demonstrate the AI life cycle in the different stages of the technology use case. This can quantify and classify the quality of service needed in order to provide global networking.

Comparison of 5G and 6G (Technology Perspective)

- 5G uses technologies like enhanced mobile broadband (eMBB) and massive machine-type communication (mMTC), whereas 6G requires eUMBB, eURLLC, UmMTC, intelligent and automated machines, extended reality (XR), augmented reality (AR), virtual reality (VR), etc.
- 5G peak data rate is up to 20 Gbps, but 6G supports 1 Tbps peak data rate.

- 5G communication network is mmWave basically, but in 6G it is AI/ML empowered, self-healing, and space air ground integrated n/w.
- 5G supports mobility up to 500 km/h with a latency of 10 ms, whereas 6G supports mobility up to 1000 km/h with a latency of less than 1 ms.
- 5G has a reliability of 99.9999%, but in 6G it is 99.99999%.
- 5G has a much lower area spectral efficiency than 6G. In 6G, it is 10 times more than in 5G.
- The energy efficiency of 6G is 100 times greater than that of 5G.
- 6G integrates powerful AI techniques, new transmission, and spectrum for community usage, which is in the initial stage of 5G.
- 5G has been designed to use different signals, like low-band sub-6 GHz for broad coverage, millimeter waves for high-band ultra-fast speed in populated areas, and mid-band for balance of speed and coverage, but 6G generally can use millimeter-wave 30 to 300 GHz signals for its global coverage and ultra-fast speed.

AI/ML Leverages in 5G Advanced/6G and Beyond Networks

In practical terms, after the globalization of 5G, the first constraint is to speed up the inclusion of AI and ML in every possible field, from network design to application and security assurance to network optimization. 5G has opened up new set of challenges like complexity in network architecture, low latency, high cost, power consumption and hybrid networking, etc., so network optimization is mostly a major problem to solve. Therefore, AI/ML and its different algorithms' inclusion in every field are required. This solves many things in a network, like self-healing capability, auto-optimization, and self-configuration of a complex network. It is called an intelligent network. Reconfigurable intelligent surface (RIS), excess AI or AI native and non-orthogonal multiple access (NOMA), are three major technologies for 6G and beyond. RIS provides information by being integrated with ML, reducing the computation cost. ML-integrated NOMA ensures control policy, security, and decision-making capability in an intelligent network. AI native makes 6G auto-operative and robust. Different AI and ML schemes have solved the network's key issues, as discussed in the literature survey section. These ML approaches are supervised, unsupervised, and reinforced. Different algorithms and ML approaches that plays vital role in 5G/6G and beyond networks are CNN, KNN, ANN, Naïve Bayes, Decision Tree, RF,PCA,Q-learning, HMM, etc. (Rekkas, Vasileios et al., 2021; Dey, S. et al., 2022).

As the technology advances from 5G to 5G advanced and to 6G and beyond in 3GPP organizations, it is estimated that AI and ML will empower the 6G completely in Release R23 in 2028–2030. The smooth progression of the releases from R14 in 2017 to R23 in 2028 gives a complete idea of how AI and ML are leveraged in next-generation wireless communication. The release of R19-R20 in the 3GPP organization (2023) has included the strength of AI in advanced 5G. This provides the roadmap to 6G and beyond networks in the upcoming releases in R21–R23 (estimated in 2028). The AI/ML technique and its excess use will give 6G a definite global network to achieve. It will support high transmission rates, full coverage of remote areas, high mobility, and a network for anyone, anywhere.

AI/ML is generally effective beyond 5G or 6G networks in many ways. Some specific cases are discussed here.

- AI/ML provides an effective solution to deal with network complexity and data rate demand.
- AI and ML can be useful for algorithm deficit problems and provide ample computational methods to resolve optimization issues.
- AI/ML-based CNNs have been very useful for modulation classification in 6G.
- AI/ML-based different algorithms provide self-healing and automated networks for 6G.

Figure 1. AI/ML approaches to 5G/6G and beyond networks

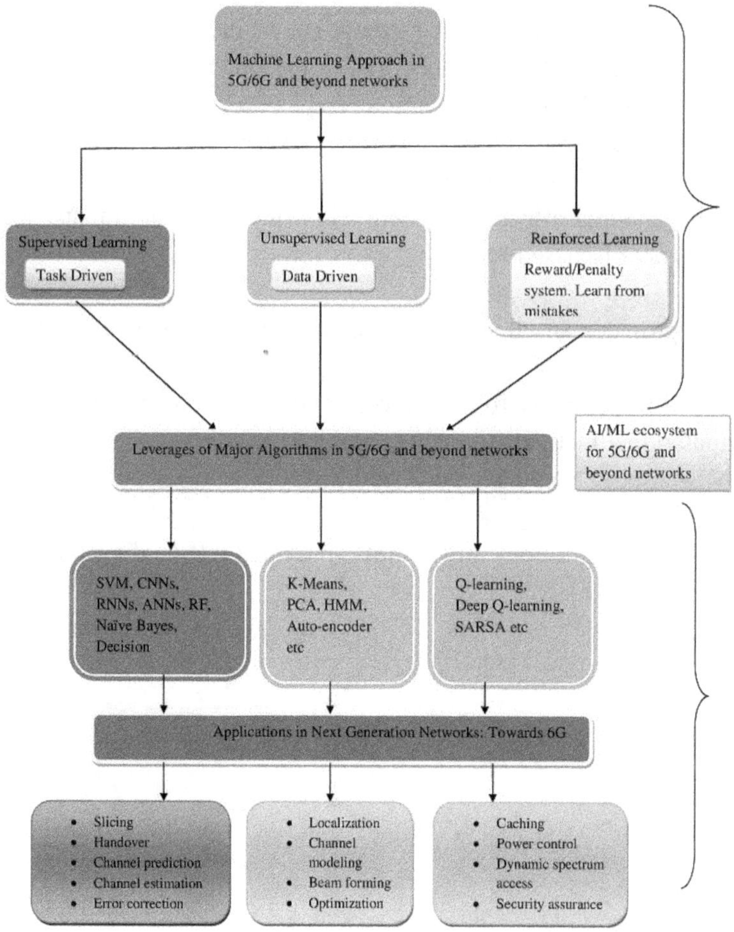

AI/ML Use Cases in Wireless Networks (UE Positioning in 3GPP): Case Study

The basic case study on AI/ML-assisted wireless networks is the user equipment or UE positioning for 3GPP beyond release 15 (Shehzad, M.K. et al., 2022). Here, various angle- and time-of-arrival based procedures are used to evaluate the UE positioning in wireless networks. Deep learning approaches and decision tree-based ML techniques are used for outdoor scenarios. These methods resolve and fix the UE position and time synchronization errors in networks. Similarly, ML-assisted proactive mobility and CSI feedback are basic case studies that show the ML use cases in wireless networks. In 3GPP, serving cell reference signal received power

(RSRP) as well as neighboring cell RSRP were used for the training of deep neural networks to get maximum accuracy.

Some important use cases of AI/ML-based 5G/6G networks are mentioned below:

- ***Ultra-smart cities*** consist of high energy efficiency, high data rates, low latency, precise sensing, and localization.
- ***Remote surgeries and telemedicines*** by the emergence of 6G networks, which include remote healthcare, by diminishing the space and time constraints
- ***Haptic communication*** by enabling humans to express information through the sense of touch, brain-controlled computer interactions, etc.
- ***Tactile Internet*** facility by transferring data, controlling, and taking feedback in real time in a highly responsive environment.
- ***Multidimensional reality*** facilities like ultra-edge human-computer interactions.

Open Challenges in 5G/6G for Deploying AI/ML

- Huge data availability and respective benchmarking for standardization
- Addressing the complexity of AI and ML algorithms and providing a set of solutions
- Stability and adaptability of AI and ML techniques for dynamic environments
- Network softwarization
- Proper AI service orchestration
- AI/ML service and infrastructure management
- Task-oriented connectivity
- Data storage management, provisioning, etc.

5G/6G SOCIAL AND ENVIRONMENTAL IMPACTS AND RESPECTIVE MITIGATION THROUGH AI/ML INTEGRATION

Impacts on Ecosystem

- The high spectrum range can lead to rapid cell mutations that cause tumors and cancers in mankind.
- The excess use of the mm wave leads to skin diseases, and it is also very harmful for birds and animals and their biological ecosystems.

- The electromagnetic radiation from 5G and 6G can cause significant health issues for both humans and the environment, like infertility and birth defects.

Mitigation Through AI/ML Integration

- 5G/6G with IoT will increase energy efficiency and reduce greenhouse gas emissions.
- Enable more use of renewable energy.
- It can reduce air and water pollution.
- Improves the percentage of decision-making capability in weather forecasting.
- Reduction and management of industrial and agricultural waste.
- AI/ML-integrated and IoT-based smart city creation by linking sensors, smart phones, and cameras through global connectivity.
- Decarbonization and recycling toxic electronic wastes.

AI Native or Excess AI

Artificial intelligence and machine learning will definitely be the ultimate and best-suited solutions to the next generation of wireless network problems. Therefore, researchers have concentrated their efforts on developing a hybrid network that will be from space to air to underwater or space-air-ground integrated networks (SAGINs) for global connectivity. It can only be possible with the inclusion of excessive AI and ML technology and its systematic leverage in 5G and 6G networks in the possible field of networking. AI native is the concept of using excessive AI principles in 5G and 6G networks. Here, AI is a natural and integral part of the functionality in terms of system design, deployment, operation, and maintenance. The ecosystem in and around the 5G/6G and beyond networks must be AI-native-driven. As 5G has already matured, it is clear that AI and ML technology will play an integral role again in 6G and beyond in every field. Except for 3GPP and air interfaces, the O-RAN alliance is also researching the AI/ML leverages in 6G and beyond to improve network orchestration. They have a unique architecture called the RAN intelligent controller (RIC), basically designed for hosting the AI-native concept for optimization. In an effort to integrate AI and ML in designing, optimizing, and maintaining the 6G network, the ecosystem around 6G will see a nativity of AI around its core. This AI native has xApps for spectral efficiency and energy efficiency improvement and rApps for leveraging AI and ML for network orchestration. In classical 5G networks, the air interface consists of human-made and designed processing units, but in 5G advanced networks, these are leveraged with AI/ML-optimized modules. Consequently, in 6G, the air interface will be completely designed by a deep neural network, and it will be optimized with an AI-native concept. The below figure shows an AI/ML-optimized 6G architecture using a deep neural network.

Figure 2. AI/ML optimized 6g architecture

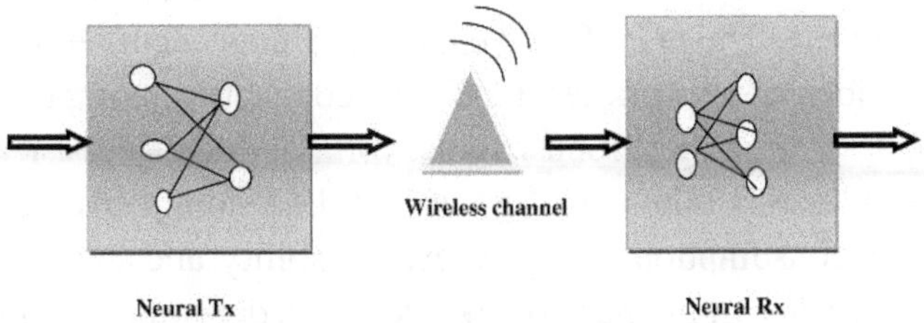

The AI native will bring four major AI services to boost the network in 6G. These are

- AI data service
- AI training service
- AI inference service
- AI verification service

Figure 3. AI native based services in 6G

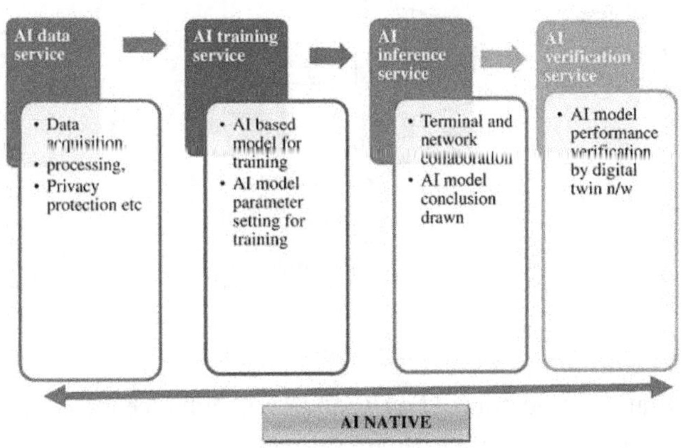

AI native in 6G ensures a complete auto-operating environment with end-to-end AI work flow. It provides resources for data, algorithms, and models on demand, fast computational ability, on-demand scheduling, data authentication, etc. for the upgrade of 6G networks. The 6G network will grace wireless communication with intelligent and inherent AI capability, and this is referred to as quality of AI services (QoAIS). To satisfy the types of AI-native services like AI data services, AI training services, AI inference services, and AI verification services, a different set of QoAIS is required each time. The 6G AI native quality is dependent on the evaluation of multiple dimensions of network resources like connection, computing, algorithm/ model, data, etc. For designing a robust 6G network, it is therefore basically dependent on the QoAIS indicators. The QoAIS indicators like storage overhead, computing overhead, energy consumption, data privacy, reusability, and algorithm security are most important for designing the network. QoAIS is definitely an important aspect of AI-native services for network orchestration, privacy control, and monitoring. The QoAIS must be mapped with QoS to make the design, computation, and connection fruitful. The basic five areas of QoS evaluations are performance, cost, safety, automation, and privacy control.

The QoAIS and QoS indicator mapping shows the overall AI services for the management plane, user plane, and control plane that guarantee security, privacy, and autonomy in the design. (J. Hoydis et al., 2021).

Figure 4. QoAIS and QoS indicator mapping

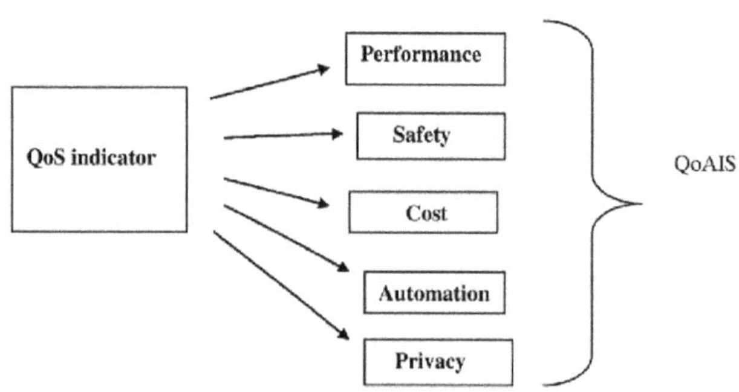

Some quantitative indicators of how QoAIS will enhance the AI training services and ultimately enhance the AI native are given below.

- Data accuracy
- Data preparation time

- Performance index
- Training time
- Computational accuracy
- Efficiency
- Bandwidth and jitter
- BER and jitter
- Reliability
- Optimization, duration, etc.

In quantifying quality assessment and assurance, these above indicators must be given emphasis because, in the 5G network, long optimization and adjustment periods, resource management, and network services are still having some issues. So quantifying the AI native and designing the end-to-end efficient QoAIS will boost the AI services in 6G networks.

AI Native Workflow

The AI native workflow life cycle, or AI life cycle, generally consists of different stages of AI operations like AI workflow generation, execution, management, evaluation, optimization, completion, deletion, etc. The AI-native workflow is shown below.

Figure 5. AI native workflow

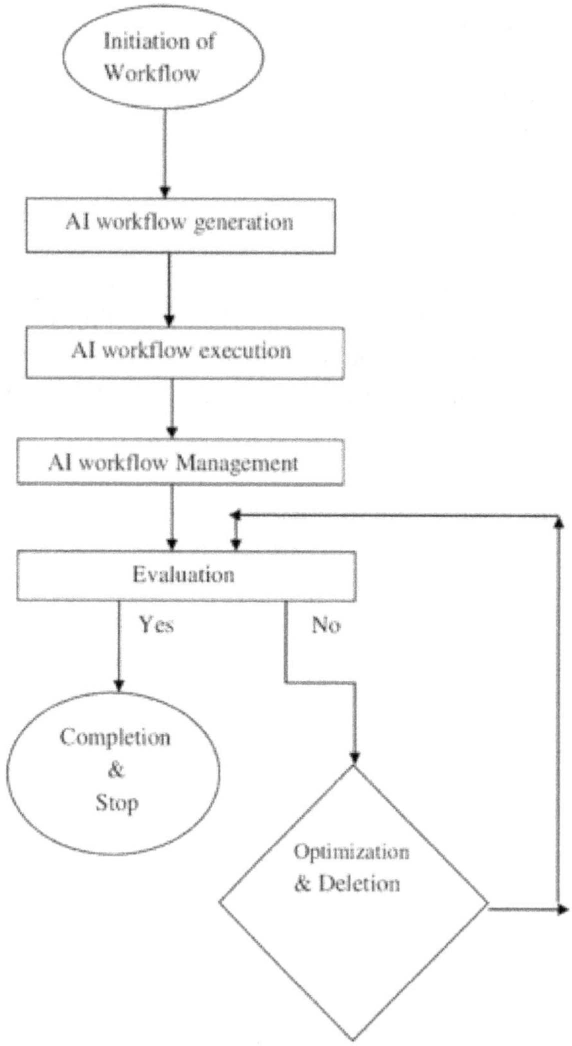

The native AI workflow also provides the complete stages of AI services in a network (J. Hoydis et al., 2021). This algorithm basically gives four main steps of the AI service that have to be completed: data model management, model learning, model validation, and model deployment. The 6G network has AI native inclusion for the complete auto-operation environment for end-to-end AI workflow to accomplish different intelligent applications, unlike the 5G network, which has offline AI workflow. So quantifying the AI services and AI workflow in 6G, we can see the proper mapping as below.

Figure 6. Quantifying the native AI workflow and AI services in 6G

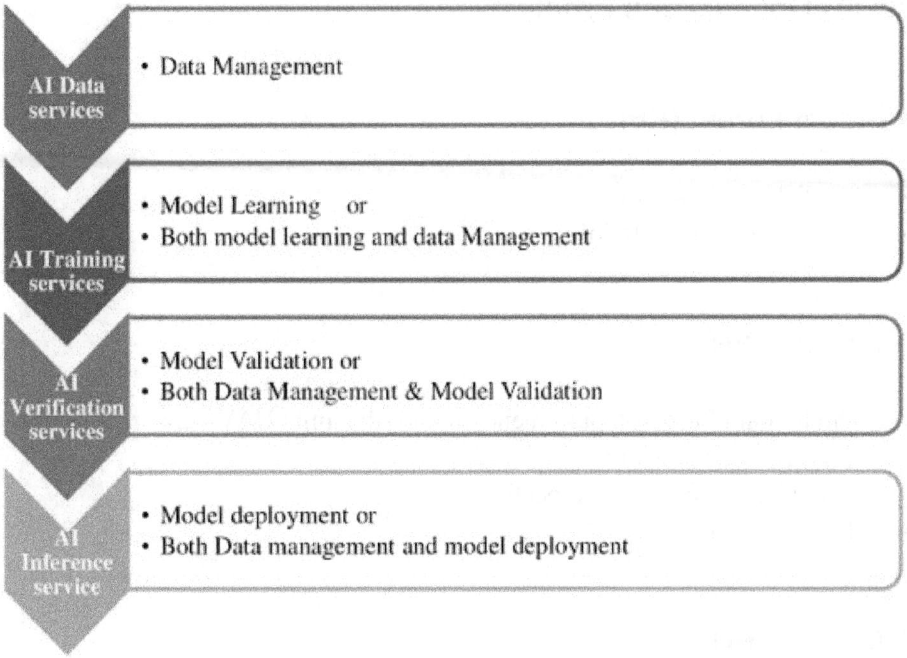

The native AI workflow includes different AI services as per requirements in 6G. For example, AI data services can only perform services related to data management, while AI training services can perform the task of model learning or both model learning and data management as well. For ensuring continuous and effective QoAIS in 6G, the AI native workflow guarantees and assigns the proper tasks for AI-related services in the network and again manages the resources like computing, algorithms, data, and connections in the network. This consequently gives intelligent orchestration management in the network. The inclusion of AI resource ingredients into 6G network design will definitely hold the key to 6G AI native. It has basically three important features of AI: algorithms, data, and computation capability.

So basic native AI services in 6G will be

- Connection: mechanism for controlling AI connections.
- Computational power: mechanism for controlling AI computation.

- Algorithm: mechanism for AI algorithm self-optimizations.
- Data: mechanism for controlling AI data collection, processing, and management.

ML Approach and Its Real Use Cases in the 6G Optimization Problem

Random Forest (RF)

- Prediction of path loss
- Optimization of path loss

DNN and RNN

- Prediction of users' requirements in high-dynamic UAV networks
- Traffic load estimation
- Optimization of power allocation
- Load balancing

SVM and ANN

- Minimization of prediction error
- Prediction of an accurate data rate
- Optimization of caching in the IOT system
- Beam selection
- Reducing false alarms
- Detects jamming
- Provides cyber security

k-Means

- Clustering with higher throughput
- Ensuring a lower SNR value

Decision Tree

- Boosting trust management
- Security assurance

LSTM

- Security assurance
- Intrusion detection
- Modulation recognition
- Channel estimation

Q-Learning and Deep Q-Learning

- Resource allocation
- Channel estimation and selection
- Energy consumption
- Power allocation
- Fault detection

ML Application in Different Layers of a 6G Network

- ML at the physical layer includes channel coding, synchronization, positioning, channel estimation, and beam forming.
- ML at medium access control layer (Predictive power management, Traffic control, predictive resource allocation)
- ML at the application layer (network performance management, ML-aided UAV control)

6G State of the Art Requirements

- Very low latency communication (10μs–100μs).
- The peak data rate is up to 1 Tb/s.
- Support high mobility (1000 km/h) for airline systems.
- Huge capacity (10^7 devices/km2) due to the IoE.

- Minimize energy consumption to introduce green communication (100x reduction compared with 5G).
- The data rate reaches 10 GB/s (*10 times 5 GB).
- Estimated Spectrum efficiency (5–10 times 5G).
- Higher 3D coverage extension, including terrestrial, aerial, space, and sea domains.

6G Design Challenges and Performance Requirements: A Research Perspective

The rapid growth of 6G has set up new design challenges and some performance requirements to satisfy. The inclusion of AI and ML has ensured a robust and automated 6G network, but end-to-end design in the data plane, service plane, and management plane is a bit complex (R. Zhong et al., 2022). Having been rooted in the principles of AI and ML, the new 6G architecture requires some key performance metrics like higher energy efficiency, SAGINs integration, peak throughput, low latency, and self-healing networking to be well addressed and mapped into it. The details of the 6G design challenges and research perspective are shown below in Table 2.

Table 2. Key performance metrics requirements and design challenges in 6G: Research perspective

Key Performance Metrics	6G Design requirements & challenges	Research direction
6G Network Architecture	• Hybrid network like space-air –ground integrated network. • Self healing network • AI/ML operated Automation • Full Connectivity and world state of art type matched networking • Anywhere anytime networking	Towards SAGINs, ML based routing and network optimization with end to end design.
Data Rates	• Expected to have theoretical data rate up to 100Gbps to 1Tbps • Across multiple gigahertz channels.	Extended reality communication, Complex System AR/VR with high speed
Energy Efficiency & Power Consumption	• Ultra dense heterogeneous network and massive MIMO for energy efficiency • ML based optimization for power utilization	ML based Resource allocation and networking. ML at infrastructure level
Data security and privacy control	• AI/ML based system for security • Quantum encryption technique for cyber security • Deployment of security within physical layer	AI/ML based cyber security
Latency	• Very low latency communication requirements • Reduction of errors	ML based application to reduce error and low latency communication

AI/ML-Based 6G Networks: Open Issues and Future Trends

AI/ML inclusion propels many research directions and opportunities in the current wireless communication system, thus speeding up the complete realization of 6G services. Some challenges and open issues are there to be resolved as quickly as possible to get the 6G services.

Issues

- ***Time Convergence:*** In high-dynamic wireless networks, optimizing the time convergence is crucial, as AI and ML can influence the performance to a great extent.
- ***E-Health Application and Resource Allocation:*** AI/ML-enabled wireless networks provide great impact on e-health applications. Proper resource allocation is essential for maintaining adequate harmony in such applications.
- ***QoS and QoE Balancing:*** For different sets of QoS and QoE requirements and balancing, action-based ML protocol design is very essential.
- ***CSI acquisition in IRS:*** Timely and accurate CSI acquisition is another important task for IRS-enhanced wireless systems. The ML and DL algorithms for CSI acquisition are crucial, then.

Future Trends

- ***Model-agnostic Meta-learning (MAML)*** is a gradient-based algorithm that learns the sensitive initialization for performing fast and adequate adaptation in sensitive environments. It also has less complexity. This is ongoing research for developing MAML that can be properly utilized in 6G networks.
- ***Social IoT Networking (S-IoT):*** S-IoT networks are virtual social networks designed by a bunch of IoT devices belonging to people having similar interests, which can be regarded as human social network support services. These networks must be geographically adjacent and capable of accessing vehicular communications.
- ***XaaS Platform:*** 6G needs a comprehensive and collaborative ecosystem where AI and ML integration can easily be deployed. AI-enabled channelization and mobility management can be done by third-party AI applications, which mean 6G can be turned into an XaaS platform providing communication and computation services. These are AI application services as a service (SaaS), infrastructure as a service (IaaS), and platform as a service (PaaS).

Towards Sustainable 6G and Ethical AI: A Socio-Economic Perspective

To design a sustainable 6G means fostering technological advancement as well as keeping an eye on the social, economical, and environmental ecosystems. No doubt AI/ML integration will provide unmatchable technological growth for present wireless communication, but according to a study by Ericsson, the balance must be kept in this ecosystem so as to ensure positive sustainability in the society in which it is scaled. Irrespective of use cases, the environmental impact must be taken into consideration when producing, operating, and recycling the 6G system. The economic perspective is to protect privacy, security control, and providing the technology at an affordable cost. Economic sustainability in 6G is about the economic advancement of society itself, capitalizing on and catering sustainable progressive development to wide areas such as job markets, productivity, infrastructure availability, and innovation. From this perspective, end-user costs are crucial for ensuring inclusive and affordable systems. Consequently, to contribute to the big extent of economic sustainability, businesses also need to be economically sound and sustainable themselves. The real value of 6G networks and downstream use cases needs to be monetized, which can carry the provisioning ecosystem of end-users, service providers, CSPs, suppliers, system integrators, and all other necessary stakeholders. Ethical AI is also another big concern that needs to be addressed. It demands the fundamental protection of human rights, privacy control, cyber security, and safety balancing in technological advancement. Ethical AI places and urges fundamental importance on ethical considerations in maintaining legitimate and illegitimate uses of AI. The 6G ecosystem must provide ethical AI that will provide clearly defined policies and review processes to ensure adherence to the strict rules, laws, and guidelines. Ethical AI is not limited to what is permissible by law; sustainability and balance in the ecosystem must be taken into consideration. Legal limits pertaining to the use of AI in the 6G ecosystem set a minimum threshold of acceptability, while ethical AI sets policies, rules and regulations that go beyond legal requirements to ensure respect for fundamental human rights and basic values.

The Global Digital Divide: A Big Concern

Since the realization of 5G/6G, a big debate around the world has been going on. This results in a digital divide. It refers to the technological and communicational gap within the globe on a demographic and regional basis towards accessing technological advancement. There are generally three types: access divide, use divide, and quality of use divide. The current scenario in the entire globe is quite a dilemma to adapt to the next generation of wireless communication, more specifically the 5G/6G incorporated with AI/ML. Some countries have raised issues about the complete realization of 5G/6G into reality over security, safety, and privacy control. Some countries raised their concerns based on economic protectionism. That will divide digital advancement across the globe on issues like cyber security, geopolitical pressures, etc. Many geographic regions are out of reach of the current advancement due to their prolonged concern about it, citing health concerns and environmental imbalances. It still takes a long time to connect to the world in a technological way, as this digital divide creates a greater impact on society. There are ample reasons for the lack of global connectivity in many remote areas of the world, such as the absence of quality infrastructure, low or average income scale, rough geography, and many more. The 6G and beyond network proposes a smart vision for breaking the digital divide, which will in turn promote sustainability alongside global connectivity. It is now roughly estimated that nearly 3.7 billion of the world population is far from the reach of internet connectivity in the current scenario. The digital divide also presents several other problems, such as socio-economic growth, education, and awareness, and inhibits the adoption of remote and advanced smart healthcare technologies. It also causes low growth in some specific sectors, like modern industrial and new-age transportation sectors. In reality, in the era of high-quality mobile internet connectivity, there are remote areas of the USA and Europe that still have low connectivity, as low as 0.4–1 Mbps.

Figure 7. Digital divide types and statistics

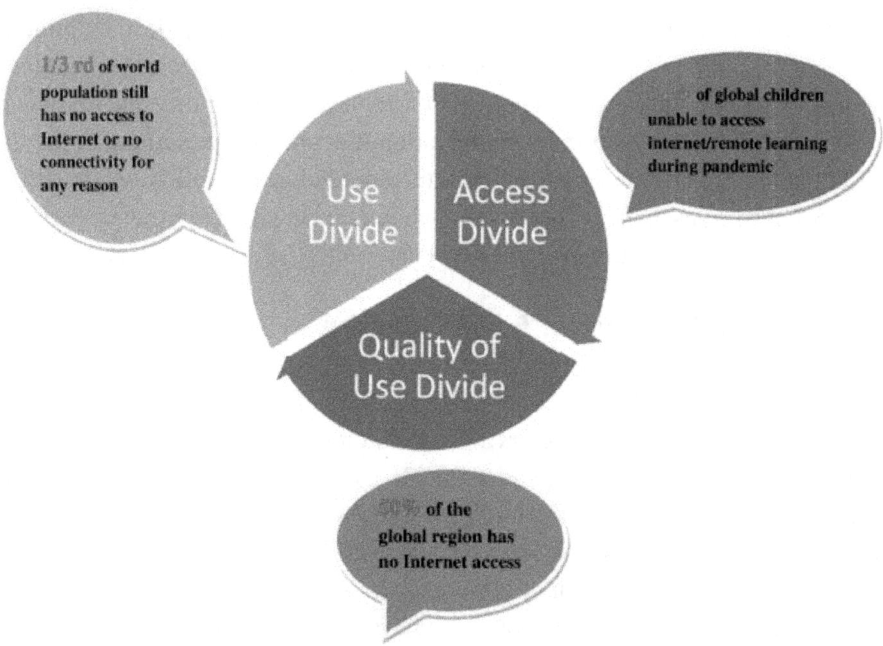

The reality is quite worse in whole Africa, Brazil, and several places in India today. Basically, if we take the digital divide into the service category, it can be divided into two major categories: the service-delivery divide and the service-fruition divide. The first one concerns the digital exclusion caused by the absence of network infrastructures necessary for Internet access and digital service delivery; the second one can be considered a measure of a person-specific divide, since an insufficient level of digital literacy or a set of physical inabilities could prevent people from enjoying the benefits deriving from the fruition of digital services.

CONCLUSION

In this chapter, we have discussed the basic state-of-the-art of requirements of 5G/6G and beyond networks and the design challenges in 6G networks. AI/ML leverages in 5G/6G and beyond networks are focused, and the basic inclusion of the AI native 6G network is also prioritized. It dives into the field of ML applications in different layers of 6G, highlighting the algorithms that are changing the 6G optimization problem in real-world use cases. Another highlight of the study is the quantification of AI and ML in the 6G network, the AI native workflow process, and different QoS services. We have properly identified the future directions for using ML in resource management, power allocations, and channel modeling. There are many ML techniques and subsequent algorithms that can be intelligently combined with the 6G wireless network to get real-world user-case solutions for 6G. Consequently, the current ML and future 6G networks will definitely ensure a solution for meeting current challenges such as latency, power allocation, privacy control, and 100 percent security at the application and infrastructure levels to enhance smart applications for smart health and smart society in a smart ecosystem. Furthermore, research will progress in meeting these requirements for 6G and beyond networks and perhaps building better SAGINs.

REFERENCES

Anton-Haro, C., & Mestre, X. (2019). Learning and data-driven beam selection for mmWave communications: An angle of arrival-based approach. *IEEE Access : Practical Innovations, Open Solutions*, *7*, 20404–20415. doi:10.1109/ACCESS.2019.2895594

Basharat, S., Hassan, S. A., Pervaiz, H., Mahmood, A., Ding, Z., & Gidlund, M. (2021). Reconfigurable Intelligent Surfaces: Potentials, Applications, and Challenges for 6G Wireless Networks. *IEEE Wireless Communications*, *28*(6), 1–8. doi:10.1109/MWC.011.2100016

Chen, M., Challita, U., Saad, W., Yin, C., & Debbah, M. (2019). Artificial neural networks-based machine learning for wireless networks: A tutorial. *IEEE Communications Surveys and Tutorials*, *21*(4), 3039–3071. doi:10.1109/COMST.2019.2926625

Dahrouj, H., Alghamdi, R., Alwazani, H., Bahanshal, S., Ahmad, A. A., Faisal, A., Shalabi, R., Alhadrami, R., Subasi, A., Alnory, M., Kittaneh, O., & Shamma, J. S. (2021). An Overview of Machine Learning-Based Techniques for Solving Optimization Problems in Communications and Signal Processing. *IEEE Access : Practical Innovations, Open Solutions*, *9*, 74908–74938. doi:10.1109/ACCESS.2021.3079639

Dang, O., Amin, O., Shihada, B., & Alouini, M.-S. (2020). What should 6g be? *Nature Electronics*, *3*(1), 2520–1131. doi:10.1038/s41928-019-0355-6

Dey, S., Khampariya, P., & Jain, M. (2022). Analysis of various AI-based algorithms and learning perspectives for 5G wireless networks. *IJMEvol.*, *7*(5), 1234–124.

Dey, S., Panda, A. K., & Pati, C. (2023). Unlocking the potential of machine learning and deep learning algorithms in recent communication: AI is the best revolutionary tool in the current era. *CPPPT*, *23*(2), 2794–2805.

Hoydis, F. A., Aoudia, F. A., Valcarce, A., & Viswanathan, H. (2021). Toward a 6G AI-native air interface. *IEEE Communications Magazine*, *59*(5), 76–81. doi:10.1109/MCOM.001.2001187

Ji, B., Han, Y., Liu, S., Tao, F., Zhang, G., Fu, Z., & Li, C. (2021). Several key technologies for 6G: Challenges and opportunities. *IEEE Commun. Stand. Mag.*, *5*(2), 44–51. doi:10.1109/MCOMSTD.001.2000038

Jiang, W., Han, B., Habibi, M. A., & Schotten, H. D. (2021). The Road Toward 6G: A Comprehensive Survey. *IEEE Open Journal of the Communications Society*, *2*, 334–366. doi:10.1109/OJCOMS.2021.3057679

Kaur, J., Khan, M. A., Iftikhar, M., Imran, M., & Haq, Q. (2021). Machine learning techniques for 5G and beyond. *IEEE Access : Practical Innovations, Open Solutions*, *9*, 1–1. doi:10.1109/ACCESS.2021.3051557

Malik, U.M., Javed, M.A., Zeadally, S., & ul Islam, S. (2021). Energy-efficient fog computing for 6G-enabled massive IoT: Recent trends and future opportunities. *IEEE Internet Things Journal.*

Mao, F., Tang, F., Kawamoto, Y., & Kato, N. (2021). Optimizing computation offloading in satellite-UAV-served 6G IoT: A deep learning approach. *IEEE Network*, *35*(4), 102–108. doi:10.1109/MNET.011.2100097

Matthaiou, M., Yurduseven, O., Ngo, H. Q., Morales-Jimenez, D., Cotton, S. L., & Fusco, V. F. (2021). The road to 6G: Ten physical layer challenges for communications engineers. *IEEE Communications Magazine*, *59*(1), 64–69. doi:10.1109/MCOM.001.2000208

Nawaz, S. J., Sharma, S. K., Wyne, S., Patwary, M. N., & Asaduzzaman, M. (2019). Quantum machine learning for 6G communication networks: State-of-the-art and vision for the future. *IEEE Access : Practical Innovations, Open Solutions*, *7*, 46317–46350. doi:10.1109/ACCESS.2019.2909490

Rekkas, V., Sotiroudis, S., Sarigiannidis, P., Wan, S., Karagiannidis, G., & Goudos, S. (2021). Machine Learning in Beyond 5G/6G Networks—State-of-the-Art and Future Trends. *Electronics (Basel)*, *10*(22), 2786. Advance online publication. doi:10.3390/electronics10222786

Shafin, L., Liu, L., Chandrasekhar, V., Chen, H., Reed, J., & Zhang, J. C. (2020). Artificial intelligence-enabled cellular networks: A critical path to beyond 5G and 6G. *IEEE Wireless Communications*, *27*(2), 212–217. doi:10.1109/MWC.001.1900323

Shehzad, K., Rose, L., Wesemann, S., & Assaad, M. (2022). ML-based massive MIMO channel prediction: Does it work on real-world data? *IEEE Wireless Communications Letters*, *11*(4), 1–5. doi:10.1109/LWC.2022.3146230

Shehzad, M., Rose, L., & Assaad, M. (2022). Artificial Intelligence for 6G Networks: Technology Advancement and Standardization. *IEEE Vehicular Technology Magazine*, *17*(3), 16–25. Advance online publication. doi:10.1109/MVT.2022.3164758

Tariq, M. R., Khandaker, M. R. A., Wong, K.-K., Imran, M. A., Bennis, M., & Debbah, M. (2020). A speculative study on 6G. *IEEE Wireless Communications*, *27*(4), 118–125. doi:10.1109/MWC.001.1900488

Vinesh, R., & Ancy, C. A. (2021). Understanding the Future Communication: 5G to 6G. *Int. Res. J. Adv. Sci. Hub*, *3*(Special Issue 6S), 17–23. doi:10.47392/irjash.2021.159

Yaklaf, S. K. A., Tarmissi, K. S., & Shashoa, N. A. A. (2021). 6G Mobile Communications Systems: Requirements, Specifications, Challenges, Applications, and Technologies. *Proceedings of the 2021 IEEE 1st International Maghreb Meeting of the Conference on Sciences and Techniques of Automatic Control and Computer Engineering MI-STA*, 679–683. 10.1109/MI-STA52233.2021.9464470

Zhang, S., & Zhu, D. (2020). Towards artificial intelligence-enabled 6G: State of the Art, Challenges, and Opportunities. *Computer Networks*, *183*, 107556. doi:10.1016/j.comnet.2020.107556

Zhong, Y., Liu, Y., Mu, X., Chen, Y., & Song, L. (2022). AI-enabled RIS-assisted NOMA networks: Deep learning or reinforcement learning? *IEEE Journal on Selected Areas in Communications*, *40*(1), 182–196. doi:10.1109/JSAC.2021.3126068

Chapter 11
Past, Present, and Future of Rate Splitting Multiple Access for Wireless Networks: A Review

Krishanu Kundu
G.L. Bajaj Institute of Technology and Management, Greater Noida, India

ABSTRACT

Rate splitting multiple access (RSMA) is a promising approach that has gained momentous attention in the field of wireless communications. With the increasing demand for higher data rates and reliable connections, traditional multiple access approaches like SDMA, NOMA have limitations in terms of interference management as well as channel state information uncertainty. The effectiveness of the RSMA has been recognized in addressing these limitations and providing improved performance in multi-antenna systems. RSMA ruptures user messages into common as well as private parts – the common part is decoded by all users to mitigate interference while the private part is decoded only by the intended user. This bridges orthogonal and non-orthogonal multiple access schemes. Benefits of RSMA include optimal performance in many scenarios, robustness against imperfect CSIT, lower feedback overhead and complexity compared to alternatives. RSMA has many potential applications in 6G use cases like mMTC, URLLC, V2X, and integrated sensing-communications. Ongoing research is still needed, especially on implementation and standardization. The current study focuses on the notion of rate splitting, the benefits of RSMA, working principle of RSMA, implementation challenges, RSMA's implementation in 5G and beyond, a comparison with other multiple access techniques, as well as its problems and future trends in wireless networks.

DOI: 10.4018/979-8-3693-0916-2.ch011

INTRODUCTION: UNDERSTANDING THE CONCEPT OF RATE SPLITTING

Traditional multiple access methods confront various obstacles in satisfying the growing needs of wireless communication. One of the most significant difficulties is the network's inadequate capacity and speed. Traditional systems are struggling to deliver optimal performance as the number of connected devices grows and the need for high-speed data transfer grows. Interference between users is also a serious concern. Multiple users broadcasting at the same time in busy wireless networks can generate interference, resulting in poor signal quality and lower overall network performance. This interference not only affects the individuals concerned, but also neighbouring users, resulting in a domino effect that reduces the overall network's performance. Rate Splitting Multiple Access (RSMA) (Y. Mao et al.,2018) is an innovative method for addressing the shortcomings of classic multiple access systems. The RSMA standard incorporates the idea of rate splitting, which divides the transmission rate into many simultaneous data streams. This one-of-a-kind feature enables each user to transmit and receive data at the same time, greatly enhancing network capacity and throughput. The rate splitting approach used in RSMA is based on the superposition coding concept. The splitting ratios are determined by a unique set of power and rate allocation characteristics provided to each user. RSMA offers more efficient utilisation of available resources by intelligently dividing the transmission rate, resulting in increased spectral efficiency and enhanced overall network performance.

Research interest in RSMA (B. Clerckx et al., 2016) has increased rapidly since 2017, with over 150 conference and journal papers published on the topic. Initial works focused on formulating the concept of rate-splitting for the MIMO broadcast channel and interference channel, and analysing theoretical performance bounds that demonstrate significant gains over SDMA and NOMA. Recent RSMA research can be categorized as: a) System Design: Optimizing RSMA system parameters like power allocation, beamforming vectors, message splitting, and decoding order. Both sum-rate maximization and max-min fairness formulations have been studied. b) Performance Analysis: Analysing performance of RSMA (B. Clerckx et al., 2020) in various wireless network scenarios like Cloud-RAN, integrated access and backhaul, Internet of Things, vehicle-to-vehicle links, and aerial platforms. RSMA demonstrates consistent gains over SDMA and NOMA in both theoretical capacity and numerical evaluations across these scenarios. c)Implementation Challenges: Key issues being addressed include optimization complexity, signalling overhead for channel state information acquisition, self-interference in full-duplex RSMA, integration with advanced PHY layer techniques like index modulation, and prototyping testbeds. d)Future Networks: Studying synergistic combinations of RSMA with promising

6G technologies like reconfigurable intelligent surfaces, cell-free Massive MIMO, THz communications, integrated sensing and communication, and machine learning.

RSMA (H. Chen et al., 2020) is a novel multiple access technology that has been identified as potential for 6G wireless networks in order to deliver the high throughput and enormous connection necessary for new applications. RSMA may efficiently handle interference and outperform traditional approaches by separating messages into common as well as private sections. To reduce interference, RSMA divides individual user's message into a common component decoded by all users and a private part decoded only by the targeted user. This bridges the gap between OMA and NOMA, in which all interference is decoded. The common-private split enables RSMA to effectively tolerate a wide range of interference levels. From an information-theoretic standpoint, RSMA (O. Dizdar et al., 2020) has been proven to reach optimal performance in a variety of contexts. RSMA outperforms 4G/5G systems in terms of spectrum efficiency, energy efficiency, coverage, reliability, quality of service, and resilience against poor channel state information (CSIT). Spectral efficiency increases in both underloaded and overloaded networks are among the primary advantages of RSMA.RSMA has resistance to faulty CSIT and user mobility. RSMA shows scalability to 6G's huge connection requirements. It has the ability to optimise for a variety of services such as eMBB and URLLC. It gives less feedback overhead and complexity compared to alternatives. Over a wide variety of network loads and channel circumstances, RSMA beats OMA and NOMA. The common-private split enables RSMA to adapt dynamically to various interference regimes. Many interesting applications for 6G and beyond are available from RSMA: a) Massive machine-to-machine communication. b) Communication with ultra-low latency. c)Networks that connect vehicles to everything (V2X). d) Sensing and communication integration. The capacity of RSMA to manage large connections effectively makes it well-suited to machine-centric and ultra-dense networks envisioned in 6G.

The chapter has been organized as follows: First, Concept of Rate Splitting has been introduced. Then, key benefits of RSMA have been discussed followed by its working and implementation challenges. Further utilization of RSMA in 5G and beyond, Comparison with other multiple access techniques, major utilization aspects as well as its future trends in wireless networks have been addressed.

Key Benefits of Rate Splitting Multiple Access

Spectral efficiency is an important statistic in wireless communication because it affects how well available spectrum is used to convey data. By leveraging the benefits of rate splitting, RSMA dramatically enhances spectral efficiency. Users are allotted a predetermined percentage of the available spectrum in classic multiple access schemes, which can result in wasteful utilisation if a user's transmission rate is much lower than the allocated bandwidth. By partitioning the available bandwidth, RSMA (A. R. Flores et al., 2020) allows users to dynamically alter their transmission rates. This adaptive rate splitting improves spectrum utilisation, maximising the network's total spectral efficiency. RSMA has numerous essential qualities and benefits that make it a game changer in wireless communication. One of its key features is its ability to reduce user intervention. RSMA assures that users may live and interact in a highly efficient and interference-free way by intelligently allocating resources and adaptively altering transmission rates. This results in quicker and more consistent connections, ultimately improving the user experience. Another distinguishing aspect of RSMA is its adaptability to diverse user requirements. The rate splitting approach enables users to tailor their transmission rates to their unique requirements, providing optimal resource allocation and network capacity. Because of its versatility, RSMA (G. Zhou et al., 2022) is appropriate for a extensive range of applications plus use cases, including multimedia streaming, Internet of Things (IoT) devices, and mission-critical communications. RSMA has the ability to revolutionise several sectors and allow new wireless communication applications. One of the primary uses is in 5G networks, where RSMA may greatly improve network capacity and performance. RSMA can play vital role in satisfying the needs of 5G networks. RSMA can be employed in wireless communication systems other than 5G, such as satellite communications, unmanned aerial vehicles (UAVs), and wireless sensor networks. RSMA's capacity to increase spectral efficiency and prevent interference makes it a potential solution for a variety of circumstances in which several users must interact at the same time.

How RSMA Works

RSMA, which relies on the rate-splitting (RS) concept, is being recognised as a viable PHY-layer transmissions model supporting non-orthogonal distribution, interference control, as well as MA methods in 6G. In contrast to the severe interference management approaches employed in SDMA and NOMA, the foremost concept behind RSMA is to partition user messages into shared as well as private sections and give the option of moderately decoding interference while also handling interference as noise. The versatility of RSMA allows it to work properly in the face of every degree of interference. Whenever there is weak or heavy interference, RSMA changes to SDMA or NOMA automatically by altering the powers. RSMA organically connects SDMA & NOMA, including complete hard switching inside them. The transmitter separates every participant's message in a shared as well as a private piece for RSMA. The aforementioned messages are encoded independently as well as linearly precoded at the transmitter. Every user begins by decoding those common streams, viewing entire private conversations as noise. Post removal of the decoded shared streaming from the signal which was received, every individual decodes their preferred private stream. This enables flexible interference control by decoding interference partially via shared stream decoding. As a result, RSMA incorporates existing MA approaches and provides a feasible PHY-layer transmissions model for non-orthogonal distribution.

Figure 1. Classification of RSMA

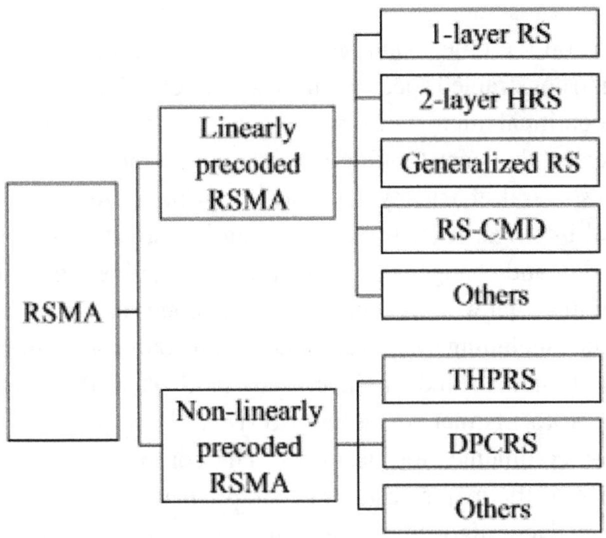

Depending on the implementation variation, RSMA may be categorized as follows.

Table 1. Characteristics of different types of linearly precoded RSMA

Strategy	1-Layer RS	2-Layer HRS	Generalized RS	RS-CMD
Complication of Encoder	Encrypt K'+1 streams	Encrypt K'+G'+1 streams	Encrypt $2^{K'}$-1 streams	Encrypt 2K' streams
Receiver complexity	Requires 1 layer of SIC at individual user.	Necessitates 2 layer of SIC at individual user.	Necessitates $2^{(k'-1)}$ -1 layer of SIC at individual user.	Necessitates k' layers of SIC at individual user.

Challenges in Implementing Rate Splitting Multiple Access

Wireless communication has become an essential aspect of our life, linking us to a wide network of information and allowing us to communicate in real time. The idea of multiple access is critical in wireless communication in order to handle various users and devices. several access systems allow several users to use the same wireless channel, allowing them to communicate at the same time. To distribute transmission resources among users, traditional multiple access practices like as TDMA and FDMA have been widely employed. These methods, however, have drawbacks. For example, TDMA assigns distinct time slots to different users, which might result in inefficient resource utilisation if the given time slots are not completely utilised. FDMA, on the other hand, assigns distinct frequency bands to different users, resulting in constrained spectrum utilisation. RSMA is a fast-expanding field of study, with continual attempts to improve its performance and investigate new applications. Researchers are looking at new coding and modulation approaches to increase the spectral efficiency of RSMA. Furthermore, machine learning as well as artificial intelligence methods are being investigated in order to optimise resource allocation and change transmission rates adaptively in real-time. RSMA implementation in actual wireless networks necessitates careful consideration of a number of aspects, including system architecture, resource allocation, and network optimisation. Researchers and engineers are hard at work designing efficient algorithms and protocols that will allow RSMA to be seamlessly integrated into existing wireless communication networks. The complexity of the rate splitting algorithms is one of the major obstacles in implementing RSMA. It is a difficult issue to create efficient algorithms that can dynamically alter transmission rates based on user requirements and network circumstances. However, continuing R&D efforts are aimed at overcoming these issues and making RSMA a viable and scalable solution for real-world wireless networks.

Rate Splitting Multiple Access in 5G Networks and Beyond

While research on Rate splitting (RS) is still ongoing, it shows great promise as a core physical layer technology for 5G and 6G networks. Specific application scenarios include: RS manages interference efficiently in very large antenna systems.RS combats beam misalignment issues in high frequency networks.RS aids joint communication and radar sensing.RS handles rapidly changing channels and interference.RS optimizes interference coordination. RS also enables enhanced mobile broadband services envisioned for 5G and 6G.

RSMA is a revolutionary multiple access approach that has emerged as an effective solution for interference control and increasing the performance of multi-antenna wireless communication systems. A brief description of RSMA for 5G networks is offered here.RSMA uses superposition coding to transmit shared plus private feeds in the exact same time-frequency channels in a non-orthogonal way. At the point of reception, users decode the shared stream first, perceiving private feeds as noise. With SIC, the common stream is eliminated from the received signal prior decoding the private stream. RSMA can control interference in a flexible manner.

Compared to existing multiple access techniques used in 5G like SDMA and NOMA, RSMA provides several key benefits: RSMA achieves optimal spatial multiplexing gain in many scenarios, outperforming SDMA and NOMA. RSMA is more robust against imperfect transmitter CSI than SDMA or NOMA, as it was designed based on imperfect CSI assumptions. RSMA reduces CSI feedback overhead compared to conventional techniques. RSMA can operate in both underloaded and overloaded regimes by adjusting common/private split. RSMA requires fewer SIC layers than NOMA, reducing receiver complexity.

RSMA achieves higher sum rate and higher spatial multiplexing gains than conventional techniques: In a 2 user MISO broadcast channel with perfect CSI, sum rates were 10-15% higher with RSMA over SDMA and NOMA. Optimal DoF achieved in several scenarios, outperforming NOMA and SDMA These spectral efficiency benefits have been confirmed in system simulations as well. In a 4 user MISO setting at 10 dB SNR, RSMA achieved ~20% higher sum rate than NOMA.

6G networks are shifting away from OMA as well as towards NOMA. In contrast to previous OMA approaches, NOMA permits users to utilise the same set of time-frequency resource unit at the same time, which is helpful for permitting numerous connections in spectrally efficient communication.

Figure 2. Different types of multiple access techniques

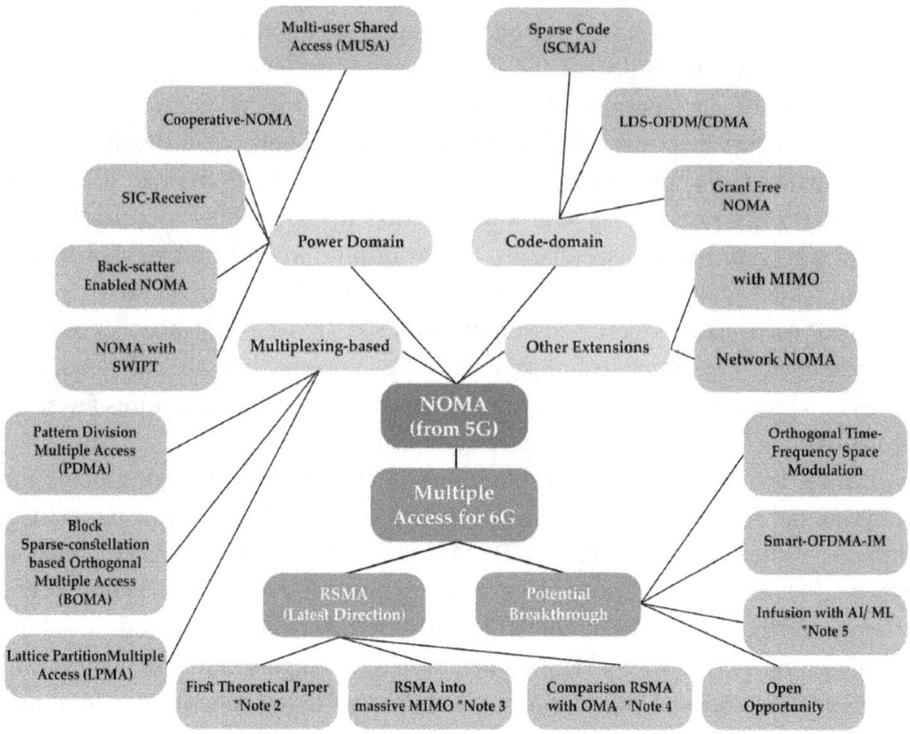

Comparison of RSMA With Other Techniques

Rate splitting (RS) is a wireless communication technique where user messages are split at the transmitter into common as well as private parts. This allows RS to smoothly transition between handling interference as noise (private messages) as well as decoding interference (common messages). RS was originally proposed in the context of interference channels, but recent research has focused on the downlink broadcast channel with multiple antennas at the transmitter.

Following are some effective multiple access practices:

- OMA: Orthogonal resources, such as time slots or frequency bands, are assigned to users. Does not take use of the entire multiplexing benefit.
- SDMA: Users share the similar time-frequency resources but are separated by spatial beams. Interference is treated as noise.
- NOMA: Employs superposition coding and SIC to overlap users. Completely decodes interference.

- RSMA: Partially decodes interference using common message decoding and treats the remainder as noise using private messages.

Table 2. Characteristics comparison of OMA, NOMA, SDMA, and RSMA

Parameter	OMA	NOMA	SDMA	RSMA
Spectral Efficiency	Low	Higher	Moderate	Higher
Reliability	Higher	Low	Moderate	Higher
Fairness	Higher	Low	Moderate	Higher
Feedback Overhead	Low	Higher	Higher	Low
Robustness to CSIT Errors	Higher	Low	Low	Higher
Complexity	Low	Higher	Moderate	Moderate

RSMA Generalises SDMA and NOMA by limiting them to special instances in suitable channel conditions. It Always produces equal or superior rate performance than SDMA and NOMA. It is More resistant to faulty CSIT, channel heterogeneity, and changing network loads. RSMA provides Lower complexity scheduling and receivers than NOMA.

So Far Major Utilization of RSMA

RSMA employs MARS techniques at the transmitter as well as SIC approach at the receivers to tenderly connect the two extreme conditions of completely decoding interference as well as addressing interference just like noise. Y. Mao et al. (2021) presented a novel multiple access strategy that employs MARS for cellular IoT. TP-RSMA as well as PP-RSMA are the two techniques presented. The former serves each of the groups of participants independently throughout orthogonal time slots, whereas the later serves the dual groups of participants in a non-orthogonal way inside the same time slot. B. Matthiesen et al. (2021) investigated the globally optimum precoder architecture enabling RSMA in Gaussian MISO downlink channels in terms of weighted sum rate as well as energy efficacy maximisation. Suggested technique addressed a situation of combined multicast and unicast beamforming, as well as multicast as well as unicast-only beamforming in particular instances. Study by J.An et al. (2020) investigated the optimisation of the RSMA sum-rate with incomplete CSI across the transmitter i.e CSIT and receivers i.e CSIR. The resilience of the RSMA technique against inadequate CSIT had been explored in prior works, but the consequences of imperfect CSIR have not been considered. L.Yin et al. (2021) investigated the possible rate efficiency of RSMA in a MU-MISO network in which mainly slow-changing statistical CSI was provided at the transmitter. O. Dizdar et al. (2021) investigated the performance of RSMA (M. Dai et al., 2016; Z. Li et al., 2020) in the presence of poor CSI at the Transmitter (CSIT) resulting from user mobility plus network delay. The suggested power allotment significantly decreases the complexity of RSMA precoder architecture. Cloud radio access networks (C-RAN) provide suitable platform for communication networks beyond the 5th generation, incorporating developments in cloud computing technology to contemporary RAN (A. Alameer Ahmad et al., 2019; A. A. Ahmad et al., 2020; A. A. Ahmad et al., 2021). Recently, it was demonstrated that RSMA can efficiently control interference in multiple-antenna communication networks. Research by A. A. Ahmad et al. (2020) investigated the use of RSMA in C-RAN. To that aim, the study addressed the topic of stochastic coordinated beamforming (SCB) optimisation in order to maximise the network's ergodic sum-rate (ESR). A.R.Flores et al.(2018) investigated the performance of Tomlinson-Harashima Precoding (THP) based Rate-Splitting (RS) in a multiple-antenna broadcast channel utilising perfect as well as imperfect CSIT. They focused on RS employing centralised and decentralised THP structures, in which just one user divides their message into a shared and private half, and developed formulations to represent the SIN ratio including sum rates related with these schemes. A.Mishra et al. (2022) addressed the precoder strategy issue for the RSMA method in downlink MU structures employing both faultless & deficient CSIT, in addition to the purpose and benefits of delivering multiple shared streams. Precoder optimisation is accomplished through the use of Sample Average Approximation (SAA) in conjunction with the suggested vectorization as well as WMMSE. The achievable sum-DoF of RSMA is calculated and its performance is confirmed using Monte Carlo simulations. Work by L.Yin et al. (2020) focused on the beamforming design challenge in multibeam satellite communications to achieve max-min fairness (MMF). A. Grant et al., (2001) proposed a unique multibeam multicast beamforming technique based on RSMA in contrast to standard linear precoding (NoRS), which depends on totally handling any residual interference as noise. N.Ha et al. (2020) investigated a double-cell downlink wireless network with two cell-center users and one cell-edge user. The authors developed coordinated RSMA, a novel multi-cell transmission method that successfully handles both inter-cell plus inter-user interference through simultaneously separating messages from many users into a collection of shared and private parts. Finally, it was demonstrated that the proposed C-RSMA achieves a bigger rate region than known multi-cell transmission systems based on orthogonal as well as NOMA. In downlink single-cell systems, RSMA has been proven to achieve higher spectrum and energy efficiency than SDMA and NOMA. Y.Mao et al. (2019) explored the best beamformer design to maximise the Weighted Sum-Rate (WSR) of all users in downlink Coordinated Multi-Point (CoMP) Joint Transmission (JT) networks under individual QoS rate limitations as well as per base station power constraints. Numerical studies revealed that in CoMP JT, RSMA outperforms SDMA and NOMA in terms of WSR across a wide range of inter-user as well as inter-cell channel strength discrepancies (D. Yu et al., 2019). Study by L.Yin et al. (2021) concentrated on the promising RSMA protocol (O. Dizdar et al.,2020) and related beamforming strategy challenge in order to attain maximum-min fairness among many co-channel multicast groups with incomplete CSIT. O. Dizdar et al. (2021) evaluated RSMA performance in circumstances involving New Radio (NR) and 6G vital core services, notably enhanced Ultra-Reliable & Low-Latency Communications (URLLC) and enhanced Mobile Broadband Communications (eMBB). They proposed optimum RSMA-based system designs aimed at short-packet, low-latency, resilient interactions with substantial throughput underneath the practical and critical setting of insufficient CSIT caused by user mobility & network latency/delay. In terms of performance, it was demonstrated that RSMA outperforms SDMA and NOMA, and that it was capable of matching the criteria for enhanced URLLC and eMBB in 6G successfully. Y. Xu et al. (2021) introduced the notion of RSMA into short-packet downlink transmission. Numerical experiments showed that RSMA could accomplish the identical transmission rate as NOMA & SDMA with smaller blocklengths (and hence reduced delay), particularly in overcrowded multi-antenna networks.Y. Mao et al. (2019) investigated the application of linearly-precoded RS in the MISO SWIPT Broadcast Channel. RS dynamically controls interference by splitting IR signals into private and shared sections & encoding the shared portions into one stream decoded by all IRs. Additional to 5G, Dual-Functional Radar Communications (DFRC) might be a viable paradigm for attaining Integrated Sensing & Communication (ISAC). C. Xu et al. (2021) suggested how RSMA (S. Naser et al., 2020) can be used as a unified, dominant multiple access approach for a multi-antenna DFRC system with double communication as well as radar competence to interact with downlink users. O. Dizdar et al. (2021) described a multi-antenna Joint Radar-Communication (JRC) system using RSMA and lower resolution DAC components. They designed the JRC system with RSMA and lower resolution DACs using a problem formulation that maximised communication sum-rate and the closeness of the generated JRC waveform to an ideal radar beampattern under an average transmit power restriction. They used the Alternating Direction approach of Multipliers (ADMM) approach to address the combined sum-rate maximisation and beampattern error reduction issue. The numerical findings demonstrated that when compared to SDMA, RSMA achieves a much greater sum-rate while maintaining the same Normalised Mean Square Error (NMSE) for the chosen radar beampattern. A. Rahmati et al. (2019) implemented RSMA and NOMA algorithms. They focused on the energy efficiency of the RSMA and NOMA systems in a mmWave downlink transmission scenario. Furthermore, they optimised the precoding vectors of both systems by explicitly considering 3GPP antenna propagation patterns. Z.Yang et al. (2020_used RSMA to investigate the resource allocation problem in a customisable intelligent surface (RIS)-assisted wireless communication system. This network's energy efficiency is maximised by regulating the phase shifts of the RIS as well as beamforming of the base station. This issue is presented as a hybrid optimisation issue involving transmit beamforming plus RIS control, with the goal of maximum energy efficiency while fulfilling the rate constraints of the users. To address this issue, an iterative technique that addresses the phase optimisation & beamforming optimisation sub issues iteratively is created.

O. Dizdar et al. (2021) examined a unique application of RSMA in MISO Broadcast Channel for cooperative communications & jamming utilising a Multi-Carrier (MC) waveform. on a platform where all users execute broadband communications utilising MC waveforms on their own networks, the anticipated aim was to communicate with Information Users (IU) whilst jamming Adversarial Users (AU) to interfere with their communications. The optimal information and jamming precoders were designed to optimise the sum-rate while adhering to jamming power constraints on the pilot subcarriers of AUs, a jamming method thought to be one of the most detrimental to MC waveforms pursuant to the system model under consideration.

The MMF problem was introduced by Zhi Wen Si et al. (2022) for a system with partial CSIT plus per-antenna power constraints at the satellite. The authors also looked into double-stage precoding with on-board processing (OBP) on the satellite level. The computational outcomes of RSMA simulations were compared to the standard linear precoding technique.

O. Dizdar et al. (2022) explored RSMA in order to provide multi-user communications with both perfect and imperfect CSIT. A unique algorithm was used to tackle the specified non-convex optimisation problem. Authors demonstrated numerically that, due to its generalised structure and enhanced interference control capabilities, RSMA achieves higher energy efficiency by deploying fewer RF chains than Space Division Multiple Access.

Using RSMA in the presence of partial CSIT, the DFRC precoders were improved to enhance the Average Weighted Sum-Rate (AWSR) under QoS constraints while reducing the DFRC beampattern Mean Squared Error (MSE) vs an ideal MIMO radar beampattern (R. Cerna-Loli et al.,2021). According to the simulation results, the RSMA-based DFRC outperforms other commonly used techniques such as SDMA and NOMA. The RSMA-based DFRC's common stream, in particular, allows for adjustable-rate splitting to assure user rate fairness utilising partial CSIT whilst also being a key contributor to the creation of a directed beampattern for effective radar detection (O. Dizdar et al., 2022).

Using recent results on the max-min fair (MMF) optimisation issue for RSMA-based multigroup multicast beamforming with imperfect CSIT, L. Yin et al. (2021)

explored the deployment of the physical layer that involve finite length polar coding, finite alphabet modulation, AMC algorithm, along with SIC receivers, among other things. In both cellular and multibeam satellite systems, LLS results verified the benefits of RSMA-based multigroup multicast transmission over SDMA based techniques.

S.Ma et al.(2021)concentrated on the best beamformer design for RSMA-assisted MISO visible light communication (VLC) networks. The authors first determined the closed-form lower limits of each user's attainable rate, which are the first theoretical constraints of achievable rate for RSMA-aided VLC networks. Second, the authors examined the best beamformer design for RSMA-aided VLC networks in order to maximise the total rate while keeping optical and electrical power limits in mind. Furthermore, it was demonstrated that the proposed RSMA-aided networks outperform SDMA and NOMA.

Sotiris A. Tegos et al. (2022) evaluated the outage probability and throughput of a two-source uplink RSMA network. The authors created closed-form formulae for the outage probability of transmitted messages, which they then utilised to compute the efficiency of sources in a random-access network employing a medium access control protocol based on slotted ALOHA and RSMA.

Future Trends in Rate Splitting Multiple Access

Several developments are predicted to affect the development and deployment of Rate Splitting Multiple Access technology in the future. The growth of Internet of Things (IoT) devices is one of the important trends. With the expansion of IoT devices in many industries such as smart homes, healthcare, transportation, and manufacturing, there is an increasing demand for wireless networks that can serve a large number of connected devices. The flexibility of RSMA to manage severe traffic increase and accept device proliferation makes it a suitable alternative for IoT applications. The arrival of the 5G era will also push the use of RSMA. 5G networks offer ultra-fast connection, minimal latency, and huge device connectivity. To fully realise the potential of 5G, new multiple access mechanisms such as RSMA will be required. The capacity of RSMA to improve spectrum efficiency, increase reliability, and enable low-latency communication is well aligned with the objectives of 5G networks. Machine Learning (ML) and Artificial Intelligence (AI) are expected to play critical roles in the development and optimisation of Rate Splitting Multiple Access technologies. ML algorithms may be employed in real-time to learn channel conditions and traffic patterns, allowing for dynamic rate and power allocation decisions. RSMA can adapt to changing network circumstances, optimise resource allocation, and increase overall network performance by integrating machine learning and artificial intelligence. Furthermore, machine learning and artificial intelligence

(AI) may be utilised to improve the security and privacy of RSMA-enabled networks. ML algorithms are capable of detecting and mitigating numerous forms of threats, including as jamming, eavesdropping, and spoofing. Anomaly detection systems driven by AI can detect aberrant network behaviour and initiate necessary responses. Rate Splitting Multiple Access has a wide range of possible applications. RSMA can provide real-time monitoring of patients in the healthcare business, enabling for remote diagnostics, telemedicine, and personalised healthcare services.

Figure 3. Future research prospectives of rate splitting

RSMA can accommodate a huge number of linked devices in smart cities, allowing effective traffic management, smart grid systems, and intelligent transportation systems. RSMA can promote machine-to-machine communication in the manufacturing industry, allowing for real-time monitoring and control of production operations. RSMA can enable high-quality video streaming, online gaming, and virtual reality experiences in the entertainment business. These are just a handful of the numerous opportunities that RSMA may open up in a variety of sectors. Rate Splitting Multiple Access (RSMA) has the ability to completely transform wireless communication. Because of its ability to divide transmission power into smaller sub-packets, it allows for simultaneous communication across several channels, enhancing efficiency and capacity in wireless networks. With the development of IoT devices and the arrival of the 5G era, RSMA is well positioned to shape the future of wireless communication. The flexibility of RSMA has generated interest in many wireless application scenarios for 6G and beyond: High-mobility communications (vehicles, drones), extreme URLLC and eMBB, massive connectivity for IoT, integrated access & backhaul, cell-free Massive MIMO, THz communications. reconfigurable intelligent surfaces However, further research is needed to address optimization challenges, signalling overhead, self-interference, and hardware prototyping before wide-scale adoption. Integrating machine learning and reconfigurable intelligent surfaces with RSMA has shown promise in overcoming some of these limitations. Overall, RSMA is poised to play a pivotal role in managing interference and guaranteeing QoS in next-generation wireless networks facing an explosion of traffic, devices, and applications. Continued research and standardization efforts are critical to unlocking the full potential of RSMA through practical implementations.

RSMA (Tegos et al., 2022) in inadequate CSIT is valid for huge MIMO (Rui Zhang et al., 2014). To deal with inter-group as well as intra-group interference, double-stage beamforming and two-layer RS are used. This structure's sum-rate outperforms standard two-stage beamforming for SDMA. Furthermore, in large MIMO (Larsson, Erik G et al., 2014), RSMA is being proven to reduce hardware weakening along with pilot contamination. The growing quantity of transmit antennas necessitates the use of more RF chains, which raises the overall expense. To decrease the number of RF links, one way is to adopt hybrid A/D beamforming. This technique, however, introduces a difficult feedback mechanism in UL for both analogue as well as digital precoders, which both necessitate feedback out of transmitters. The higher coverage needs of 6G (Chowdhury, Mostafa Zaman et al., 2020) necessitate the adoption of additional infrastructure to enable network coverage, such as aerial networks that employ UAVs (Gabor Vaszkun et al.,2015) as BS to provide coverage. When RSMA is used on UAVs and BSs, the maximum sum-rate enhances. It efficiently handles inter-beam interference in overburdened multibeam satellite communication environments with perfect as well as imperfect CSIT. When compared to time-sharing, RSMA attains superior sum-rate & DoF, and a latency enhancement could be accomplished with restricted CSIT quality. RSMA is a great option for usage in Intelligent Reflecting Surfaces (IRS)-enabled systems (Rui Zhang et al., 2021) because it is resistant to poor CSIT and can deal with IRS issues more effectively than other technologies. Because of RSMA's enhanced rate performance, its application in VLC at higher SNR levels seems promising. Because it is intended to function in the presence of interference, RSMA is an obvious option for usage in Cognitive Radio networks.

CONCLUSION

In summary, rate splitting is an emerging and broadly applicable technique to manage interference in 5G and 6G networks. By bridging between decoding interference and treating interference as noise, RS provides spectral efficiency, reliability, and complexity improvements. While research continues, RS shows immense promise as a core physical layer technology for next generation wireless systems. Practical implementation and standardization efforts will be vital to unlocking its full potential.

REFERENCES

Ahmad, A. A., Dahrouj, H., Chaaban, A., Sezgin, A., & Alouini, M. S. (2019). Interference mitigation via rate-splitting and common message decoding in cloud radio access networks. *IEEE Access : Practical Innovations, Open Solutions*, *7*, 80350–80365. doi:10.1109/ACCESS.2019.2921626

Ahmad, A. A., Mao, Y., Sezgin, A., & Clerckx, B. (2020, August). Rate splitting multiple access in C-RAN. In *2020 IEEE 31st Annual International Symposium on Personal, Indoor and Mobile Radio Communications* (pp. 1-6). IEEE. 10.1109/PIMRC48278.2020.9217249

Ahmad, A. A., Mao, Y., Sezgin, A., & Clerckx, B. (2021). Rate splitting multiple access in C-RAN: A scalable and robust design. *IEEE Transactions on Communications*, *69*(9), 5727–5743. doi:10.1109/TCOMM.2021.3085343

Ahmad, A. A., Matthiesen, B., Sezgin, A., & Jorswieck, E. (2020, June). Energy efficiency in C-RAN using rate splitting and common message decoding. In *2020 IEEE International Conference on Communications Workshops (ICC Workshops)* (pp. 1-6). IEEE. 10.1109/ICCWorkshops49005.2020.9145076

An, J., Dizdar, O., Clerckx, B., & Shin, W. (2021). Rate-splitting multiple access for multi-antenna broadcast channel with imperfect CSIT and CSIR. *arXiv preprint arXiv:2102.08738.*

Chen, H., Mi, D., Chu, Z., Xiao, P., Xu, Y., & He, D. (2020, August). Link-level performance of rate-splitting based downlink multiuser MISO systems. In *2020 IEEE 31st Annual International Symposium on Personal, Indoor and Mobile Radio Communications* (pp. 1-5). IEEE. 10.1109/PIMRC48278.2020.9217103

Chowdhury, M. Z., Shahjalal, M., Ahmed, S., & Jang, Y. M. (2020). 6G wireless communication systems: Applications, requirements, technologies, challenges, and research directions. *IEEE Open Journal of the Communications Society*, *1*, 957–975. doi:10.1109/OJCOMS.2020.3010270

Clerckx, B., Joudeh, H., Hao, C., Dai, M., & Rassouli, B. (2016). Rate splitting for MIMO wireless networks: A promising PHY-layer strategy for LTE evolution. *IEEE Communications Magazine*, *54*(5), 98–105. doi:10.1109/MCOM.2016.7470942

Clerckx, B., Mao, Y., Schober, R., & Poor, H. V. (2019). Rate-splitting unifying SDMA, OMA, NOMA, and multicasting in MISO broadcast channel: A simple two-user rate analysis. *IEEE Wireless Communications Letters*, *9*(3), 349–353. doi:10.1109/LWC.2019.2954518

Dai, M., Clerckx, B., Gesbert, D., & Caire, G. (2016). A rate splitting strategy for massive MIMO with imperfect CSIT. *IEEE Transactions on Wireless Communications*, *15*(7), 4611–4624. doi:10.1109/TWC.2016.2543212

Dizdar, O., & Clerckx, B. (2021, September). Rate splitting multiple access for multi-antenna multi-carrier joint communications and jamming. In 2021 Sensor Signal Processing for Defence Conference (SSPD) (pp. 1-5). IEEE. doi:10.1109/SSPD51364.2021.9541482

Dizdar, O., & Clerckx, B. (2022). Rate-splitting multiple access for communications and jamming in multi-antenna multi-carrier cognitive radio systems. *IEEE Transactions on Information Forensics and Security*, *17*, 628–643. doi:10.1109/TIFS.2022.3149145

Dizdar, O., Kaushik, A., Clerckx, B., & Masouros, C. (2021, June). *Rate-splitting multiple access for joint radar-communications with low-resolution DACs. In 2021 IEEE international conference on communications workshops (ICC workshops).* IEEE.

Dizdar, O., Kaushik, A., Clerckx, B., & Masouros, C. (2022). Energy efficient dual-functional radar-communication: Rate-splitting multiple access, low-resolution DACs, and RF chain selection. *IEEE Open Journal of the Communications Society*, *3*, 986–1006. doi:10.1109/OJCOMS.2022.3183950

Dizdar, O., Mao, Y., & Clerckx, B. (2021). Rate-splitting multiple access to mitigate the curse of mobility in (massive) MIMO networks. *IEEE Transactions on Communications*, *69*(10), 6765–6780. doi:10.1109/TCOMM.2021.3098695

Dizdar, O., Mao, Y., Han, W., & Clerckx, B. (2020, August). Rate-splitting multiple access for downlink multi-antenna communications: Physical layer design and link-level simulations. In *2020 IEEE 31st Annual International Symposium on Personal, Indoor and Mobile Radio Communications* (pp. 1-6). IEEE.

Dizdar, O., Mao, Y., Han, W., & Clerckx, B. (2020, November). Rate-splitting multiple access: A new frontier for the PHY layer of 6G. In *2020 IEEE 92nd vehicular technology conference (VTC2020-Fall)* (pp. 1-7). IEEE.

Dizdar, O., Mao, Y., Xu, Y., Zhu, P., & Clerckx, B. (2021, September). Rate-splitting multiple access for enhanced URLLC and eMBB in 6G. In *2021 17th International Symposium on Wireless Communication Systems (ISWCS)* (pp. 1-6). IEEE.

Flores, A. R., Clerckx, B., & De Lamare, R. C. (2018, August). Tomlinson-harashima precoded rate-splitting for multiuser multiple-antenna systems. In *2018 15th International Symposium on Wireless Communication Systems (ISWCS)* (pp. 1-6). IEEE. 10.1109/ISWCS.2018.8491245

Flores, A. R., de Lamare, R. C., & Clerckx, B. (2020). Linear precoding and stream combining for rate splitting in multiuser MIMO systems. *IEEE Communications Letters*, *24*(4), 890–894. doi:10.1109/LCOMM.2020.2969158

Grant, A. J., Rimoldi, B., Urbanke, R. L., & Whiting, P. A. (2001). Rate-splitting multiple access for discrete memoryless channels. *IEEE Transactions on Information Theory*, *47*(3), 873–890. doi:10.1109/18.915637

Gupta, L., Jain, R., & Vaszkun, G. (2015). Survey of important issues in UAV communication networks. *IEEE Communications Surveys and Tutorials*, *18*(2), 1123–1152. doi:10.1109/COMST.2015.2495297

Ha, N., Shin, W., Vaezi, M., & Poor, H. V. (2020, November). Coordinated rate splitting multiple access for multi-cell downlink networks. In *2020 54th Asilomar Conference on Signals, Systems, and Computers* (pp. 996-1001). IEEE. 10.1109/IEEECONF51394.2020.9443398

Larsson, E. G., Edfors, O., Tufvesson, F., & Marzetta, T. L. (2014). Massive MIMO for next generation wireless systems. *IEEE Communications Magazine*, *52*(2), 186–195. doi:10.1109/MCOM.2014.6736761

Li, Z., Ye, C., Cui, Y., Yang, S., & Shamai, S. (2020). Rate splitting for multi-antenna downlink: Precoder design and practical implementation. *IEEE Journal on Selected Areas in Communications*, *38*(8), 1910–1924. doi:10.1109/JSAC.2020.3000824

Loli, R. C., Dizdar, O., & Clerckx, B. (2022). Rate-splitting multiple access for multi-antenna joint radar and communications with partial CSIT: Precoder optimization and link-level simulations. *arXiv preprint arXiv:2201.10621.*

Lu, L., Li, G. Y., Swindlehurst, A. L., Ashikhmin, A., & Zhang, R. (2014). An overview of massive MIMO: Benefits and challenges. *IEEE Journal of Selected Topics in Signal Processing*, *8*(5), 742–758. doi:10.1109/JSTSP.2014.2317671

Ma, S., Zhang, G., Zhang, Z., & Gu, R. (2021). Robust beamforming design for rate splitting multiple access-aided MISO visible light communications. *arXiv preprint arXiv:2108.07014.*

Mao, Y., Clerckx, B., & Li, V. O. (2018). Rate-splitting multiple access for downlink communication systems: Bridging, generalizing, and outperforming SDMA and NOMA. *EURASIP Journal on Wireless Communications and Networking*, *2018*(1), 1–54. doi:10.1186/s13638-018-1104-7 PMID:30996723

Mao, Y., Clerckx, B., & Li, V. O. (2019, July). Rate-splitting for multi-user multi-antenna wireless information and power transfer. In *2019 IEEE 20th International Workshop on Signal Processing Advances in Wireless Communications (SPAWC)* (pp. 1-5). IEEE. 10.1109/SPAWC.2019.8815494

Mao, Y., Clerckx, B., & Li, V. O. (2019, May). Rate-splitting multiple access for coordinated multi-point joint transmission. In *2019 IEEE International Conference on Communications Workshops (ICC Workshops)* (pp. 1-6). IEEE. 10.1109/ICCW.2019.8756668

Mao, Y., Piovano, E., & Clerckx, B. (2021). Rate-splitting multiple access for overloaded cellular internet of things. *IEEE Transactions on Communications*, *69*(7), 4504–4519. doi:10.1109/TCOMM.2021.3067642

Matthiesen, B., Mao, Y., Popovski, P., & Clerckx, B. (2021, June). Globally optimal beamforming for rate splitting multiple access. In *ICASSP 2021-2021 IEEE International Conference on Acoustics, Speech and Signal Processing (ICASSP)* (pp. 4775-4779). IEEE. 10.1109/ICASSP39728.2021.9413616

Mishra, A., Mao, Y., Dizdar, O., & Clerckx, B. (2021). Rate-splitting multiple access for downlink multiuser MIMO: Precoder optimization and PHY-layer design. *IEEE Transactions on Communications*, *70*(2), 874–890. doi:10.1109/TCOMM.2021.3138437

Naser, S., Sofotasios, P. C., Bariah, L., Jaafar, W., Muhaidat, S., Al-Qutayri, M., & Dobre, O. A. (2020). Rate-splitting multiple access: Unifying NOMA and SDMA in MISO VLC channels. *IEEE Open Journal of Vehicular Technology*, *1*, 393–413. doi:10.1109/OJVT.2020.3031656

Rahmati, A., Yapici, Y., Rupasinghe, N., Guvenc, I., Dai, H., & Bhuyan, A. (2019, May). Energy efficiency of RSMA and NOMA in cellular-connected mmWave UAV networks. In *2019 IEEE International Conference on Communications Workshops (ICC Workshops)* (pp. 1-6). IEEE. 10.1109/ICCW.2019.8756699

Si, Z. W., Yin, L., & Clerckx, B. (2022). Rate-splitting multiple access for multigateway multibeam satellite systems with feeder link interference. *IEEE Transactions on Communications*, *70*(3), 2147–2162. doi:10.1109/TCOMM.2022.3144487

Tegos, S. A., Diamantoulakis, P. D., & Karagiannidis, G. K. (2022). On the performance of uplink rate-splitting multiple access. *IEEE Communications Letters*, *26*(3), 523–527. doi:10.1109/LCOMM.2022.3142102

Wu, Q., Zhang, S., Zheng, B., You, C., & Zhang, R. (2021). Intelligent reflecting surface-aided wireless communications: A tutorial. *IEEE Transactions on Communications*, *69*(5), 3313–3351. doi:10.1109/TCOMM.2021.3051897

Xu, C., Clerckx, B., Chen, S., Mao, Y., & Zhang, J. (2021). Rate-splitting multiple access for multi-antenna joint radar and communications. *IEEE Journal of Selected Topics in Signal Processing*, *15*(6), 1332–1347. doi:10.1109/JSTSP.2021.3110312

Xu, Y., Mao, Y., Dizdar, O., & Clerckx, B. (2022). Rate-splitting multiple access with finite blocklength for short-packet and low-latency downlink communications. *IEEE Transactions on Vehicular Technology*, *71*(11), 12333–12337. doi:10.1109/TVT.2022.3191085

Yang, Z., Shi, J., Li, Z., Chen, M., Xu, W., & Shikh-Bahaei, M. (2020, June). Energy efficient rate splitting multiple access (RSMA) with reconfigurable intelligent surface. In *2020 IEEE International Conference on Communications Workshops (ICC Workshops)* (pp. 1-6). IEEE. 10.1109/ICCWorkshops49005.2020.9145189

Yin, L., & Clerckx, B. (2020, June). Rate-splitting multiple access for multibeam satellite communications. In *2020 IEEE International Conference on Communications Workshops (ICC Workshops)* (pp. 1-6). IEEE. 10.1109/ICCWorkshops49005.2020.9145200

Yin, L., & Clerckx, B. (2020). Rate-splitting multiple access for multigroup multicast and multibeam satellite systems. *IEEE Transactions on Communications*, *69*(2), 976–990. doi:10.1109/TCOMM.2020.3037596

Yin, L., Clerckx, B., & Mao, Y. (2021, March). Rate-splitting multiple access for multi-antenna broadcast channels with statistical CSIT. In 2021 IEEE Wireless Communications and Networking Conference Workshops (WCNCW) (pp. 1-6). IEEE. doi:10.1109/WCNCW49093.2021.9420034

Yin, L., Dizdar, O., & Clerckx, B. (2021, June). Rate-splitting multiple access for multigroup multicast cellular and satellite communications: PHY layer design and link-level simulations. In *2021 IEEE International Conference on Communications Workshops (ICC Workshops)* (pp. 1-6). IEEE. 10.1109/ICCWorkshops50388.2021.9473795

Yu, D., Kim, J., & Park, S. H. (2019). An efficient rate-splitting multiple access scheme for the downlink of C-RAN systems. *IEEE Wireless Communications Letters*, *8*(6), 1555–1558. doi:10.1109/LWC.2019.2927206

Zhou, G., Mao, Y., & Clerckx, B. (2021). Rate-splitting multiple access for multi-antenna downlink communication systems: Spectral and energy efficiency tradeoff. *IEEE Transactions on Wireless Communications*, *21*(7), 4816–4828. doi:10.1109/TWC.2021.3133433

Compilation of References

. Khan, F., & Cho, S. H. (2017). Hand based Gesture Recognition inside a car through IR-UWB Radar. 대한전자공학회 학술대회, 154-157.

. Kim, B. S., Jin, Y., Lee, J., & Kim, S. (2022). FMCW radar estimation algorithm with high resolution and low complexity based on reduced search area. *Sensors, 22*(3), 1202.

. Kothapudi, V. K., & Kumar, V. (2019). Vertical polarized 1 × 3 series-fed linear array with gain and front-to-back ratio enhancement for airborne X-band SAR applications. *Progress in Electromagnetic Research (PIER-M), 80,* 169-179.

. Phaisiri, K., Boonpoonja, A., Kittiyanpunya, C., Kaemarungsi, K., & Chudpooti, N. (2023, August). Through-Wall Human Sensing Using mmWave FMCW Radar. In *2023 Research, Invention, and Innovation Congress: Innovative Electricals and Electronics (RI2C)* (pp. 1-4). IEEE.

Abbas, A., Sufian, M. A., Awan, W. A., Hussain, N., Lee, J., & Kim, N. (2023, March). A Reconfigurable Dual Sharp Notch-band UWB Antenna. In *2023 17th European Conference on Antennas and Propagation (EuCAP)* (pp. 1-5). IEEE. 10.23919/EuCAP57121.2023.10133304

Abdalla & Ibrahim. (2013). Compact and Closely Spaced Metamaterial MIMO Antenna With High Isolation for Wireless Applications. IEEE Antennas and Wireless Propagation Letters, 12.

Abdalla, M., Abdelnaby, U., & Mitkees, A. A. (2012). Compact and Triple Band Meta-material Antenna for All WiMAX Applications. *Proceedings of ISAP2012.*

Abdulhussein, A. J., Farhan, M. J., & Ali, G. M. (2023). Design and implementation of a frequency reconfigurable antenna using PIN switch for sub-6 GHz applications. *Open Engineering*, *13*(1), 20220453. doi:10.1515/eng-2022-0453

Abdulraheem, Y. I., Abdullah, A. S., Mohammed, H. J., Mohammed, B. A., & Abd-Alhameed, R. A. (2014). Design of radiation pattern-reconfigurable 60-GHz antenna for 5G applications. *Journal of Telecommunications*, *52*(1), 1–5.

Abdulsahib. (2023). Reverse polarity optical Orthogonal frequency Division Multiplexing for High-Speed visible light communications system. *Egyptian Informatics Journal, 24*(4). . doi:10.1016/j.eij.2023.100407

Abdulzahra, D. H., Alnahwi, F. M., & Abdullah, A. S. (2023). Design of Compact Wideband/ Bi-Band Frequency Reconfigurable Antenna for IoT Applications. *Iraqi Journal for Electrical and Electronic Engineering*, *19*(2), 100–109. doi:10.37917/ijeee.19.2.12

Abraham, J. K., Vinoy, K. J., Tellakula, R. A., & Varadan, V. K. (2003). Experimental Results of Planar Meta Materials. *IEEE-APS International Symposium.*

Adam, F. (2023). Selected Aspects of Nonorthogonal Multiple Access for Future Wireless Communications. Mathematics in Computer Science. doi:10.1007/s11786-023-00561-y

Ahmad, A. A., Mao, Y., Sezgin, A., & Clerckx, B. (2020, August). Rate splitting multiple access in C-RAN. In *2020 IEEE 31st Annual International Symposium on Personal, Indoor and Mobile Radio Communications* (pp. 1-6). IEEE. 10.1109/PIMRC48278.2020.9217249

Ahmad, A. A., Dahrouj, H., Chaaban, A., Sezgin, A., & Alouini, M. S. (2019). Interference mitigation via rate-splitting and common message decoding in cloud radio access networks. *IEEE Access : Practical Innovations, Open Solutions*, *7*, 80350–80365. doi:10.1109/ACCESS.2019.2921626

Ahmad, A. A., Mao, Y., Sezgin, A., & Clerckx, B. (2021). Rate splitting multiple access in C-RAN: A scalable and robust design. *IEEE Transactions on Communications*, *69*(9), 5727–5743. doi:10.1109/TCOMM.2021.3085343

Ahmad, A. A., Matthiesen, B., Sezgin, A., & Jorswieck, E. (2020, June). Energy efficiency in C-RAN using rate splitting and common message decoding. In *2020 IEEE International Conference on Communications Workshops (ICC Workshops)* (pp. 1-6). IEEE. 10.1109/ICCWorkshops49005.2020.9145076

Ahmad, A., Lee, G. O., & Choi, D. Y. (2023). Design and Performance Evaluation of a Compact Frequency-Reconfigurable Coplanar-Waveguide-Fed Slotted Patch Antenna for Multi-Band Wireless Communication. *Electronics (Basel)*, *12*(18), 3889. doi:10.3390/electronics12183889

Ahmad, W. A., Wessel, J., Ng, H. J., & Kissinger, D. (2020, December). IoT-ready millimeter-wave radar sensors. In *2020 IEEE Global Conference on Artificial Intelligence and Internet of Things (GCAIoT)* (pp. 1-5). IEEE.

Ahmed, A., Zhang, Y. D., & Hassanien, A. (2021). Joint Radar-Communications Exploiting Optimized OFDM Waveforms. *Remote Sensing (Basel)*, *13*(21), 4376. doi:10.3390/rs13214376

Ahmed, S., & Cho, S. H. (2020). Hand gesture recognition using an IR-UWB radar with an inception module-based classifier. *Sensors (Basel)*, *20*(2), 564. doi:10.3390/s20020564 PMID:31968587

Ahmed, S., & Cho, S. H. (2023). Machine learning for healthcare radars: Recent progresses in human vital sign measurement and activity recognition. *IEEE Communications Surveys and Tutorials.*

Ahmed, S., Kallu, K. D., Ahmed, S., & Cho, S. H. (2021). Hand gestures recognition using radar sensors for human-computer-interaction: A review. *Remote Sensing (Basel)*, *13*(3), 527. doi:10.3390/rs13030527

Aijaz, A. (2020). Private 5G: The Future of Industrial Wireless. *IEEE Industrial Electronics Magazine*, *14*(4), 136–145. doi:10.1109/MIE.2020.3004975

Alieldin, A., Eid, A. M., Salama, A. A., & El-Akhdar, A. M. (2023, February). A Reconfigurable Cosecant-Squared/Pencil Beam Antenna Array for Radar Systems. In *2023 International Microwave and Antenna Symposium (IMAS)* (pp. 50-52). IEEE. 10.1109/IMAS55807.2023.10066936

Allanic, R., Le Berre, D., Quendo, C., Ball, E. A., Ng, J. S., Huang, G., ... Merlet, T. (2023, November). Monolithic Frequency-Reconfigurable Antenna on Silicon Carbide for Constrained Environments. In *2023 IEEE Conference on Antenna Measurements and Applications (CAMA)* (pp. 649-652). IEEE. 10.1109/CAMA57522.2023.10352802

Al-Mahdi, A., & Al-Hassan, O. (2022). Reconfigurable Intelligent Surfaces: A game changer for future wireless systems. *Journal of Electromagnetic Waves and Applications*, *36*(13), 1789–1803.

Almekhlafi, M. H., Arfaoui, M. A., Assi, C., & Ghrayeb, A. (2023, May). A Low Complexity Passive Beamforming Design for Reconfigurable Intelligent Surface (RIS) in 6G Networks. *IEEE Transactions on Vehicular Technology*, *72*(5), 6309–6321. Advance online publication. doi:10.1109/TVT.2022.3233469

Alsharif, F., & Kurnaz, C. (2023, July). Design and analysis frequency reconfigurable coplanar waveguide antenna for S/C bands satellite applications. In *2023 46th International Conference on Telecommunications and Signal Processing (TSP)* (pp. 70-73). IEEE.

Al-Yasir, Y. I. A., Abdullah, A. S., Ojaroudi Parchin, N., Abd-Alhameed, R. A., & Noras, J. M. (2018). A new polarization-reconfigurable antenna for 5G applications. *Electronics (Basel)*, *7*(11), 293. doi:10.3390/electronics7110293

Al-Yasir, Y., Abdullah, A., Ojaroudi Parchin, N., Abd-Alhameed, R., & Noras, J. (2018). A new polarization-reconfigurable antenna for 5G wireless communications. *Proceedings of the 9th International Conference on Broadband Communications, Networks, and Systems.*

Amar, R., Alaee-Kerahroodi, M., & Shankar, M. B. (2021, June). FMCW-FMCW interference analysis in mm-wave radars; an indoor case study and validation by measurements. In *2021 21st International Radar Symposium (IRS)* (pp. 1-11). IEEE.

Amri, M., & Gharsallah, A. (2023, July). Design of Frequency-Pattern Reconfigurable Patch Array Antenna for Radar Application. In *2023 3rd International Conference on Electrical, Computer, Communications and Mechatronics Engineering (ICECCME)* (pp. 1-7). IEEE. 10.1109/ICECCME57830.2023.10253299

An, J., Dizdar, O., Clerckx, B., & Shin, W. (2021). Rate-splitting multiple access for multi-antenna broadcast channel with imperfect CSIT and CSIR. *arXiv preprint arXiv:2102.08738.*

Anagnostou, E., Torres, D., & Sepulveda, N. (2017). Vanadium dioxide switch for a reconfigurable bandwidth antenna. *Proceedings of the Loughborough Antennas & Propagation Conference (LAPC 2017).*

Anand, S., Raj, R. K., Sinha, S., Upadhyay, D., & Mishra, G. K. (2016). Bandwidth reconfigurable patch antenna for next generation wireless communication system applications. *Proceedings of the International Conference on Emerging Trends in Communication Technologies (ETCT)*. 10.1109/ETCT.2016.7882991

Anton-Haro, C., & Mestre, X. (2019). Learning and data-driven beam selection for mmWave communications: An angle of arrival-based approach. *IEEE Access : Practical Innovations, Open Solutions*, *7*, 20404–20415. doi:10.1109/ACCESS.2019.2895594

Anwar, M. S. (2023). *Microfluidics Based 3D Printed Reconfigurable Antenna for K-Band Applications* (Doctoral dissertation).

Apparao, M., & Karunakar, G. (2023). Design and analysis of frequency reconfigurable antenna in S-band. *Bulletin of Electrical Engineering and Informatics*, *12*(5), 2878–2886. doi:10.11591/eei.v12i5.4052

Arbab-Zavar, B., Palacios-Garcia, E., Vasquez, J., & Guerrero, J. (2019). Smart Inverters for Microgrid Applications: A Review. *Energies*, *12*(5), 840. doi:10.3390/en12050840

Arzykulov, S., Celik, A., Nauryzbayev, G. M., & Eltawil, A. (2023). Artificial noise and RIS-aided physical layer security: Optimal RIS partitioning and power control. *IEEE Wireless Communications Letters*, *12*(6), 992–996. doi:10.1109/LWC.2023.3256001

Ashraf, M., Tan, B., Moltchanov, D., Thompson, J. S., & Valkama, M. (2023). Joint Optimization of Radar and Communications Performance in 6G Cellular Systems. *IEEE Transactions on Green Communications and Networking*, *7*(1), 522–536. doi:10.1109/TGCN.2023.3234258

Attygalle, N. T., Leiva, L. A., Kljun, M., Sandor, C., Plopski, A., Kato, H., & Čopič Pucihar, K. (2021). No interface, no problem: Gesture recognition on physical objects using radar sensing. *Sensors (Basel)*, *21*(17), 5771. doi:10.3390/s21175771 PMID:34502662

AWR1843 data sheet, product information and support | TI.com. (n.d.). https://www.ti.com/product/AWR1843

Bahrami, S., Lee, D., Hong, W., & Song, H. J. (2023). Analysis of On-Antenna Power Combining for Realization of Pattern-Reconfigurable Multifeed Microstrip Antennas. *IEEE Transactions on Antennas and Propagation*, *71*(4), 3091–3103. doi:10.1109/TAP.2023.3240073

Bai, T., Pan, C., Deng, Y., Elkashlan, M., Nallanathan, A., & Hanzo, L. (2020). Latency Minimization for Intelligent Reflecting Surface Aided Mobile Edge Computing. *IEEE Journal on Selected Areas in Communications*, *38*(11), 2666–2682. doi:10.1109/JSAC.2020.3007035

Bai, Y. Y., Xiao, S., Tang, M. C., Ding, Z. F., & Wang, B. Z. (2011). Wide-angle scanning phased array with pattern reconfigurable elements. *IEEE Transactions on Antennas and Propagation*, *59*(11), 4071–4076. doi:10.1109/TAP.2011.2164176

Balanis, C. A. (1998). *Antenna Theory Analysis and Design* (4th ed.). John Wiley & Sons.

Bammidi, D., & Kota, C. B. R. (2023). Design and validation of frequency reconfigurable multiband antenna using varied current distribution method. *Measurement. Sensors*, *29*, 100844. doi:10.1016/j.measen.2023.100844

Bao, X. L., Ruvio, G., & Ammann, M. J. (2007). Low-profile dual-frequency GPS patch antenna enhanced with dual-band EBG structure. *Microwave and Optical Technology Letters*, *49*(11), 2630–2634. doi:10.1002/mop.22815

Basar E., Di Renzo M., De Rosny J., Debbah M., Alouini M.-S., & Zhang R., (2019). Wireless communications through reconfigurable intelligent surfaces. *IEEE Access, 7*.

Basharat, S., Hassan, S. A., Pervaiz, H., Mahmood, A., Ding, Z., & Gidlund, M. (2021). Reconfigurable Intelligent Surfaces: Potentials, Applications, and Challenges for 6G Wireless Networks. *IEEE Wireless Communications*, *28*(6), 1–8. doi:10.1109/MWC.011.2100016

Basharat, S., Khan, M., Iqbal, M., Hashmi, U. S., Zaidi, S. A. R., & Robertson, I. (2022). Exploring reconfigurable intelligent surfaces for 6G: State-of-the-art and the road ahead. *IET Communications*, *16*(13), 1458–1474. doi:10.1049/cmu2.12364

Battaglia, G. M., Isernia, T., Palmeri, R., & Morabito, A. F. (2023). Four-Beams-Reconfigurable Circular-Ring Array Antennas for Monopulse Radar Applications. *Radio Science*, *58*(9).

Behdad, N., & Sarabandi, K. (2006). Dual-band reconfigurable antenna with a very wide tunability range. *IEEE Transactions on Antennas and Propagation*, *54*(2), 409–416. doi:10.1109/TAP.2005.863412

Bekali, Y. K., Zugari, A., El Bhiri, B., Ennasar, M. A., Elmajid, H., Khalladi, M., & Elghandour, A. (2023). A New Design for a Novel Compact Frequency Reconfigurable Antenna. *International Journal of Microwave & Optical Technology*, *18*(4).

Bellapukonda, V. K., Muduli, A., & Nayak, D. K. (2023, April). Reconfigurable T-shaped discrete tunable Antenna for Cognitive Radio and Natural Sciences Applications. In *2023 International Conference on Recent Advances in Electrical, Electronics, Ubiquitous Communication, and Computational Intelligence (RAEEUCCI)* (pp. 1-4). IEEE. 10.1109/RAEEUCCI57140.2023.10134509

Benlakehal, M. E., Hocini, A., Khedrouche, D., Temmar, M. N., & Denidni, T. A. (2022). Design and analysis of a 1 × 2 microstrip patch antenna array based on photonic crystals with a graphene load in THz. *Journal of Optics*. Advance online publication. doi:10.1007/s12596-022-01006-8

Benny, S., & Sahoo, S. (2023, June). Polarization switching frequency reconfigurable dual polarization antenna element for phased array weather applications. In *2023 IEEE Wireless Antenna and Microwave Symposium (WAMS)* (pp. 1-5). IEEE. 10.1109/WAMS57261.2023.10242976

Berhab, S., Annou, A., & Chebbara, F. (2023). Reconfigurable Low-Profile Antenna-Based Metamaterial for on/off Body Communications. In *Handbook of Research on Emerging Designs and Applications for Microwave and Millimeter Wave Circuits* (pp. 166–200). IGI Global. doi:10.4018/978-1-6684-5955-3.ch007

Bernhard, J. T. (2005). *Reconfigurable Antennas*. Wiley.

Bernhard, J. T. (2007, August). Reconfigurable multifunction antennas: Next steps for the future. In *2007 International Symposium on Microwave, Antenna, Propagation and EMC Technologies for Wireless Communications* (pp. K2-1). IEEE. 10.1109/MAPE.2007.4393804

Bhagat, P. P. (2012). Analysis and Synthesis of Microstrip Patch Antenna using Artificial Neural Networks. *IEEE Asia-Pacific Conference on Antennas and Propagation*. 10.1109/APCAP.2012.6333177

Bhartia, P., & Bahl, I. J. (1982). Frequency agile microstrip antennas. *Microwave Journal*, *25*(1), 67–70.

Bhatnagar, M. R., & Ghassemlooy, Z. (2016). Performance analysis of Gamma–Gamma fading FSO MIMO links with pointing errors. *Journal of Lightwave Technology*, *34*(9), 2158–2169. doi:10.1109/JLT.2016.2526053

Bikrat, M., & Bri, S. (2023). A bandwidth reconfigurable antenna for devices in low UWB-applications. *TELKOMNIKA*, *21*(2), 272–279. doi:10.12928/telkomnika.v21i2.24256

Bildik, S., Dieter, S., Fritzsch, C., Menzel, W., & Jakoby, R. (2014). Reconfigurable folded reflectarray antenna based upon liquid crystal technology. *IEEE Transactions on Antennas and Propagation*, *63*(1), 122–132. doi:10.1109/TAP.2014.2367491

Bitchikh, M., Rili, W., & Mokhtar, M. (2018). An UWB to narrow band and BI-bands reconfigurable octagonal antenna. *Progress in Electromagnetics Research Letters*, *74*, 69–75. doi:10.2528/PIERL18011023

Bjornson, E., Ozdogan, O., & Larsson, E. G. (2020). Reconfigurable intelligent surfaces: Three myths and two critical questions. *IEEE Communications Magazine*, *58*(12), 90–96. doi:10.1109/MCOM.001.2000407

Björnson, E., Sanguinetti, L., Wymeersch, H., Hoydis, J., & Marzetta, T. L. (2019). Massive MIMO is a reality—What is next? *Digital Signal Processing*, *94*, 3–20. doi:10.1016/j.dsp.2019.06.007

Borzov, A. B., Likhoedenko, K. P., Seregin, G. M., & Suchkov, V. B. (2017, August). Applications of short-range radars on the base of ultrashort pulse single chip. In *2017 3rd IEEE International Conference on Control Science and Systems Engineering (ICCSSE)* (pp. 461-464). IEEE. 10.1109/CCSSE.2017.8088041

Boukarkar, A., Lin, X. Q., Jiang, Y., Chen, Y. J., Nie, L. Y., & Mei, P. (2018). Compact mechanically frequency and pattern reconfigurable patch antenna. *IET Microwaves, Antennas & Propagation*, *12*(12), 1864–1869. doi:10.1049/iet-map.2017.0917

Britto, E. C., Danasegaran, S. K., & Sagadevan, K. (2023). Performance Analysis of SSRR in High-Speed Terahertz Antenna for Biomedical Applications. In S. Mehta & A. Abougreen (Eds.), *Metamaterial Technology and Intelligent Metasurfaces for Wireless Communication Systems* (pp. 180–199). IGI Global. doi:10.4018/978-1-6684-8287-2.ch008

Brown, M., & Li, C. (2023). Advancement of PMCW Radar and Its Board-Level Prototyping. *2023 IEEE 16th Dallas Circuits and Systems Conference (DCAS)*, 1–4. 10.1109/DCAS57389.2023.10130242

Brown, E. R. (1998). RF-MEMS switches for reconfigurable integrated circuits. *IEEE Transactions on Microwave Theory and Techniques*, *46*(11), 1868–1880. doi:10.1109/22.734501

Bruder, J. A. (2013). IEEE Radar standards and the radar systems panel. *IEEE Aerospace and Electronic Systems Magazine*, *28*(7), 19–22. doi:10.1109/MAES.2013.6559377

Bshara, O. (2023). *Millimeter wave coarse beamforming using outband sub-6GHz reconfigurable antennas* (U.S. Patent No. 11,843,434). United States Patent and Trademark Office.

Buzzi, S., Grossi, E., Lops, M., & Venturino, L. (2022). Foundations of MIMO radar detection aided by reconfigurable intelligent surfaces. *IEEE Transactions on Signal Processing*, *70*, 1749–1763. doi:10.1109/TSP.2022.3157975

Cai, Y., Guo, Y. J., & Bird, T. S. (2012). A frequency reconfigurable printed Yagi-Uda dipole antenna for cognitive radio applications. *IEEE Transactions on Antennas and Propagation*, *60*(6), 2905–2912. doi:10.1109/TAP.2012.2194654

Cai, Y.-M., Gao, S., Yin, Y., Li, W., & Luo, Q. (2016). Compact-size low-profile wideband circularly polarized omnidirectional patch antenna with reconfigurable polarizations. *IEEE Transactions on Antennas and Propagation*, *64*(5), 2016–2021. doi:10.1109/TAP.2016.2535502

Chacko, P., Augustin, G., & Denidni, T. A. (2015). Electronically reconfigurable uniplanar antenna for cognitive radio applications. *IET Microwaves, Antennas & Propagation*, *9*(3), 213–216.

Chakrabarti, S., & Chakraborty, A. (2024). A Pattern Reconfigurable Antenna for Ka-band Application. *Journal of the Institution of Electronics and Telecommunication Engineers*, 1–13. doi:10.1080/03772063.2024.2303994

Chamok, N. H., Yılmaz, M. H., Arslan, A., & Ali, M. (2016). High-gain pattern reconfigurable MIMO antenna array for wireless handheld terminals. *IEEE Transactions on Antennas and Propagation*, *64*(10), 4306–4315. doi:10.1109/TAP.2016.2598201

Chen, H., Mi, D., Chu, Z., Xiao, P., Xu, Y., & He, D. (2020, August). Link-level performance of rate-splitting based downlink multiuser MISO systems. In *2020 IEEE 31st Annual International Symposium on Personal, Indoor and Mobile Radio Communications* (pp. 1-5). IEEE. 10.1109/PIMRC48278.2020.9217103

Chen, A., Ning, X., Wang, L., & Zhang, Z. (2017). A design of radiation pattern and polarization reconfigurable antenna using metasurface. *Proceedings of the 2017 IEEE Asia Pacific Microwave Conference (APMC)*. 10.1109/APMC.2017.8251389

Cheng, B., Cui, Z., Lu, B., Qin, Y., Liu, Q., Chen, P., He, Y., Jiang, J., He, X., Deng, X., Zhang, J., & Zhu, L. (2018). 340-GHz 3-D imaging radar with 4Tx-16Rx MIMO array. *IEEE Transactions on Terahertz Science and Technology*, *8*(5), 509–519. doi:10.1109/TTHZ.2018.2853551

Chen, M. S., Zhang, Z. X., & Wu, X. L. (2018). Design of frequency-and polarization-reconfigurable antenna based on the polarization conversion metasurface. *IEEE Antennas and Wireless Propagation Letters*, *17*(1), 78–81. doi:10.1109/LAWP.2020.3040509

Chen, M., Challita, U., Saad, W., Yin, C., & Debbah, M. (2019). Artificial neural networks-based machine learning for wireless networks: A tutorial. *IEEE Communications Surveys and Tutorials*, *21*(4), 3039–3071. doi:10.1109/COMST.2019.2926625

Chen, M., Zhang, Z. Q., Li, X. D., Liu, H. L., Chen, R. S., Shao, Q., & Huang, G. L. (2023, May). A Beam Reconfigurable Antenna Array Based on Highly Conductive Graphene-Assembled Films. In *2023 International Conference on Microwave and Millimeter Wave Technology (ICMMT)* (pp. 1-3). IEEE. 10.1109/ICMMT58241.2023.10276949

Chen, S. L., Qin, P. Y., Lin, W., & Guo, Y. J. (2018). Pattern-reconfigurable antenna with five switchable beams in elevation plane. *IEEE Antennas and Wireless Propagation Letters*, *17*(3), 454–457. doi:10.1109/LAWP.2018.2794990

Chen, S.-Y., Chu, Q.-X., & Shinohara, N. (2016). A bandwidth reconfigurable planar antenna for WLAN/WiMAX applications. *Proceedings of the Asia-Pacific Microwave Conference (APMC)*. 10.1109/APMC.2016.7931467

Chen, W., & Lu, Y. (2023). Reconfigurable intelligent surfaces: Unleashing potential for energy-efficient communication. *Journal of Wireless Systems*, *44*(3), 299–312.

Chen, Y., Guo, H., Liu, Y., Li, J., Zhan, Y., Wu, Q., & Li, M. (2023). An L-Slot Frequency Reconfigurable Antenna Based on MEMS Technology. *Micromachines*, *14*(10), 1945. doi:10.3390/mi14101945 PMID:37893382

Chen, Z., Tang, J., Zhang, X. Y., Wu, Q., Wang, Y., Jin, S., & Wong, K.-K. (2022). Offset learning-based channel estimation for intelligent reflecting surface-assisted indoor communication. *IEEE Journal of Selected Topics in Signal Processing*, *16*(1), 41–55. doi:10.1109/JSTSP.2021.3129350

Chidurala, S., & Rao, A. P. (2023, July). Frequency Reconfigurable Antenna with Partial Ground for WLAN, WiMAX, C-band, and Sub-6GHz Wireless Communications. In *2023 14th International Conference on Computing Communication and Networking Technologies (ICCCNT)* (pp. 1-4). IEEE.

Chi, L., Weng, Z., Qi, Y., & Drewniak, J. L. (2020). A 60 GHz PCB Wideband Antenna-in-Package for 5G/6G Applications. *IEEE Antennas and Wireless Propagation Letters*, *1225*(c), 1–1. doi:10.1109/LAWP.2020.3006873

Choi, D., Shin, C., Kim, N., & Shin, H. (2005). Design and SAR Analysis of Broadband PIFA with Triple Band. *Progress in Electromagnetics Research Symposium*, *1*(3), 290–293. doi:10.2529/PIERS041212211650

Choudhury, S. (2023). *Reconfigurable Antennas: Trends and applications*. IoP Series in Electromagnetics and Metamaterials. doi:10.1088/978-0-7503-5457-8

Chowdhury, M. Z., Shahjalal, M., Ahmed, S., & Jang, Y. M. (2020). 6G wireless communication systems: Applications, requirements, technologies, challenges, and research directions. *IEEE Open Journal of the Communications Society*, *1*, 957–975. doi:10.1109/OJCOMS.2020.3010270

Christodoulou, C. G., Tawk, Y., Lane, S. A., & Erwin, S. R. (2012). Reconfigurable antennas for wireless and space applications. *Proceedings of the IEEE*, *100*(7), 2250–2261. doi:10.1109/JPROC.2012.2188249

Clemente, A., Dussopt, L., Sauleau, R., Potier, P., & Pouliguen, P. (2012). 1-Bit reconfigurable unit cell based on PIN diodes for transmit-array applications in X -Band. *IEEE Transactions on Antennas and Propagation*, *60*(5), 2260–2269. doi:10.1109/TAP.2012.2189716

Clerckx, B., Joudeh, H., Hao, C., Dai, M., & Rassouli, B. (2016). Rate splitting for MIMO wireless networks: A promising PHY-layer strategy for LTE evolution. *IEEE Communications Magazine*, *54*(5), 98–105. doi:10.1109/MCOM.2016.7470942

Clerckx, B., Mao, Y., Schober, R., & Poor, H. V. (2019). Rate-splitting unifying SDMA, OMA, NOMA, and multicasting in MISO broadcast channel: A simple two-user rate analysis. *IEEE Wireless Communications Letters*, *9*(3), 349–353. doi:10.1109/LWC.2019.2954518

Cohen, D., Mishra, K. V., & Eldar, Y. C. (2018). Spectrum Sharing Radar: Coexistence via Xampling. *IEEE Transactions on Aerospace and Electronic Systems*, *54*(3), 1279–1296. doi:10.1109/TAES.2017.2780599

Correas-Serrano, D., Gomez-Diaz, J. S., Alù, A., & Melcón, A. Á. (2015). Electrically and magnetically biased graphene-based cylindrical waveguides: Analysis and applications as reconfigurable antennas. *IEEE Transactions on Terahertz Science and Technology*, *5*(6), 951–960. doi:10.1109/TTHZ.2015.2472985

Cummings, N. P. (2003). *Active Antenna Bandwidth Control Using Reconfigurable Antenna Elements* [Ph.D. Thesis]. Virginia Tech, Bradley Department of Electrical and Computer Engineering, Blacksburg, VA, USA.

Dadgarpour, A., Zarghooni, B., Virdee, B., & Denidni, T. A. (2014). Beam Tilting Antenna Using Integrated Metamaterial Loading. *IEEE Transactions on Antennas and Propagation*, *62*(May), 2874–2879. doi:10.1109/TAP.2014.2308516

Dahrouj, H., Alghamdi, R., Alwazani, H., Bahanshal, S., Ahmad, A. A., Faisal, A., Shalabi, R., Alhadrami, R., Subasi, A., Alnory, M., Kittaneh, O., & Shamma, J. S. (2021). An Overview of Machine Learning-Based Techniques for Solving Optimization Problems in Communications and Signal Processing. *IEEE Access : Practical Innovations, Open Solutions*, *9*, 74908–74938. doi:10.1109/ACCESS.2021.3079639

Dai, M., Clerckx, B., Gesbert, D., & Caire, G. (2016). A rate splitting strategy for massive MIMO with imperfect CSIT. *IEEE Transactions on Wireless Communications*, *15*(7), 4611–4624. doi:10.1109/TWC.2016.2543212

Danasegaran, S. K., Britto, E. C., Sagadevan, K., Paranthaman, M., Poonguzhali, S., & Krishnakumar, M. (2024). Design and Investigation of Photonic Crystal Antenna Performance for Expeditious Data Rate in Wireless Communication. *Brazilian Journal of Physics*, *54*(1), 31. doi:10.1007/s13538-023-01408-4

Dang, O., Amin, O., Shihada, B., & Alouini, M.-S. (2020). What should 6g be? *Nature Electronics*, *3*(1), 2520–1131. doi:10.1038/s41928-019-0355-6

Dekker, B., Jacobs, S., Kossen, A. S., Kruithof, M. C., Huizing, A. G., & Geurts, M. (2017, October). Gesture recognition with a low power FMCW radar and a deep convolutional neural network. In *2017 European Radar Conference (EURAD)* (pp. 163-166). IEEE. 10.23919/EURAD.2017.8249172

Delaune, D., Guan, N., & Ito, K. (2008). *Simple Multiband Antenna for Mobile Phone Application Based on a Dual-arm Monopole Structure*. PIERS Online, 4(1).

Deng, H., Guo, S., Jin, L., Cao, X., & Xu, X. (2022, April). A human gesture recognition method for through-the-wall MIMO radar. In *2022 7th International Conference on Computer and Communication Systems (ICCCS)* (pp. 239-244). IEEE. 10.1109/ICCCS55155.2022.9846748

Deshmukh, R., & Patel, S. (2021). A critical review of Reconfigurable Intelligent Surfaces in modern communication. *Journal of Electromagnetic Research and Applications*, *13*(2), 60–73.

Dey, S., Khampariya, P., & Jain, M. (2022). Analysis of various AI-based algorithms and learning perspectives for 5G wireless networks. *IJMEvol.*, *7*(5), 1234–124.

Dey, S., Panda, A. K., & Pati, C. (2023). Unlocking the potential of machine learning and deep learning algorithms in recent communication: AI is the best revolutionary tool in the current era. *CPPPT*, *23*(2), 2794–2805.

Dhouibi, A., Burokur, S. N., Lustrac, A., & Priou, A. (2012). Compact Metamaterial-Based Substrate-Integrated Luneburg Lens Antenna. *IEEE Transactions On Antennas And Propagation, VOL.*, *11*, 1504–1507.

Dixit, A. S., Shevada, L. K., Raut, H. D., Malekar, R. R., & Kumar, S. (2020). Fifth generation antennas: A bibliometric survey and future research directions. *Library Philosophy and Practice*, *2020*, 1–24.

Dizdar, O., & Clerckx, B. (2021, September). Rate splitting multiple access for multi-antenna multi-carrier joint communications and jamming. In 2021 Sensor Signal Processing for Defence Conference (SSPD) (pp. 1-5). IEEE. doi:10.1109/SSPD51364.2021.9541482

Dizdar, O., Mao, Y., Han, W., & Clerckx, B. (2020, August). Rate-splitting multiple access for downlink multi-antenna communications: Physical layer design and link-level simulations. In *2020 IEEE 31st Annual International Symposium on Personal, Indoor and Mobile Radio Communications* (pp. 1-6). IEEE.

Dizdar, O., Mao, Y., Han, W., & Clerckx, B. (2020, November). Rate-splitting multiple access: A new frontier for the PHY layer of 6G. In *2020 IEEE 92nd vehicular technology conference (VTC2020-Fall)* (pp. 1-7). IEEE.

Dizdar, O., Mao, Y., Xu, Y., Zhu, P., & Clerckx, B. (2021, September). Rate-splitting multiple access for enhanced URLLC and eMBB in 6G. In *2021 17th International Symposium on Wireless Communication Systems (ISWCS)* (pp. 1-6). IEEE.

Dizdar, O., & Clerckx, B. (2022). Rate-splitting multiple access for communications and jamming in multi-antenna multi-carrier cognitive radio systems. *IEEE Transactions on Information Forensics and Security*, *17*, 628–643. doi:10.1109/TIFS.2022.3149145

Dizdar, O., Kaushik, A., Clerckx, B., & Masouros, C. (2021, June). *Rate-splitting multiple access for joint radar-communications with low-resolution DACs. In 2021 IEEE international conference on communications workshops (ICC workshops)*. IEEE.

Dizdar, O., Kaushik, A., Clerckx, B., & Masouros, C. (2022). Energy efficient dual-functional radar-communication: Rate-splitting multiple access, low-resolution DACs, and RF chain selection. *IEEE Open Journal of the Communications Society*, *3*, 986–1006. doi:10.1109/OJCOMS.2022.3183950

Dizdar, O., Mao, Y., & Clerckx, B. (2021). Rate-splitting multiple access to mitigate the curse of mobility in (massive) MIMO networks. *IEEE Transactions on Communications*, *69*(10), 6765–6780. doi:10.1109/TCOMM.2021.3098695

Dokhanchi, S. H., Shankar, M. R. B., Stifter, T., & Ottersten, B. (2018). OFDM-based automotive joint radar-communication system. *2018 IEEE Radar Conference (RadarConf18)*, 902–907. 10.1109/RADAR.2018.8378680

Dokhanchi, S. H., Mysore, B. S., Mishra, K. V., & Ottersten, B. (2019). A mmWave Automotive Joint Radar-Communications System. *IEEE Transactions on Aerospace and Electronic Systems*, *55*(3), 1241–1260. doi:10.1109/TAES.2019.2899797

Donthu, N., Kumar, S., Mukherjee, D., Pandey, N., & Lim, W. M. (2021). How to conduct a bibliometric analysis: An overview and guidelines. *Journal of Business Research*, *133*, 285–296. doi:10.1016/j.jbusres.2021.04.070

Drinan, D. P., Zale, A. V., Webb, M. A. H., Taper, M. L., Shepard, B. B., & Kalinowski, S. T. (2012). Evidence of Local Adaptation in Westslope Cutthroat Trout. *Transactions of the American Fisheries Society*, *141*(4), 872–880. doi:10.1080/00028487.2012.675907

Duan. (2020). Evolution and innovation of antenna systems for beyond 5G and 6G. *Frontiers of Information Technology & Electronic Engineering, 21*(1), 1–3. doi:10.1631/FITEE.2010000

Dussopt, L., & Rebeiz, G. M. (2003, April). Intermodulation distortion and power handling in RF MEMS switches, varactors, and tunable filters. *IEEE Transactions on Antennas and Propagation*, *51*(4), 1247–1256.

Dwivedi. (2022). Design and Analysis of Metamaterial Waveguide Antenna for broadband applications. *ICT4SD*. . doi:10.1007/978-981-19-5224-1_42

Dwivedi, S. (2022). *ReconfigurableArray antenna*. Intech Open. doi:10.5772/intechopen.106343

Dwivedi, S. (2023). *Metamaterials-BasedAntennafor5GandBeyond*. IGI Global. doi:10.4018/978-1-6684-8287-2.ch001

Dyogtyev, A.V., Sukhoivanov, I.A., & De La Rue, R.M. (2010). Photonic band-gap maps for different two dimensionally periodic photonic crystal structures. *Journal of Applied Physics, 107*(1), 1-7.

Eduardo, V. P. (2022). FORMAT: A Reconfigurable Tile-Based Antenna Array System for 5G and 6G Millimeter-Wave Testbeds. *IEEE Systems Journal, 16*(3), 4489–4500. Advance online publication. doi:10.1109/JSYST.2022.3146360

Elhemaily, H. (2023). Reconfigurable Monopole Antenna Using Graphene Material for Wireless Applications. *Journal of Advanced Engineering Trends, 42*(2), 173–179. doi:10.21608/jaet.2022.111686.1134

Elijah, O., Abdul Rahim, S. K., New, W. K., Leow, C. Y., Cumanan, K., & Kim Geok, T. (2022). Intelligent Massive MIMO Systems for Beyond 5G Networks: An Overview and Future Trends. *IEEE Access : Practical Innovations, Open Solutions, 10*, 102532–102563. doi:10.1109/ACCESS.2022.3208284

Elsner, J., Blumenau, A., Frauenheim, T., Jones, R., & Heggie, M. (2000). Structural and Electronic Properties of Line Defects in GaN. *Materials Research Society Internet Journal of Nitride Semiconductor Research, 5*(1), 481–492. doi:10.1557/S1092578300004671

Emara, M. K., Stuhec-Leonard, S. K., Tomura, T., Hirokawa, J., & Gupta, S. (2020). Laser-Drilled All- Dielectric Huygens' Transmit-Arrays as 120 GHz Band Beamformers. *IEEE Access : Practical Innovations, Open Solutions, 8*, 153815–153825. doi:10.1109/ACCESS.2020.3018297

Erdil, E., Topalli, K., Unlu, M., Civi, O. A., & Akin, T. (2007, April). Frequency tunable patch antenna using RF MEMs technology. *IEEE Transactions on Antennas and Propagation, 55*(4), 1193–1196. doi:10.1109/TAP.2007.893426

Fang, Z., & Lu, C. (2023, November). Design of a 1-Bit Reconfigurable Low-Profile UWB Antenna Element. In *2023 IEEE MTT-S International Microwave Workshop Series on Advanced Materials and Processes for RF and THz Applications (IMWS-AMP)* (pp. 1-2). IEEE. 10.1109/IMWS-AMP57814.2023.10381461

Fang, Z., Yang, H., Gao, Y., Zhai, F., Wu, J. W., Cheng, Q., & Cui, T. J. (2023). Design of a 2-Bit Reconfigurable UWB Planar Antenna Array for Beam Scanning Application. *IEEE Open Journal of Antennas and Propagation, 4*, 91–96. doi:10.1109/OJAP.2023.3234541

Farahani, H. S., Veysi, M., Kamyab, M., & Tadjalli, A. (2010). Mutual coupling reduction in patch antenna arrays using a UC-EBG superstrate. *IEEE Antennas and Wireless Propagation Letters, 9*, 57–59. doi:10.1109/LAWP.2010.2042565

Farshad, R. (2023). *Analytical Characterization of Coverage Regions for STAR-RIS-aided NOMA/ OMA Communication Systems*. ArXiv.org, doi: /arXiv.2306.00404. doi:10.48550

Fathy, A. E., Rosen, A., Owen, H. S., McGinty, F., McGee, D. J., Taylor, G. C., Amenta, R., Swain, P. K., Perlow, S. M., & ElSherbiny, M. (2003). Silicon-based reconfigurable antennas-concepts, analysis, implementation, and feasibility. *IEEE Transactions on Microwave Theory and Techniques*, *51*(6), 1650–1661. doi:10.1109/TMTT.2003.812559

Feng, Z., Fang, Z., Wei, Z., Chen, X., Quan, Z., & Ji, D. (2020). Joint radar and communication: A survey. *China Communications*, *17*(1), 1–27. doi:10.23919/JCC.2020.01.001

Fernandes, C. A., Lima, E. B., & Costa, J. R. (2016). Dielectric lens antennas in Handbook of Antenna Technologies. Springer.

Fernandez, G. L., & Montes, R. H. (2022). Integrating advanced antenna technologies for sustainable wireless networks. *IEEE Transactions on Sustainable Computing*, *7*(2), 230–241.

Fhafhiem, N., Krachodnok, P., & Wongsan, R. (2012). The Circularly Polarized Resonator Antenna using Double Polarizing Metallic EBG. *IEEE Asia-Pacific Conference on Antennas and Propagation*. 10.1109/APCAP.2012.6333133

Fitri, I., & Akbart, A. A. (2018). A new gridded parasitic patch stacked microstrip antenna for enhanced wide bandwidth in 60 GHz Band. *2017 International Conference on Broadband Communication, Wireless Sensors and Powering, BCWSP 2017,* 1–4. 10.1109/BCWSP.2017.8272571

Flores, A. R., Clerckx, B., & De Lamare, R. C. (2018, August). Tomlinson-harashima precoded rate-splitting for multiuser multiple-antenna systems. In *2018 15th International Symposium on Wireless Communication Systems (ISWCS)* (pp. 1-6). IEEE. 10.1109/ISWCS.2018.8491245

Flores, A. R., de Lamare, R. C., & Clerckx, B. (2020). Linear precoding and stream combining for rate splitting in multiuser MIMO systems. *IEEE Communications Letters*, *24*(4), 890–894. doi:10.1109/LCOMM.2020.2969158

Foroozesh, A., & Shafai, L. (2006). Size reduction of a microstrip antenna with dielectric superstrate using meta-materials: Artificial magnetic conductors versus magneto- dielectrics. *Proc. IEEE Antennas Propag. Soc. Int. Symp.*, 11–14. 10.1109/APS.2006.1710439

Frist, Y., Elhabchi, M., & Srifi, M. N. (2023, October). A Novel Frequency Reconfigurable Antenna Based on Metamaterial for Millimeter-Wave 5G Applications. In *2023 10th International Conference on Wireless Networks and Mobile Communications (WINCOM)* (pp. 1-5). IEEE.

Fujimoto, K., & Nakamura, T. (2023). Emerging trends in antenna technology for 6G and beyond. *Future Wireless Systems Journal*, *6*(1), 9–24.

Fu, X., Yang, F., Liu, C., Wu, X., & Cui, T. J. (2020). Terahertz beam steering technologies: From phased arrays to field-programmable metasurfaces. *Advanced Optical Materials*, *8*(3), 1900628. doi:10.1002/adom.201900628

Galvan, L. A. H. (2023). *Simple Mechanically Reconfigurable Patch Antennas* [Doctoral dissertation]. The University of Texas Rio Grande Valley.

Gao, X., Xing, G., Roy, S., & Liu, H. (2019, November). Experiments with mmWave automotive radar test-bed. In *2019 53rd Asilomar conference on signals, systems, and computers* (pp. 1-6). IEEE. 10.1109/IEEECONF44664.2019.9048939

Ge, L., Dong, P., Zhang, H., Wang, J. B., & You, X. (2020). Joint beamforming and trajectory optimization for intelligent reflecting surfaces-assisted UAV communication. *IEEE Access : Practical Innovations, Open Solutions*, *8*, 78702–78712. doi:10.1109/ACCESS.2020.2990166

Gençoğlan, D. N., Çolak, Ş., & Palandöken, M. (2023). Spiral-resonator-based frequency reconfigurable antenna design for sub-6 GHz applications. *Applied Sciences (Basel, Switzerland)*, *13*(15), 8719. doi:10.3390/app13158719

Gençoğlan, D. N., Palandöken, M., & Çolak, Ş. (2023). Novel Frequency-Reconfigurable Antennas with Ring Resonators and RF Switches: Enhancing Versatility and Adaptability in Wireless Communication Systems. *Applied Sciences (Basel, Switzerland)*, *13*(18), 10237. doi:10.3390/app131810237

Ghani, U., & Rahman, F. (2022). A comparative study of traditional vs. Metamaterial Antennas for IoT devices. *IoT Systems and Applications*, *7*(3), 210–224.

Ghimire, J., Khan, D., & Choi, D. Y. (2023). Microstrip to Slot-Line-Fed Microstrip Patch Antenna with Radiation Pattern Diversity for X-Band Application. *Electronics (Basel)*, *12*(17), 3672. doi:10.3390/electronics12173672

Ghorbanian, M., Dolatabadi, S. H., Masjedi, M., & Siano, P. (2019). Communication in Smart Grids: A Comprehensive Review on the Existing and Future Communication and Information Infrastructures. *IEEE Systems Journal*, *13*(4), 4001–4014. doi:10.1109/JSYST.2019.2928090

Giacomo, Benoni, Rocca, Salucci, & Massa. (2023). *RIS Dynamic Control Strategies in Smart EM Environments for 6G Networks*. . doi:10.23919/EuCAP57121.2023.10133521

Gök, Ç., & Alatan, L. (2023, March). A Compact Pattern Reconfigurable Antenna with 4 Different Radiation Modes. In *2023 17th European Conference on Antennas and Propagation (EuCAP)* (pp. 1-5). IEEE.

Gopalakrishnan, R., & Gunasekaran, N. (2005). Design of equilateral triangular microstrip anten using artificial neural networks. *IEEE International Workshop on Antenna Technology: Small Antennas and Novel Metamaterials*, 246 - 249.

Grant, A. J., Rimoldi, B., Urbanke, R. L., & Whiting, P. A. (2001). Rate-splitting multiple access for discrete memoryless channels. *IEEE Transactions on Information Theory*, *47*(3), 873–890. doi:10.1109/18.915637

Grau, A., Romeu, J., Ming-Jer Lee, Blanch, S., Jofre, L., & De Flaviis, F. (2010, January). A dual linearly polarized MEMS-reconfigurable antenna for narrow band MIMO communication systems. *IEEE Transactions on Antennas and Propagation*, *58*(1), 4–16. doi:10.1109/TAP.2009.2036197

Griffiths, H., Cohen, L., Watts, S., Mokole, E., Baker, C., Wicks, M., & Blunt, S. (2015). Radar Spectrum Engineering and Management: Technical and Regulatory Issues. *Proceedings of the IEEE, 103*(1), 85–102. doi:10.1109/JPROC.2014.2365517

Griguer, H. (2009). Patch Antenna Bandwidth Enhancement through the Use of Metamaterials. *International Conference on Telecommunications*, 323-327.

Gross, F. B. (2015). *Smart antennas with MATLAB*. McGraw-Hill Education.

Gu, C., Gao, S., Liu, H., Luo, Q., Loh, T.-H., Sobhy, M., ... Wei, G. (2015). Compact smart antenna with electronic beam-switching and reconfigurable polarizations. *IEEE Transactions on Antennas and Propagation, 63*(12), 5325–5333. doi:10.1109/TAP.2015.2490239

Gunasekaran, V., & Harmantzis, F. C. (2007). Emerging wireless technologies for developing countries. *Technology in Society, 29*(1), 23–42. doi:10.1016/j.techsoc.2006.10.001

Guo, L., Huang, F., & Tang, X. (2014). A novel integrated MEMS helix antenna for terahertz applications. *Optik (Stuttgart), 125*(1), 101–103. doi:10.1016/j.ijleo.2013.06.016

Gupta, L., Jain, R., & Vaszkun, G. (2015). Survey of important issues in UAV communication networks. *IEEE Communications Surveys and Tutorials, 18*(2), 1123–1152. doi:10.1109/COMST.2015.2495297

Ha, N., Shin, W., Vaezi, M., & Poor, H. V. (2020, November). Coordinated rate splitting multiple access for multi-cell downlink networks. In *2020 54th Asilomar Conference on Signals, Systems, and Computers* (pp. 996-1001). IEEE. 10.1109/IEEECONF51394.2020.9443398

Haider, N., Caratelli, D., & Yarovoy, A. (2013). Recent developments in reconfigurable and multiband antenna technology. *International Journal of Antennas and Propagation, 2013*, 1–14. doi:10.1155/2013/869170

Han, L., Wang, C., Zhang, W., Ma, R., & Zeng, Q. (2018). Design of frequency- and pattern-reconfigurable wideband slot antenna. *International Journal of Antennas and Propagation, 2018*, 1–7. doi:10.1155/2018/3678018

Hannula, J.-M., Saarinen, T., Holopainen, J., & Viikari, V. (2017). Frequency reconfigurable multiband handset antenna based on a multichannel transceiver. *IEEE Transactions on Antennas and Propagation, 65*(9), 65. doi:10.1109/TAP.2017.2725384

Hao, Z., Wang, R., Peng, J., & Dang, X. (2023). Static Hand Gesture Recognition Based on Millimeter-Wave Near-Field FMCW-SAR Imaging. *Electronics (Basel), 12*(19), 4013. doi:10.3390/electronics12194013

HarzingA. W. (2007). *Publish or Perish*. Available from http://www.harzing.com/pop.htm

Hasan, L., Nassr, M., Anbar, M., & Alasadi, H. A. A. (2023). Inverted U-shaped Frequency Reconfigurable Microstrip patch antenna for 5G communication systems. *Óptica Pura y Aplicada, 56*(3).

Hassan Mirzaei & Eleftheriades. (2011). Frequency Reconfigurable Metamaterial Inspired Antenna. IEEE Antennas and Wireless Propagation Letters, 10.

Hassanien, A., Amin, M. G., Zhang, Y. D., & Ahmad, F. (2016). Dual-Function Radar-Communications: Information Embedding Using Sidelobe Control and Waveform Diversity. *IEEE Transactions on Signal Processing*, *64*(8), 2168–2181. doi:10.1109/TSP.2015.2505667

Haupt, R. L., & Lanagan, M. (2013). Reconfigurable antennas. *IEEE Antennas & Propagation Magazine*, *55*(1), 49–61. doi:10.1109/MAP.2013.6474484

Hazra, S., & Santra, A. (2018). Robust gesture recognition using millimetric-wave radar system. *IEEE Sensors Letters*, *2*(4), 1–4. doi:10.1109/LSENS.2018.2882642

Hernandez Galvan, L. A. (2023). Simple Mechanically Reconfigurable Patch Antennas. *IEEE Antennas and Wireless Propagation Letters*, *17*(7), 1186–1189.

Hinsz, L., & Braaten, B. D. (2014). A frequency reconfigurable transmitter antenna with autonomous switching capabilities. *IEEE Transactions on Antennas and Propagation*, *62*(7), 3809–3813. doi:10.1109/TAP.2014.2316298

Homod, R. Z., Mohammed, H. I., Abderrahmane, A., Alawi, O. A., Khalaf, O. I., Mahdi, J. M., Guedri, K., Dhaidan, N. S., Albahri, A. S., Sadeq, A. M., & Yaseen, Z. M. (2023, December). Deep clustering of Lagrangian trajectory for multi-task learning to energy saving in intelligent buildings using cooperative multi-agent. *Applied Energy*, *351*(1), 121843. doi:10.1016/j.apenergy.2023.121843

Hong. (2022). Review of antenna technology using metasurfaces. *Electromagnet. Tech. KIEES*, *33*, 20–53.

Hong, I.-P. (2023). Reviews Based on the Reconfigurable Intelligent Surface Technical Issues. *Electronics (Basel)*, *12*(21), 4489. doi:10.3390/electronics12214489

Hong, W., Jiang, Z. H., Yu, C., Zhou, J., Chen, P., Yu, Z., Zhang, H., Yang, B., Pang, X., Jiang, M., Cheng, Y., Al-Nuaimi, M. K. T., Zhang, Y., Chen, J., & He, S. (2017, December). Multibeam antenna technologies for 5G wireless communications. *IEEE Transactions on Antennas and Propagation*, *65*(12), 6231–6249. doi:10.1109/TAP.2017.2712819

Horestani, A. K., Shaterian, Z., Naqui, J., Martín, F., & Fumeaux, C. (2016). Reconfigurable and tunable S-shaped split-ring resonators and application in band-notched UWB antennas. *IEEE Transactions on Antennas and Propagation*, *64*(9), 3766–3776. doi:10.1109/TAP.2016.2585183

Hossain, M. A., Rahman, M. A., Murshed, A. H., Nishiyama, E., & Toyoda, I. (2023). A simple design and fabrication of polarization reconfigurable antenna for industrial scientific and medical-band applications. International. *Journal of Electrical and Computer Engineering*, *13*(2), 1580.

Hoydis, F. A., Aoudia, F. A., Valcarce, A., & Viswanathan, H. (2021). Toward a 6G AI-native air interface. *IEEE Communications Magazine*, *59*(5), 76–81. doi:10.1109/MCOM.001.2001187

Huang, H.-C. (2018). Overview of antenna designs and considerations in 5G cellular phones. *Proc. Int. Workshop Antenna Technol. (iWAT),* 1–4. 10.1109/IWAT.2018.8379253

Huang, Liu, & Gong. (2015). Multiband Metamaterial-Loaded Monopole Antenna for WLAN/WiMAX Applications. *IEEE Antennas and Wireless Propagation Letters, 14*, 662-665.

Huang, R., Luo, Y., Zhao, S., Yan, N., An, W., & Ma, K. (2023, August). A Compact Anti-Interference Pattern Reconfigurable Quasi-Yagi Antenna. In *2023 International Applied Computational Electromagnetics Society Symposium (ACES-China)* (pp. 1-3). IEEE. 10.23919/ACES-China60289.2023.10250083

Huang, Y. C., Ma, X., Pan, W., & Luo, X. (2014). A low profile polarization reconfigurable dipole antenna using tunable electromagnetic band-gap surface. *Microwave and Optical Technology Letters*, *56*(6), 1281–1285. doi:10.1002/mop.29074

Huff, G. H., Bahukudumbi, P. B., Everett, W. N., Beskok, A., Bevan, M. A., Lagoudas, D., & Ounaies, Z. (2007). Microfluidic reconfiguration of antennas. *Proc. Antenna Appl. Symp.,* 241-258.

Huff, G. H., Feng, J., Zhang, S., & Bernhard, J. T. (2003). A novel radiation pattern and frequency reconfigurable single turn square spiral microstrip antenna. *IEEE Microwave and Wireless Components Letters*, *13*(2), 57–59. doi:10.1109/LMWC.2003.808714

Hum, S. V., & Perruisseau-Carrier, J. (2013). Reconfigurable reflect arrays and array lenses for dynamic antenna beam control: A review. *IEEE Transactions on Antennas and Propagation*, *62*(1), 183–198. doi:10.1109/TAP.2013.2287296

Huseien, G. F., & Shah, K. W. (2022). A review on 5G technology for smart energy management and smart buildings in Singapore. *Energy and AI*, *7*, 100116. doi:10.1016/j.egyai.2021.100116

Hussain, M., Awan, W. A., Alzaidi, M. S., Hussain, N., Ali, E. M., & Falcone, F. (2023). Metamaterials and their application in the performance enhancement of reconfigurable antennas: A review. *Micromachines*, *14*(2), 349. doi:10.3390/mi14020349 PMID:36838049

Hussain, R., Khan, M. U., & Sharawi, M. S. (2018). An integrated dual MIMO antenna system with dual-function GND-plane frequency-agile antenna. *IEEE Antennas and Wireless Propagation Letters*, *17*(1), 142–145. doi:10.1109/LAWP.2017.2778182

Hussain, R., & Sharawi, M. S. (2015). A cognitive radio reconfigurable MIMO and sensing antenna system. *IEEE Antennas and Wireless Propagation Letters*, *14*, 257–260. doi:10.1109/LAWP.2014.2361450

Hussain, R., & Sharawi, M. S. (2015). Integrated reconfigurable multiple-input–multiple-output antenna system with an ultra-wideband sensing antenna for cognitive radio platforms. *IET Microwaves, Antennas & Propagation*, *9*(9), 940–947. doi:10.1049/iet-map.2014.0605

Hussain, R., & Sharawi, M. S. (2017). An integrated frequency reconfigurable antenna for cognitive radio application. *Wuxiandian Gongcheng*, *26*(3), 746–754.

Hussain, R., Sharawi, M. S., & Shamim, A. (2018). An integrated four-element slot-based MIMO and a UWB sensing antenna system for CR platforms. *IEEE Transactions on Antennas and Propagation*, *66*(2), 978–983. doi:10.1109/TAP.2017.2781220

Instruments, T. (2020). *The fundamentals of millimeter wave radar sensors*. Academic Press.

Iqbal, A., & Khan, Z. (2021). Sustainability in wireless networks: Role of advanced antenna systems. *Green Communications Review*, *12*(4), 245–260.

Iqbal, A., Smida, A., Abdulrazak, L. F., Saraereh, O. A., Mallat, N. K., Elfergani, I., & Kim, S. (2019). Low-profile frequency reconfigurable antenna for heterogeneous wireless systems. *Electronics (Basel)*, *8*(9), 976. doi:10.3390/electronics8090976

Islam, M. Z., & Hossain, M. S., ul Islam, R., & Andersson, K. (2019, May). Static hand gesture recognition using convolutional neural network with data augmentation. In *2019 Joint 8th International Conference on Informatics, Electronics & Vision (ICIEV) and 2019 3rd International Conference on Imaging, Vision & Pattern Recognition (icIVPR)* (pp. 324-329). IEEE.

Ismail, M. M., Elwi, T. A., & Salim, A. J. (2023). Reconfigurable composite right/left-handed transmission line antenna-based Hilbert/Minkowski stepped impedance resonator for wireless applications. *Radioelectronic and Computer Systems*, (1), 77–91. doi:10.32620/reks.2023.1.06

Ito, H., & Tanaka, S. (2020). Challenges and breakthroughs in next-gen wireless communication: A focus on antenna technologies. *Communications Research Letters*, *15*(4), 657–665.

Jaafar, Kadhim, & Hasan. (2023). Design and Implementation Unit Cell for 6G Reconfigurable Intelligent Surface Application. *International Journal of Online Engineering*. doi:10.3991/ijoe.v19i05.37585

Jaafar, H., Sidek, O., Miskam, A., & Abdul-Rahman, R. (2013). High-isolation and low-loss RF MEMS shunt switches. ACES J.-. *Appl. Comput. Electromagnet. Soc.*, *25*, 780–786.

Jabbar, A., Jamshed, M. A., Abbasi, Q., Imran, M. A., & Ur-Rehman, M. (2023). Leveraging the Role of Dynamic Reconfigurable Antennas in Viewpoint of Industry 4.0 and Beyond. *Research*, *6*, 0110.

Janagouda, M., Badagavi, P., & Iyer, S. (2023, October). Comparative Study of Origami Enabled Helical and Curlicue Reconfigurable Antennas. In *2023 IEEE International Conference on Distributed Computing, VLSI, Electrical Circuits and Robotics (DISCOVER)* (pp. 269-274). IEEE. 10.1109/DISCOVER58830.2023.10316702

Jankiraman, M. (2018). *FMCW radar design*. Artech House.

Jhaung, Y. C., Lin, Y. M., Zha, C., Leu, J. S., & Köppen, M. (2022). Implementing a hand gesture recognition system based on range-doppler map. *Sensors (Basel)*, *22*(11), 4260. doi:10.3390/s22114260 PMID:35684880

Jiang. (2014). A Low-Profile High-Gain Substrate-Integrated Waveguide Slot Antenna Enabled by an Ultrathin Anisotropic Zero-Index Metamaterial Coating. *IEEE Transactions On Antennas And Propagation, 62*(March), 1173–1184.

Jiang, M., Cezanne, J., Sampath, A., Shental, O., Wu, Q., Koymen, O., Bedewy, A., & Li, J. (2022). Wireless front haul for 5G and future radio access networks: Challenges and enabling technologies. *IEEE Wireless Communications, 29*(2), 108–114. doi:10.1109/MWC.003.2100482

Jiang, R., Fei, Z., Huang, S., Wang, X., Wu, Q., & Ren, S. (2023). Bivariate Pilot Optimization for Compressed Channel Estimation in RIS-Assisted Multiuser MISO-OFDM Systems. *IEEE Transactions on Vehicular Technology, 72*(7), 9115–9130. doi:10.1109/TVT.2023.3250252

Jiang, W., Han, B., Habibi, M. A., & Schotten, H. D. (2021). The Road Toward 6G: A Comprehensive Survey. *IEEE Open Journal of the Communications Society, 2*, 334–366. doi:10.1109/OJCOMS.2021.3057679

Ji, B., Han, Y., Liu, S., Tao, F., Zhang, G., Fu, Z., & Li, C. (2021). Several key technologies for 6G: Challenges and opportunities. *IEEE Commun. Stand. Mag., 5*(2), 44–51. doi:10.1109/MCOMSTD.001.2000038

Jin, G., Li, M., Liu, D., & Zeng, G. (2018). A simple planar pattern-reconfigurable antenna based on arc dipoles. *IEEE Antennas and Wireless Propagation Letters, 17*(9), 1664–1668. doi:10.1109/LAWP.2018.2862624

Johnson, S. G., Fan, S., Villeneuve, P. R., Joannopoulos, J. D., & Kolodziejski, L. A. (1999). Guided modes in photonic crystal slabs. *Physical Review B: Condensed Matter, 60*(8), 5751–5758. doi:10.1103/PhysRevB.60.5751

Joshi, J. G., Pattnaik, S. S., & Devi, S. (2011). *Metamaterial loaded dual band microstrip patch antenna. Antenna Week.*

Joshi, J. G., Pattnaik, S. S., Devi, S., Lohokare, M. R., & Vidyasagar, C. (2009). Offset fed diamond shaped split ring (DSSR) planar metamaterial antenna. *IEEE Conference on Applied Electromagnetics*, 1-4. 10.1109/AEMC.2009.5430686

Jouvaud, C., de Rosny, J., & Ourir, A. (2013). Adaptive metamaterial antenna using coupled tunable split-ring resonators. *Electronics Letters, 49*(8), 518–519. doi:10.1049/el.2013.0398

JungSook, Khalid, Lee, Lee, Noh, & Yu. (2024). Overview of RIS-enabled secure transmission in 6G wireless networks. *Digital Communications and Networks*. doi:10.1016/j.dcan.2024.02.005

Kamran Shereen, M., Khattak, M. I., Basit, A., & Ahmad, G. (2023). Intelligent Frequency, Radiation Pattern and Polarization Reconfigurable Antenna for 5G Applications. *Journal of Electrical Engineering & Technology*, 1–15. doi:10.1007/s42835-023-01718-2

Kandasamy, K., Majumder, B., Mukherjee, J., & Ray, K. P. (2015). Low-RCS and polarization-reconfigurable antenna using cross-slot-based metasurface. *IEEE Antennas and Wireless Propagation Letters, 14*, 1638–1641. doi:10.1109/LAWP.2015.2415585

Kang, M. J., Kim, T. H., Choi, Y. S., & Lee, W. S. (2023, July). Low-Profile Beam-Reconfigurable Antenna Using Multiple Parasitic Elements for UAV Applications. In *2023 IEEE International Symposium on Antennas and Propagation and USNC-URSI Radio Science Meeting (USNC-URSI)* (pp. 621-622). IEEE. 10.1109/USNC-URSI52151.2023.10238150

Karra, J. M., & Eroglu, A. (2023, July). Performance Improvement of Wideband Reconfigurable Planar Antennas. In *2023 IEEE International Symposium on Antennas and Propagation and USNC-URSI Radio Science Meeting (USNC-URSI)* (pp. 1075-1076). IEEE. 10.1109/USNC-URSI52151.2023.10237838

Karthika, K., Anusha, K., Kavitha, K., & Geetha, D. M. (2024). Optimization Algorithms for Reconfigurable Antenna Design: A Review. *Advances in Microwave Engineering*, 85-103.

Karthika, K., & Kavitha, K. (2023). Design and Development of Parasitic Elements Loaded Quadband Frequency and Pattern Reconfigurable Antenna. *International Journal of RF and Microwave Computer-Aided Engineering, 2023*, 2023. doi:10.1155/2023/4034241

Karthikeyan, M., Pradeep, J., Harikrishnan, M., Sitharthan, R., Mishra, N., Rajesh, M., & Sivanesan, T. (2023). Recent developments in flexible and printed reconfigurable antennas for medical and Internet of things applications. *Advances in Flexible and Printed Electronics: Materials, fabrication, and applications*, 9-1.

Karuppanan, S., Vinothini, V. R., & Chenguttuvan, E. (2023). Reconfigured antenna with switchable polarization for S Band Wireless applications. *El-Cezeri Journal of Science and Engineering, 10*(2), 231–239. doi:10.31202/ecjse.1137882

Kataja, J., & Nikoskinen, K. (2011). The Parametric Optimization of Wire Dipole Antenna. *IEEE Transactions on Antennas and Propagation, 59*(2), 350–356. doi:10.1109/TAP.2010.2096181

Kaur, J., Khan, M. A., Iftikhar, M., Imran, M., & Haq, Q. (2021). Machine learning techniques for 5G and beyond. *IEEE Access : Practical Innovations, Open Solutions, 9*, 1–1. doi:10.1109/ACCESS.2021.3051557

Kaushik, A., Masouros, C., & Liu, F. (2021). Hardware Efficient Joint Radar-Communications with Hybrid Precoding and RF Chain Optimization. *ICC 2021 - IEEE International Conference on Communications*, 1–6. 10.1109/ICC42927.2021.9500661

Khan, F., Leem, S. K., & Cho, S. H. (2017). Hand-based gesture recognition for vehicular applications using IR-UWB radar. *Sensors (Basel), 17*(4), 833. doi:10.3390/s17040833 PMID:28398267

Khidre, A., Lee, K. F., Yang, F., & Elsherbeni, A. Z. (2012). Circular polarization reconfigurable wideband E-shaped patch antenna for wireless applications. *IEEE Transactions on Antennas and Propagation, 61*(2), 960–964. doi:10.1109/TAP.2012.2223436

Kim, Y., & Toomajian, B. (2017, March). Application of Doppler radar for the recognition of hand gestures using optimized deep convolutional neural networks. In *2017 11th European conference on antennas and propagation (EUCAP)* (pp. 1258-1260). IEEE. 10.23919/EuCAP.2017.7928465

Kim. (2008). The Design of a Small Monopole Type Antenna using Negative Permittivity Metamaterial Structure. *Antenna Technology: Small Antennas and Novel Metamaterials, iWAT 2008. International Workshop on*, 275-278.

Kim, G. W., Lee, S. W., Son, H. Y., & Choi, K. W. (2023). A study on 3D human pose estimation using through-wall IR-UWB radar and transformer. *IEEE Access : Practical Innovations, Open Solutions*, *11*, 15082–15095. doi:10.1109/ACCESS.2023.3244017

Kim, S. Y., Han, H. G., Kim, J. W., Lee, S., & Kim, T. W. (2017). A hand gesture recognition sensor using reflected impulses. *IEEE Sensors Journal*, *17*(10), 2975–2976. doi:10.1109/JSEN.2017.2679220

Kim, Y., & Toomajian, B. (2016). Hand gesture recognition using micro-Doppler signatures with convolutional neural network. *IEEE Access : Practical Innovations, Open Solutions*, *4*, 7125–7130. doi:10.1109/ACCESS.2016.2617282

Kingsly, S., Thangarasu, D., Kanagasabai, M., Alsath, M. G., Thipparaju, R. R., Palaniswamy, S. K., & Sambandam, P. (2018). Multiband reconfigurable filtering monopole antenna for cognitive radio applications. *IEEE Antennas and Wireless Propagation Letters*, *17*(8), 1416–1420. doi:10.1109/LAWP.2018.2848702

Knight, J. C., Birks, T. A., Russell, P. J., & Sandro, J. P. (1998). Properties of photonic crystal fiber and the effective index model. *Journal of the Optical Society of America. A, Optics, Image Science, and Vision*, *15*(3), 748–752. doi:10.1364/JOSAA.15.000748

Kothapudi, V. K., & Kumar, V. (2017). A Single Layer S/X-Band Series-Fed Shared Aperture Antenna for SAR Applications. *Progress In Electromagnetics Research (PIER-C), 76,* 207–219.

Kothapudi, V. K., & Kumar, V. (2018). A 6-port Two Dimensional 3x3 Series-fed Planar Array Antenna for Dual Polarized X-band Airborne Synthetic Aperture Radar Applications. *IEEE Access, 6,* 12001-12007.

Kothapudi, V. K., & Kumar, V. (2018). Compact 1 × 2 and 2 × 2 Dual Polarized Series-Fed Antenna Array for X-Band Airborne Synthetic Aperture Radar Applications. *Journal of Electromagnetic Engineering and Science*, *18*(2), 117–128. doi:10.26866/jees.2018.18.2.117

Kothapudi, V. K., & Kumar, V. (2019). SFCFOS Uniform and Chebyshev amplitude distribution linear array antenna for K-band Radar applications. *Journal of Electromagnetic Engineering and Science*, *19*(1), 64–70. doi:10.26866/jees.2019.19.1.64

Kothapudi, V. K., & Kumar, V. (2021). Hybrid-fed shared aperture antenna array for X/K-band airborne synthetic aperture radar applications. *IET Microwaves, Antennas & Propagation*, *15*(1), 93–103. doi:10.1049/mia2.12024

Kounta, C. A. K. A., Arnaud, L., Kamsu-Foguem, B., & Tangara, F. (2022). Review of ai-based methods for chatter detection in machining based on bibliometric analysis. *International Journal of Advanced Manufacturing Technology*, *122*(5-6), 2161–2186. doi:10.1007/s00170-022-10059-9

Kovitz, J. M., Rajagopalan, H., & Rahmat-Samii, Y. (2015). Design and implementation of broadband MEMS RHCP/LHCP reconfigurable arrays using rotated E-shaped patch elements. *IEEE Transactions on Antennas and Propagation*, *63*(6), 2497–2507. doi:10.1109/TAP.2015.2417892

Kumari, P., Choi, J., Gonzalez-Prelcic, N., & Heath, R. W. (2018). IEEE 802.11ad-Based Radar: An Approach to Joint Vehicular Communication-Radar System. *IEEE Transactions on Vehicular Technology*, *67*(4), 3012–3027. doi:10.1109/TVT.2017.2774762

Kumar, M. S., & Choukiker, Y. K. (2017). Frequency and polarization reconfigurable antenna using BLC feed network. *Proceedings of the 2017 IEEE International Conference on Antenna Innovations & Modern Technologies for Ground, Aircraft and Satellite Applications (iAIM)*. 10.1109/IAIM.2017.8402550

Kumar, P. R., Sunitha, P., & Prasad, M. V. (2023). Compact Reconfigurable Patch Antenna for Wireless Applications. *Progress in Electromagnetics Research, C*, 138, 161–174.

Kumar, R., & Malhotra, J. (2023). The interplay between Beamforming and Massive MIMO: A comprehensive analysis. *Journal of Advanced Wireless Systems*, *5*(2), 120–134.

Kumar, S., Rayavarapu, N., & Naidu, P. V. (2023). A novel frequency reconfigurable antenna for smart grid applications in TV white space band. *Iranian Journal of Electrical and Computer Engineering*, *13*(1), 611–618. doi:10.11591/ijece.v13i1.pp611-618

Labadi, M., Jamel, B. T., & Choubani, F. (2011). New design of antenna array using Left Handed Meta-Material "LHM" based on circular Split Ring Resonator "SRR". *Mediterranean Microwave Symposium (MMS)*, 52-56.

Lalithakumari, S., Danasegaran, S. K., Rajalakshmi, G., Pandian, R., & Britto, E. C. (2023). Analysis of Wave Propagation in Hybrid Metamaterial Structure for Terahertz Applications. *Brazilian Journal of Physics*, *53*(5), 140. doi:10.1007/s13538-023-01351-4

Larsson, E. G., Edfors, O., Tufvesson, F., & Marzetta, T. L. (2014). Massive MIMO for next generation wireless systems. *IEEE Communications Magazine*, *52*(2), 186–195. doi:10.1109/MCOM.2014.6736761

Lee, Y., Ga, D., & Choi, J. (2011). Design of a MIMO Antenna with Improved Isolation Using MNG Metamaterial. *International workshop on Antenna Technology (iWAT)*, 231 – 234.

Lee, C. H. L., & Itoh, T. (2006). Compact dual-band antenna using an anisotropic metamaterial. *Proceedings of the 36th European Microwave Conference*, 1044-1047. 10.1109/EUMC.2006.281112

Lee, J. H., & Kim, D. W. (2022). An overview of 5G and beyond: Challenges and advancements. *IEEE Journal on Wireless Communication Systems*, *39*(7), 2450–2463.

Leem, S. K., Khan, F., & Cho, S. H. (2019). Detecting mid-air gestures for digit writing with radio sensors and a CNN. *IEEE Transactions on Instrumentation and Measurement*, *69*(4), 1066–1081. doi:10.1109/TIM.2019.2909249

Lee, S., & Park, Y. (2021). Innovations in antenna technology: A pathway to robust wireless communication. *Advanced Communication Reviews*, *14*(5), 486–499.

Letaief, K. B., Chen, W., Shi, Y., Zhang, J., & Zhang, Y. J. A. (2019, August). The Roadmap to 6G: AI Empowered Wireless Networks. *IEEE Communications Magazine*, *57*(8), 84–90. doi:10.1109/MCOM.2019.1900271

Li, G., Zhang, Z., Yang, H., Pan, J., Chen, D., & Zhang, J. (2020, March). Capturing human pose using mmWave radar. In *2020 IEEE International Conference on Pervasive Computing and Communications Workshops (PerCom Workshops)* (pp. 1-6). IEEE. 10.1109/PerComWorkshops48775.2020.9156151

Li. (2013). Design of Electrically Small Metamaterial Antenna With ELC and EBG Loading. *IEEE Antennas and Wireless Propagation Letters, 12*, 678-681.

Liang, B., Sanz-Izquierdo, B., Parker, E. A., & Batchelor, J. C. (2015). A frequency and polarization reconfigurable circularly polarized antenna using active EBG structure for satellite navigation. *IEEE Transactions on Antennas and Propagation*, *63*(1), 33–40. doi:10.1109/TAP.2014.2367537

Liang, C., Li, Y., Hu, X., Wang, Y., Zhang, L., Wang, M., & Guo, J. (2023). Coherent-on-Receive Synthesis Using Dominant Scatterer in Millimeter-Wave Distributed Coherent Aperture Radar. *Remote Sensing (Basel)*, *15*(6), 1505. doi:10.3390/rs15061505

Li, H., Liu, Y., Liao, G., & Chen, Y. (2023). Joint Radar and Communications Waveform Design Based on Complementary Sequence Sets. *Remote Sensing (Basel)*, *15*(3), 645. doi:10.3390/rs15030645

Li, H., Shen, S., & Clerckx, B. (2023). Beyond diagonal reconfigurable intelligent surfaces: From transmitting and reflecting modes to single-, group-, and fully-connected architectures. *IEEE Transactions on Wireless Communications*, *22*(4), 2311–2324. doi:10.1109/TWC.2022.3210706

Li, L.-W. (2008). Design of a Novel Rectangular Patch Antenna with Planar Metamaterial Patterned Substrat. *International Workshop on Antenna Technology: Small Antennas and Novel Metamaterials,* 124-126.

Li, M., Li, Q. L., Wang, B., Zhou, C. F., & Cheung, S. W. (2018). A low-profile dual-polarized dipole antenna using wideband AMC reflector. *IEEE Transactions on Antennas and Propagation*, *66*(5), 2610–2615. doi:10.1109/TAP.2018.2806424

Lin, T. C., & Hung, C. M. (2023). Reconfigurable RF front end and antenna arrays for radar mode switching (U.S. Patent No. 11,835,645). *United States Patent and Trademark Office.*

Lin, W., Chen, S. L., Ziolkowski, R. W., & Guo, Y. J. (2018). Reconfigurable, wideband, low-profile, circularly polarized antenna and array enabled by artificial magnetic conductor ground. *IEEE Transactions on Antennas and Propagation*, *66*(3), 1564–1569. doi:10.1109/TAP.2018.2790437

Lin, W., & Wong, H. (2015). Polarization reconfigurable wheel-shaped antenna with conical-beam radiation pattern. *IEEE Transactions on Antennas and Propagation*, *63*(2), 491–499. doi:10.1109/TAP.2014.2381263

Lin, W., Wong, H., & Ziolkowski, R. W. (2017). Wideband pattern-reconfigurable antenna with switchable broadside and conical beams. *IEEE Antennas and Wireless Propagation Letters, 16*, 2638–2641. doi:10.1109/LAWP.2017.2738101

Lin, W., Wong, H., & Ziolkowski, R. W. (2018). Circularly-polarized antenna with reconfigurable broadside and conical beams facilitated by a mode switchable feed network. *IEEE Transactions on Antennas and Propagation, 66*(2), 996–1001. doi:10.1109/TAP.2017.2784452

Li, P. K., Shao, Z. H., Wang, Q., & Cheng, Y. J. (2015). Frequency- and pattern reconfigurable antenna for multistandard wireless applications. *IEEE Antennas and Wireless Propagation Letters, 14*, 333–336. doi:10.1109/LAWP.2014.2359196

Liu, C. C., & Wu, C. C. (2014). Analysis of defect mode in a dielectric photonic crystal containing ITO defect. *Optik (Stuttgart), 125*(24), 7140–7142. doi:10.1016/j.ijleo.2014.07.120

Liu, D. (2023, June). Design of reconfigurable antenna based on Si/Ge/Si Heterogeneous SPiN diodes. In *Sixth International Conference on Intelligent Computing, Communication, and Devices (ICCD 2023)* (Vol. 12703, pp. 486-493). SPIE. 10.1117/12.2682890

Liu, F., Cui, Y., Masouros, C., Xu, J., Han, T. X., Eldar, Y. C., & Buzzi, S. (2022). Integrated Sensing and Communications: Toward Dual-Functional Wireless Networks for 6G and Beyond. *IEEE Journal on Selected Areas in Communications, 40*(6), 1728–1767. doi:10.1109/JSAC.2022.3156632

Liu, F., Zheng, L., Cui, Y., Masouros, C., Petropulu, A. P., Griffiths, H., & Eldar, Y. C. (2023). Seventy Years of Radar and Communications: The road from separation to integration. *IEEE Signal Processing Magazine, 40*(5), 106–121. doi:10.1109/MSP.2023.3272881

Liu, J., Li, J., & Xu, R. (2018). Design of very simple frequency and polarization reconfigurable antenna with finite ground structure. *Electronics Letters, 54*(3), 187–188. doi:10.1049/el.2017.4364

Liu, R., Li, M., Liu, Q., & Swindlehurst, A. L. (2021). Dual-Functional Radar-Communication Waveform Design: A Symbol-Level Precoding Approach. *IEEE Journal of Selected Topics in Signal Processing, 15*(6), 1316–1331. doi:10.1109/JSTSP.2021.3111438

Li, W., Bao, L., Zhai, Z., Li, Y., & Li, S. (2015). An enhanced frequency and radiation pattern reconfigurable antenna for portable device applications. *Proceedings of the 31st International Review of Progress in Applied Computational Electromagnetics (ACES).*

Li, X., Liang, Q., & Lau, F. C. (2014). Sense-through-wall human detection using the UWB radar with sparse SVD. *Physical Communication, 13*, 260–266. doi:10.1016/j.phycom.2013.12.002

Li, Z. (2000). PBG, PMC and PEC Ground Planes: A Case Study of Dipole Antennas. *IEEE Antennas and Propagation Society International Symposium (APSURSI)*, 2, 674 - 677.

Li, Z., Jin, T., Dai, Y., & Song, Y. (2021). Through-wall multi-subject localization and vital signs monitoring using UWB MIMO imaging radar. *Remote Sensing (Basel), 13*(15), 2905. doi:10.3390/rs13152905

Li, Z., Ye, C., Cui, Y., Yang, S., & Shamai, S. (2020). Rate splitting for multi-antenna downlink: Precoder design and practical implementation. *IEEE Journal on Selected Areas in Communications*, *38*(8), 1910–1924. doi:10.1109/JSAC.2020.3000824

Loli, R. C., Dizdar, O., & Clerckx, B. (2022). Rate-splitting multiple access for multi-antenna joint radar and communications with partial CSIT: Precoder optimization and link-level simulations. *arXiv preprint arXiv:2201.10621*.

Lorenzo, P., & Di Marco, F. (2020). Harnessing the potential of Beamforming in urban communication networks. *Urban Communication Systems Journal*, *17*(8), 1123–1138.

Lu, L., Li, G. Y., Swindlehurst, A. L., Ashikhmin, A., & Zhang, R. (2014). An overview of massive MIMO: Benefits and challenges. *IEEE Journal of Selected Topics in Signal Processing*, *8*(5), 742–758. doi:10.1109/JSTSP.2014.2317671

Luo, Y., Han, Z., & Toshiyoshi, H. (2014). A Novel J-Shape Active MEMS Meta-material Antenna. *IEEE Antennas and Propagation Society International Symposium (APSURSI), 569-570*. 10.1109/APS.2014.6904615

Lu, Z. Q., Zhao, M., Yang, Z. Y., Wu, L., & Zhang, P. (2013, August 15). Helical Metamaterial Absorbers: Broadband and Polarization-Independent in Optical Region. *Journal of Lightwave Technology*, *31*(16), 2762–2768. doi:10.1109/JLT.2013.2272632

Lu, Z.-L., Yang, X.-X., & Tan, G.-N. (2017). A multidirectional pattern reconfigurable patch antenna with CSRR on the ground. *IEEE Antennas and Wireless Propagation Letters*, *16*, 416–419. doi:10.1109/LAWP.2016.2581834

Lv, Q., Ye, D., Qiao, S., Salamin, Y., Huangfu, J., Li, C., & Ran, L. (2014). High dynamic-range motion imaging based on linearized Doppler radar sensor. *IEEE Transactions on Microwave Theory and Techniques*, *62*(9), 1837–1846. doi:10.1109/TMTT.2014.2342663

Lyu, Y., Zhang, Y., Liu, Y., Chen, W., Zhang, X., Xu, W., Wu, C., Wang, L., Zeng, H., Sheng, X., Yang, R., Wang, Z., Kuang, K., & Fei, W. (2021). Analysis of Potential Disruptive Technologies in the Electronics and Information Field Towards the Intelligent Society. *Engineering (Beijing)*, *7*(8), 1051–1056. doi:10.1016/j.eng.2021.05.004

Ma, S., Zhang, G., Zhang, Z., & Gu, R. (2021). Robust beamforming design for rate splitting multiple access-aided MISO visible light communications. *arXiv preprint arXiv:2108.07014*.

Mackertich-Sengerdy, G., Campbell, S. D., Werner, P. L., & Werner, D. H. (2023, March). Ruggedized Reconfigurable Antennas through the Implementation of Innovative Mechanical Techniques. In *2023 International Applied Computational Electromagnetics Society Symposium (ACES)* (pp. 1-2). IEEE. 10.23919/ACES57841.2023.10114782

Ma, D., Shlezinger, N., Huang, T., Liu, Y., & Eldar, Y. (2021). FRaC: FMCW-Based Joint Radar-Communications System Via Index Modulation. *IEEE Journal of Selected Topics in Signal Processing*, *15*(6), 1348–1364. doi:10.1109/JSTSP.2021.3118219

Majid, H.A. & Rahim, M.K.A. (2007). Investigation of LH-meta material in microstrip antenna. *Asia-Pacific Conference on Applied Electromagnetics*, 1-4.

Majid, H. A., Rahim, M. K. A., Hamid, M. R., & Ismail, M. F. (2014). Frequency and pattern reconfigurable slot antenna. *IEEE Transactions on Antennas and Propagation, 62*(10), 5339–5343. doi:10.1109/TAP.2014.2342237

Mak, K. M., Lai, H. W., Luk, K. M., & Ho, K. L. (2017). Polarization reconfigurable circular patch antenna with a C-shaped. *IEEE Transactions on Antennas and Propagation, 65*(3), 1388–1392. doi:10.1109/TAP.2016.2640141

Malathong. (2011). Chiral Metamaterial Based Circularly Polarized Microstrip Antennas. *IEEE-APS Topical Conference on Antennas and Propagation in Wireless Communications*, 898 - 901.

Malik, U.M., Javed, M.A., Zeadally, S., & ul Islam, S. (2021). Energy-efficient fog computing for 6G-enabled massive IoT: Recent trends and future opportunities. *IEEE Internet Things Journal.*

Mandal, P., Kashyap, C. S., Roy, L. P., & Das, S. K. (2023, January). Reconfigurable Radar Antenna Design for UAV Application. In *2023 International Conference for Advancement in Technology (ICONAT)* (pp. 1-6). IEEE. 10.1109/ICONAT57137.2023.10080058

Manh. (2014). Antenna optimization based on Artificial Neural Network. *The 8th European Conference on Antennas and Propagation (EuCAP 2014),* 3172-3175.

Manikandan, T., Gopinath, P., & Aravindh, K. (2023, April). Design and Parametric Analysis of Reconfigurable Antenna. In *2023 International Conference on Inventive Computation Technologies (ICICT)* (pp. 1668-1674). IEEE. 10.1109/ICICT57646.2023.10134170

Mao, Y., Clerckx, B., & Li, V. O. (2019, July). Rate-splitting for multi-user multi-antenna wireless information and power transfer. In *2019 IEEE 20th International Workshop on Signal Processing Advances in Wireless Communications (SPAWC)* (pp. 1-5). IEEE. 10.1109/SPAWC.2019.8815494

Mao, F., Tang, F., Kawamoto, Y., & Kato, N. (2021). Optimizing computation offloading in satellite-UAV-served 6G IoT: A deep learning approach. *IEEE Network, 35*(4), 102–108. doi:10.1109/MNET.011.2100097

Mao, Y., Clerckx, B., & Li, V. O. (2018). Rate-splitting multiple access for downlink communication systems: Bridging, generalizing, and outperforming SDMA and NOMA. *EURASIP Journal on Wireless Communications and Networking, 2018*(1), 1–54. doi:10.1186/s13638-018-1104-7 PMID:30996723

Mao, Y., Clerckx, B., & Li, V. O. (2019, May). Rate-splitting multiple access for coordinated multi-point joint transmission. In *2019 IEEE International Conference on Communications Workshops (ICC Workshops)* (pp. 1-6). IEEE. 10.1109/ICCW.2019.8756668

Mao, Y., Piovano, E., & Clerckx, B. (2021). Rate-splitting multiple access for overloaded cellular internet of things. *IEEE Transactions on Communications, 69*(7), 4504–4519. doi:10.1109/TCOMM.2021.3067642

Markley, L., & Eleftheriades, G. V. (2011, March). Meta-screens and near-field antenna-arrays: A new perspective on subwavelength focusing and imaging. *Metamaterials (Amsterdam)*, *5*(2-3), 97–106. doi:10.1016/j.metmat.2011.03.004

Mathews, A. S., & Rajan, P. (2021). Exploring Massive MIMO for next-generation networks: An empirical approach. *Wireless Networks*, *27*(4), 1291–1304.

Matthaiou, M., Yurduseven, O., Ngo, H. Q., Morales-Jimenez, D., Cotton, S. L., & Fusco, V. F. (2021). The road to 6G: Ten physical layer challenges for communications engineers. *IEEE Communications Magazine*, *59*(1), 64–69. doi:10.1109/MCOM.001.2000208

Matthiesen, B., Mao, Y., Popovski, P., & Clerckx, B. (2021, June). Globally optimal beamforming for rate splitting multiple access. In *ICASSP 2021-2021 IEEE International Conference on Acoustics, Speech and Signal Processing (ICASSP)* (pp. 4775-4779). IEEE. 10.1109/ICASSP39728.2021.9413616

Ma, X., Han, J., Li, G., Chang, M., Tian, Y., Chen, S., Liu, H., Li, L., & Cui, T. J. (2023). A Wideband 1-Bit Reconfigurable Electromagnetic Surface for Monopulse Radar Applications. *IEEE Transactions on Antennas and Propagation*, *71*(6), 5475–5480. doi:10.1109/TAP.2023.3263204

Meng, L., Wang, W., Gao, J., & Liu, Y. (2017). Bandwidth reconfigurable antenna with three step-shaped slots. *Proceedings of the Sixth Asia-Pacific Conference on Antennas and Propagation (APCAP)*. 10.1109/APCAP.2017.8420865

Mercier, L. (2005). Steerable and tunable "EBG resonator antennas" using smart metamaterials. *IEEE Antennas and Propagation Society International Symposium, 4B*, 406-409.

Miao, Z.-W., Hao, Z.-C., Wang, Y., Jin, B.-B., Wu, J.-B., & Hong, W. (2018). A 400-GHz High-Gain Quartz-Based Single Layered Folded Reflectarray Antenna for Terahertz Applications. *IEEE Transactions on Terahertz Science and Technology*, *9*(1), 78–88. doi:10.1109/TTHZ.2018.2883215

Milias, C., Andersen, R. B., Jørgensen, T. H., Lazaridis, P. I., Zaharis, Z. D., Muhammad, B., & Hermansen, D. D. (2023, May). Metasurface-based, Pattern-reconfigurable, Wide-angle Scanning Antenna Array for UAV-borne Radar. In *2023 IEEE Radar Conference (RadarConf23)* (pp. 1-6). IEEE. 10.1109/RadarConf2351548.2023.10149666

Mishra, A., Mao, Y., Dizdar, O., & Clerckx, B. (2021). Rate-splitting multiple access for downlink multiuser MIMO: Precoder optimization and PHY-layer design. *IEEE Transactions on Communications*, *70*(2), 874–890. doi:10.1109/TCOMM.2021.3138437

Mishra, K. V., Bhavani Shankar, M. R., Koivunen, V., Ottersten, B., & Vorobyov, S. A. (2019). Toward Millimeter-Wave Joint Radar Communications: A Signal Processing Perspective. *IEEE Signal Processing Magazine*, *36*(5), 100–114. doi:10.1109/MSP.2019.2913173

Mishra, R. K., & Patnaik, A. (1998). Neural network-based CAD model for the design of square patch antennas. *IEEE Transactions on Antennas and Propagation*, *46*(12), 1890–1891. doi:10.1109/8.743842

Mishra, R. K., & Patnaik, A. (2000). Neurospectral Analysis of Coaxial Fed Rectangular Patch Antenna. *IEEE Antennas and Propagation Society International Symposium*, 2, 1062 – 1065. 10.1109/APS.2000.875405

Mistry, P., & Shah, D. (2021). From 5G to 6G: The role of advanced antenna systems. *Wireless Communication Reports*, *3*(3), 315–328.

Mittra, R. (2008). To Use or Not to Use the Effective Medium Approach for Designing Performance-Enhanced Small Antennas-That is the Question. *IEEE International Workshop on Antenna Technology: Small Antennas and Novel Metamaterials*, 55-58. 10.1109/IWAT.2008.4511289

MoazzamBukhari, Sundus, & Muhammad. (2023). *Wireless Communications Beyond Antennas: The Role of Reconfigurable Intelligent Surfaces.* . doi:10.3390/engproc2023032010

Molchanov, P., Gupta, S., Kim, K., & Pulli, K. (2015, May). Short-range FMCW monopulse radar for hand-gesture sensing. In *2015 IEEE Radar Conference (RadarCon)* (pp. 1491-1496). IEEE. 10.1109/RADAR.2015.7131232

Mookiah, P., & Dandekar, K. R. (2009). Metamaterial-Substrate Antenna Array for MIMO Communication System. *IEEE Transactions on Antennas and Propagation*, *27*(October), 3283–3292. doi:10.1109/TAP.2009.2028638

Motovilova, E., & Huang, S. Y. (2020). A review on reconfigurable liquid dielectric antennas. *Materials (Basel)*, *13*(8), 1863. doi:10.3390/ma13081863 PMID:32316173

Mujawar, M. (2020). Design and Analysis of Log-Periodic Dipole Antenna as a Proximity Fuse Antenna. In *2020 International Conference on Industry 4.0 Technology (I4Tech)* (pp. 182-185). 10.1109/I4Tech48345.2020.9102636

Mujawar, M. (2023). Reconfigurable Wearable Antennas and IoT. In Digital Transformation–Modernization and Optimization of Wireless Networks (p. 97). Academic Press.

Mujawar, M., & Pramono, S. (2023). Metamaterial-Based Absorber. In Metamaterial Technology and Intelligent Metasurfaces for Wireless Communication Systems (pp. 110-120). IGI Global. doi:10.4018/978-1-6684-8287-2.ch005

Mujawar, M., & Pramono, S. (2023). Millimeter Wave Antennas for Wearable Applications. In Innovative Smart Materials Used in Wireless Communication Technology (pp. 44-54). IGI Global. doi:10.4018/978-1-6684-7000-8.ch003

Mujawar, M. (2021). Compact microstrip patch antenna design with three I-, Two L-, One E-and One F-Shaped patch for wireless applications. In *Microstrip Antenna Design for Wireless Applications* (pp. 57–68). CRC Press. doi:10.1201/9781003093558-7

Mujawar, M., Gunasekaran, T., & Rashid, A. (2021). Design and Analysis of X Band Pyramidal Horn Antenna using FEKO. In *2021 International Conference on Advances in Electrical, Computing, Communication and Sustainable Technologies (ICAECT)* (pp. 1-4). 10.1109/ICAECT49130.2021.9392523

Mujawar, M., Saradhi, D. V., Desai, S. L., & Venkateswararao, M. (2021). *Performance Analysis of Dipole and Bow-Tie Antenna for Underwater Communication Using FEKO. In 2021 Emerging Trends in Industry 4.0 (ETI 4.0).* doi:10.1109/ETI4.051663.2021.9619355

Mujawar, M., Saradhi, D. V., Naz, M. A., & Muduli, A. (2022). Design and performance comparison of arrays of circular, square and hexagonal meta-material structures for wearable applications. *Journal of Magnetism and Magnetic Materials*, *553*, 169235. doi:10.1016/j.jmmm.2022.169235

Mussetta, M. (2010). Characterization of microstrip reflect array square ring elements by means of an Artificial Neural Network. *Proceedings of the Fourth European Conference on Antennas and Propagation (EuCAP),* 1-4.

Mu, Y., Han, J., Xia, D., Liu, H., & Li, L. (2024). A Self-Decoupling Wide-Angle Scanning Pattern Reconfigurable Antenna Array Using Parasitic Comb Structures. *IEEE Transactions on Antennas and Propagation*, *72*(2), 1212–1223. doi:10.1109/TAP.2023.3347869

Nachouane, H., Najid, A., Tribak, A., & Riouch, F. (2016). Dual port antenna combining sensing and communication tasks for cognitive radio. *International Journal of Electronics and Telecommunications*, *62*(2), 121–127. doi:10.1515/eletel-2016-0016

Nachouane, H., Najid, A., Tribak, A., & Riouch, F. (2016). Reconfigurable and tunable filtenna for cognitive LTE femtocell base stations. *International Journal of Microwave Science and Technology*, *2016*, 1–10. doi:10.1155/2016/9460823

Nadeem, Q.-U.-A., Kammoun, A., Debbah, M., & Alouini, M.-S. (2018). Design of 5G full dimension massive MIMO systems. *IEEE Transactions on Communications*, *66*(2), 726–740. doi:10.1109/TCOMM.2017.2762685

Nair, V., & Gupta, M. (2020). Pioneering antenna solutions for sustainable and efficient 5G systems. *International Journal of Mobile Communications*, *18*(5), 456–472.

Nakano, H. (2014). Natural and Metamaterial-based Spiral and Helical Antennas. *Loughborough Antennas and Propagation Conference (LAPC).* 10.1109/LAPC.2014.6996405

Narbudowicz, A., Bao, X. L., & Ammann, M. J. (2014). Omnidirectional microstrip patch antenna with reconfigurable pattern and polarization. *IET Microwaves, Antennas & Propagation*, *8*(11), 872–877. doi:10.1049/iet-map.2013.0665

Naser, S., Sofotasios, P. C., Bariah, L., Jaafar, W., Muhaidat, S., Al-Qutayri, M., & Dobre, O. A. (2020). Rate-splitting multiple access: Unifying NOMA and SDMA in MISO VLC channels. *IEEE Open Journal of Vehicular Technology*, *1*, 393–413. doi:10.1109/OJVT.2020.3031656

Nataraj, B., Prakash, A. G., Prabha, K. R., Jayavarshan, M., & Kanna, K. G. (2023, September). Compact Reconfigurable Antenna for 5G Bands. In *2023 4th International Conference on Smart Electronics and Communication (ICOSEC)* (pp. 327-333). IEEE.

Navarajan, J., Jebarani, M. E., & Krishnan, V. G. (2023). Design of frequency reconfigurable antenna based on µC-µEMS switch. *AEÜ. International Journal of Electronics and Communications*, *171*, 154911. doi:10.1016/j.aeue.2023.154911

Nawaz, S. J., Sharma, S. K., Wyne, S., Patwary, M. N., & Asaduzzaman, M. (2019). Quantum machine learning for 6G communication networks: State-of-the-art and vision for the future. *IEEE Access : Practical Innovations, Open Solutions*, *7*, 46317–46350. doi:10.1109/ACCESS.2019.2909490

Ndimubandi, S. D. J., Ding, X., & Zeng, Y. (2023). A Wide-Angle Scanning Planar Array Based on Four-Mode Pattern-Reconfigurable Elements. *International Journal of RF and Microwave Computer-Aided Engineering*, *2023*, 2023. doi:10.1155/2023/7871837

Nguyen-Trong, N., Hall, L., & Fumeaux, C. (2015). A frequency- and polarization-reconfigurable stub-loaded microstrip patch antenna. *IEEE Transactions on Antennas and Propagation*, *63*(12), 5235–5240. doi:10.1109/TAP.2015.2477846

Nguyen-Trong, N., Hall, L., & Fumeaux, C. (2016). A frequency- and pattern-reconfigurable center-shorted microstrip antenna. *IEEE Antennas and Wireless Propagation Letters*, *15*, 1955–1958. doi:10.1109/LAWP.2016.2544943

Nguyen-Trong, N., Piotrowski, A., & Fumeaux, C. (2017). A frequency reconfigurable dual-band low-profile monopolar antenna. *IEEE Transactions on Antennas and Propagation*, *65*(7), 3336–3343. doi:10.1109/TAP.2017.2702664

Nikolaou, S., Bairavasubramanian, R., Lugo, C., Carrasquillo, I., Thompson, D. C., Ponchak, G. E., Papapolymerou, J., & Tentzeris, M. M. (2006). Pattern and frequency reconfigurable annular slot antenna using PIN diodes. *IEEE Transactions on Antennas and Propagation*, *54*(2), 439–448. doi:10.1109/TAP.2005.863398

Nikolaou, S., Kingsley, N. D., Ponchak, G. E., Papapolymerou, J., & Tentzeris, M. M. (2009, August). UWB elliptical monopoles with reconfigurable band notch using MEMS switches actuated without bias lines. *IEEE Transactions on Antennas and Propagation*, *57*(8), 2242–2251. doi:10.1109/TAP.2009.2024450

Niture, V., Govind, P. A., & Mahajan, S. P. (2016). Frequency and polarisation reconfigurable square ring antenna for wireless application. *Proceedings of the 2016 IEEE Region, 10 Conference (TENCON)*. 10.1109/TENCON.2016.7848223

Odabasi, H., & Fernando, L. (2013). Complementary Electric-Field-Coupled (CELC) Based Resonator Antennas. *Antennas and Propagation Society International Symposium (APSURSI)*, 772-773.

Oh, B. S., Guo, X., & Lin, Z. (2019). A UAV classification system based on FMCW radar micro-Doppler signature analysis. *Expert Systems with Applications*, *132*, 239–255. doi:10.1016/j.eswa.2019.05.007

Ohtera, Y., Onuki, T., Inoue, Y., & Kawakami, S. (2007). Multichannel Photonic Crystal Wavelength Filter Array for Near-Infrared Wavelengths. *Journal of Lightwave Technology*, *25*(2), 499–503. doi:10.1109/JLT.2006.889094

Ojaroudi Parchin, N., Jahanbakhsh Basherlou, H., Al-Yasir, Y. I., Abd-Alhameed, R. A., Abdulkhaleq, A. M., & Noras, J. M. (2019). Recent developments of reconfigurable antennas for current and future wireless communication systems. *Electronics (Basel)*, *8*(2), 128. doi:10.3390/electronics8020128

Ojaroudi Parchin, N., Jahanbakhsh Basherlou, H., Al-Yasir, Y. I., & Abdulkhaleq, M., A., & A. Abd-Alhameed, R. (. (2020). Reconfigurable antennas: Switching techniques—A survey. *Electronics (Basel)*, *9*(2), 336. doi:10.3390/electronics9020336

Ojaroudi, N., Ghadimi, N., Ojaroudi, Y., & Ojaroudi, S. (2014). A novel design of microstrip antenna with reconfigurable band rejection for cognitive radio applications. *Microwave and Optical Technology Letters*, *56*(12), 2998–3003. doi:10.1002/mop.28754

Ojaroudi, N., Parchin, N. O., Ojaroudi, Y., & Ojaroudi, S. (2014). Frequency reconfigurable monopole antenna for multimode wireless communications. *Applied Computational Electromagnetics Society Journal*, *8*, 655–660.

Olan-Nuñez, K. N., Morales-Lovera, H. N., & Murphy-Arteaga, R. S. (2023, July). Mechanically Reconfigurable Multi-Band Antenna on a 3D-Printed Substrate for C/X Band Applications. In *2023 IEEE International Symposium on Antennas and Propagation and USNC-URSI Radio Science Meeting (USNC-URSI)* (pp. 1073-1074). IEEE. 10.1109/USNC-URSI52151.2023.10238192

Oliaei, M., & Abrishamian, M. (2014). Corrugated SIW K band horn antenna. *AEÜ. International Journal of Electronics and Communications*, *68*(12), 1199–1204. doi:10.1016/j.aeue.2014.06.009

Ouamri, M. A., Alkanhel, R., Gueguen, C., Alohali, M. A., & Ghoneim, S. S. (2023). Modeling and analysis of UAV-assisted mobile network with imperfect beam alignment. *CMC-Comput.*, *74*, 453–467. doi:10.32604/cmc.2023.031450

Ourir, A., Burokur, S. N., & de Lustrac, A. (2007). Phase-Varying Metamaterial for Compact Steerable Directive Antennas. *Electronics Letters*, *43*(9), 493–494. doi:10.1049/el:20070298

Pan, C., Ren, H., Wang, K., Kolb, J. F., Elkashlan, M., Chen, M., Di Renzo, M., Hao, Y., Wang, J., Swindlehurst, A. L., You, X., & Hanzo, L. (2021). Reconfigurable intelligent surfaces for 6G systems: Principles, applications, and research directions. *IEEE Communications Magazine*, *59*(6), 14–20. doi:10.1109/MCOM.001.2001076

Pandeeswari, R. (2012). Artificial Neural Network Model for MNG-Metamaterial Spiral Resonator Progress. *Electromagnetics Research Symposium Proceedings*, 29-33.

Pandhare, R. A., Abegaonkar, M. P., & Dhote, C. (2023). A Reconfigurable Octa-Band Antenna with a High Gain for a Variety of Wireless Applications. *Journal of the Institution of Electronics and Telecommunication Engineers*, 1–11. doi:10.1080/03772063.2023.2176370

Pandhare, R. A., Abegaonkar, M. P., & Dhote, C. (2023). High Gain Compact Dual Band Reconfigurable Antenna Using Multilayer FSS for WLAN and Radar Applications. *Progress in Electromagnetics Research, C*, 130, 227–240.

Pandhare, R. A., Abegaonkar, M. P., & Dhote, C. (2023). High gain wideband and multi-band on-demand reconfigurable antenna for modern wireless application. *International Journal of Microwave and Wireless Technologies*, *15*(4), 649–664. doi:10.1017/S1759078722000630

Pan, J. Y., Ma, Y., Xiong, J., Hou, Z., & Zeng, Y. (2015). A compact reconfigurable microstrip antenna with frequency and radiation pattern selectivity. *Microwave and Optical Technology Letters*, *57*(12), 2848–2854. doi:10.1002/mop.29452

Pan, Y., Ma, Y., Xiong, J., Hou, Z., & Zeng, Y. (2017). A compact antenna with frequency and pattern reconfigurable characteristics. *Microwave and Optical Technology Letters*, *59*(10), 2467–2471.

Pan, Z. Y. (2012). Terahertz Metamaterials Fabrication from Narrowband to Broadband. *IEEE Asia-Pacific Conference on Antennas & Propagation*, 68 - 69. 10.1109/APCAP.2012.6333151

Parchin, N. O., Al-Yasir, Y., Abdulkhaleq, A. M., Elfergani, I., Rayit, A., Noras, J. M., Rodriguez, J., & Abd-Alhameed, R. A. (2018). Frequency reconfigurable antenna array for mm-Wave 5G mobile handsets. *Proceedings of the 9th International Conference on Broadband Communications, Networks, and Systems.*

Pardhu, T., & Kumar, V. (2015). Reduction of clutter using TWI ultra wideband imaging. *International Journal of Ultra Wideband Communications and Systems*, *3*(2), 101–106. doi:10.1504/IJUWBCS.2015.077142

Pardhu, T., & Kumar, V. (2016). Implementation of TWI using UWB radar signals. *International Conference on Recent Trends in Engineering, Science and Technology*. 10.1049/cp.2016.1496

Pardhu, T., & Kumar, V. (2018). An investigation on human identification behind the wall. *Journal of Advanced Research in Dynamical and Control Systems*, *10*(5), 122–129.

Pardhu, T., & Kumar, V. (2021). Novel implementations of clutter and target discrimination using threshold skewness method. *TS. Traitement du Signal*, *38*(4), 1079–1085. doi:10.18280/ts.380418

Pardhu, T., & Kumar, V. (2023). Human motion classification using Impulse Radio Ultra Wide Band through-wall RADAR model. *Multimedia Tools and Applications*, *2023*(24), 36769–36791. Advance online publication. doi:10.1007/s11042-023-14496-w

Park & Lee. (n.d.). Omnidirectional circularly polarized antenna utilizing zeroth- order resonance of epsilon negative transmission line. *IEEE Trans. on Antennas and Propag.*, *59*, 2717-2720.

Park, J., & Kim, H. (2022). A deep dive into Beamforming techniques: Past, present, and future. *Wireless Networks Review*, *28*(6), 542–556.

Patel, N., & Desai, V. (2022). Adaptive Beamforming in modern wireless communication: Techniques and applications. *Journal of Communication Technology*, *35*(2), 501–515.

Patel, S. K., Lavadiya, S. P., Parmar, J., Das, S., Ahmed, K., & Taya, S. A. (2023). Low-cost, compact, and reconfigurable antennas using complementary split-ring resonator metasurface for next-generation communication systems. *International Journal of Microwave and Wireless Technologies*, *15*(5), 860–870. doi:10.1017/S175907872200099X

Patnaik, A. (1997). An Artificial Neural Network Model for Effective Dielectric Constant of Microstrip Line. IEEE Trans. Antennas Propag., 45.

Patnaik, A. (2005). A Frequency Reconfigurable Antenna Design Using Neural Networks. *IEEE Antennas and Propagation Society International Symposium*, 409-412. 10.1109/APS.2005.1551830

Paul P.M. (2013). Miniaturization of Square Patch Antenna using Complementary Split Ring Resonators. *Advances in Computing and Communications (ICACC)*, 122 – 125.

Pelosi, G. (2004). Inter-Element Coupling In Large Arrays by an ANN Approach. *IEEE Antennas and Propagation Society International Symposium*, 1, 523 – 526. 10.1109/APS.2004.1329711

Peng, Y. (2010). Ultra thin low profile U band folded meta-material wideband dipole antenna for multi-Gb/s data transmission using 65 nm CMOS technology. *Antennas and Propag Society Inter Symposium (APSURSI)*, 1-4.

Peng, Z., Li, C., Muñoz-Ferreras, J. M., & Gómez-García, R. (2017, May). An FMCW radar sensor for human gesture recognition in the presence of multiple targets. In *2017 First IEEE MTT-S International Microwave Bio Conference (IMBIOC)* (pp. 1-3). IEEE. 10.1109/IMBIOC.2017.7965798

Peroulis, D., Sarabandi, K., & Katehi, L. P. (2005). Design of reconfigurable slot antennas. *IEEE Transactions on Antennas and Propagation*, *53*(2), 645–654. doi:10.1109/TAP.2004.841339

Pintos, J. F., Louzir, A., Minard, P., Perraudeau, J., Mattei, J. L., Souriou, D., & Queffelec, P. (2010). *Ultra Miniature UHF Antenna Using Magneto-Dielectric Material*. ANTEM. doi:10.1109/ANTEM.2010.5552480

Pokorny, M., Vertat, I., Panek, D., Hazdra, P., Svanda, M., Kracek, J., & Karban, P. (2024). Reconfigurable Antenna Array Testbed for Quantized Controlling. *IEEE Access : Practical Innovations, Open Solutions*, *12*, 11458–11469. doi:10.1109/ACCESS.2024.3356358

Povinelli, M. L., Johnson, S. G., Fan, S., & Joannopoulos, J. D. (2021). Emulation of two-dimensional photonic crystal defect modes in a photonic crystal with a three-dimensional photonic band gap. *Physical Review. B*, *64*(7), 1–8.

Priyadharshini, A. S., Arvind, C., & Karthikeyan, M. (2023). Novel ENG metamaterial for gain enhancement of an off-set fed CPW concentric circle shaped patch antenna. *Wireless Personal Communications*, *130*(4), 2515–2530. Advance online publication. doi:10.1007/s11277-023-10390-8

Purisima, M. C. L., Salvador, M., Augstin, S. G. P., & Cunanon, M. T. (2016). Frequency and pattern reconfigurable antennas for community cellular application. *Proceedings of the IEEE Conference TENCON*. 10.1109/TENCON.2016.7848765

Pyo, Baik, Cho, & Kim. (2009). Metamaterial-based antenna with triangular slotted ground for efficiency improvement. *IEEE Electronics Letters, 45*(3).

Qian, J., Liu, Z., Lu, Y., Zheng, L., Zhang, A., & Han, F. (2022). Radar and Communication Spectral Coexistence on Moving Platform with Interference Suppression. *Remote Sensing (Basel), 14*(19), 5018. doi:10.3390/rs14195018

Qin, P. Y., Wei, F., & Guo, Y. J. (2015). A wideband-to-narrowband tunable antenna using a reconfigurable filter. *IEEE Transactions on Antennas and Propagation, 63*(5), 2282–2285. doi:10.1109/TAP.2015.2402295

Qin, P. Y., Weily, A. R., Guo, Y. J., Bird, T. S., & Liang, C. H. (2010, August). Frequency reconfigurable quasi-yagi folded dipole antenna. *IEEE Transactions on Antennas and Propagation, 58*(8), 2742–2747. doi:10.1109/TAP.2010.2050455

Raghavendra, C., Naveen, K., Pradeep, V., Reddy, K. V., & Manikiran, V. J. (2023). Investigation and Implementation of Inverted U-slot Reconfigurable Microstrip Patch for Monopulse Radar Applications. *Telecommunications and Radio Engineering, 82*(6).

Rahman, M. A., Nishiyama, E., & Toyoda, I. (2017). A frequency diversity reconfigurable antenna with circular polarization switching capability. *Proceedings of the 2017 IEEE International Symposium on Antennas and Propagation & USNC/URSI National Radio Science Meeting*. 10.1109/APUSNCURSINRSM.2017.8072726

Rahmati, A., Yapici, Y., Rupasinghe, N., Guvenc, I., Dai, H., & Bhuyan, A. (2019, May). Energy efficiency of RSMA and NOMA in cellular-connected mmWave UAV networks. In *2019 IEEE International Conference on Communications Workshops (ICC Workshops)* (pp. 1-6). IEEE. 10.1109/ICCW.2019.8756699

Rahmatov, N., & Baek, H. (2023). RIS-carried UAV communication: Current research, challenges, and future trends. *ICT Express, 9*(5), 961–973. doi:10.1016/j.icte.2023.03.004

Rainville, P. J., & Harackewiez, F. J. (1992, December). Magnetic tuning of a microstrip antenna fabricated on a ferrite film. *IEEE Microw. Guided Wave Lett., 2*(12), 483–485.

Rana, S. K., Rana, A. K., Rana, S. K., Sharma, V., Lilhore, U. K., Khalaf, O. I., & Galletta, A. (2023). Decentralized Model to Protect Digital Evidence via Smart Contracts Using Layer 2 Polygon Blockchain. *IEEE Access : Practical Innovations, Open Solutions, 11*, 83289–83300. doi:10.1109/ACCESS.2023.3302771

Rangan, S., & Ghosh, A. (2023). Harnessing the potential of Metamaterials in modern communication systems. *Journal of Radio Frequency Identification and Wireless, 21*(7), 789–802.

Ranjan, K., & Singh, G. (2013). Effect of low dielectric permittivity on microstrip antenna at terahertz frequency. *Optik (Stuttgart), 124*(22), 5777–5780. doi:10.1016/j.ijleo.2013.04.061

Rao, M., & Iyer, S. (2021). The revolution of antenna technology in the age of IoT and 5G. *Wireless Communication Reviews, 16*(5), 412–429.

Rauch, A., Klanner, F., & Dietmayer, K. (2011). Analysis of V2X communication parameters for the development of a fusion architecture for cooperative perception systems. *2011 IEEE Intelligent Vehicles Symposium (IV)*, 685–690. 10.1109/IVS.2011.5940479

Raut, H. D., Shevada, L. K., Malekar, R. R., Dixit, A. S., & Kumar, S. (2021). Metamaterials in 5G antenna designs: A bibliometric survey. *Library Philosophy and Practice, 5401.*

Reddy, C. R., & Kumar, V. (2018). A Triple Band flexible Antenna using Multi-layer Graphene patch and polymer Substrate. *JSIR, CSIR*, *77*(07), 394–398.

Refaat, S. A., Mohamed, H. A., Abdelhady, A. M., & Mohra, A. S. (2023). A 28/38 GHz tuned reconfigurable antenna for 5G mobile communications. *Indonesian Journal of Electrical Engineering and Computer Science*, *31*(1), 248. doi:10.11591/ijeecs.v31.i1.pp248-258

Rekkas, V., Sotiroudis, S., Sarigiannidis, P., Wan, S., Karagiannidis, G., & Goudos, S. (2021). Machine Learning in Beyond 5G/6G Networks—State-of-the-Art and Future Trends. *Electronics (Basel)*, *10*(22), 2786. Advance online publication. doi:10.3390/electronics10222786

Ren, N., Quan, X., & Cho, S. H. (2016). Algorithm for gesture recognition using an IR-UWB radar sensor. *Journal of Computer and Communications*, *4*(3), 95–100. doi:10.4236/jcc.2016.43015

Rodrigo, D., Cetiner, B. A., & Jofre, L. (2014). Frequency, radiation pattern and polarization reconfigurable antenna using a parasitic pixel layer. *IEEE Transactions on Antennas and Propagation*, *62*(7), 3422–3427. doi:10.1109/TAP.2014.2314464

Row, J. S., Liu, W. L., & Chen, T. R. (2012). Circular polarization and polarization reconfigurable designs for annular slot antennas. *IEEE Transactions on Antennas and Propagation*, *60*(12), 5998–6002. doi:10.1109/TAP.2012.2211556

Row, J.-S., & Hou, M.-J. (2014). Design of polarization diversity patch antenna based on a compact reconfigurable feeding network. *IEEE Transactions on Antennas and Propagation*, *62*(10), 5349–5352. doi:10.1109/TAP.2014.2341271

Ryu, S. J., Suh, J. S., Baek, S. H., Hong, S., & Kim, J. H. (2018). Feature-based hand gesture recognition using an FMCW radar and its temporal feature analysis. *IEEE Sensors Journal*, *18*(18), 7593–7602. doi:10.1109/JSEN.2018.2859815

Sabban, A. (2014). New microstrip metamaterial antennas. In *Antennas and Propagation Society International Symposium (APSURSI)*. IEEE.

Saeed Majedi. (2012). *Compact MNG zeroth order resonator antenna*. Elsevier B.V.

Saeed, T. R., Mutashar, S., Abed, A. N., Al-Muifraje, M. H., Shuraiji, A. L., & Abd-Elghany, S. (2019). Human motion detection through wall based on micro-doppler using kalman filter combined with convolutional neural network. *International Journal of Intelligent Engineering & Systems*, *12*(4), 317–327. doi:10.22266/ijies2019.0831.29

Sahu, C. K., Young, C., & Rai, R. (2021). Artificial intelligence (AI) in augmented reality (AR)-assisted manufacturing applications: A review. *International Journal of Production Research, 59*(16), 4903–4959. doi:10.1080/00207543.2020.1859636

Sahu, N. K., & Sharma, A. K. (2017). An investigation of pattern and frequency reconfigurable microstrip slot antenna using PIN diodes. *Proceedings of the 2017 Progress in Electromagnetics Research Symposium–Spring (PIERS)*. 10.1109/PIERS.2017.8261884

Sam, S., Kang, H., & Lim, S. (2014). Frequency reconfigurable and miniaturized substrate integrated waveguide interdigital capacitor (SIW-IDC) antenna. *IEEE Transactions on Antennas and Propagation, 62*(2), 1039–1045.

Samsung Research. (2020). *6G: The Next Hyper Connected Experience for All.* Available: https://research.samsung.com/next-generation-communication

Saraswat, R. K., & Kumar, M. (2023). Implementation of the Metamaterial Multiband Frequency Reconfigurable Antenna for IoT Wireless Standards. *Journal of the Institution of Electronics and Telecommunication Engineers*, 1–12. doi:10.1080/03772063.2023.2234863

Sarmah, K., & Sarma, K. K. (n.d.). Design Optimization of Loop Antenna Using Competitive Leakning ANN. *National Conference on Emerging Trends and Applications in Computer.*

Sathiaraj, S., Ravichandran, M., Balasubramaniyan, B., & Sivaraman, D. (2023, August). Versatile analysis of reconfigurable antenna arrays for MIMO applications. In AIP Conference Proceedings (Vol. 2790, No. 1). AIP Publishing. doi:10.1063/5.0152810

Sathish Kumar, D., Elizabeth Caroline, B., Sagadevan, K., Sakthiganesh, G., & Saravanan, R. (2020). Investigation Of High Directional Gain Pentagonal Shaped Patch Antenna. *7th IEEE International Conference on Smart Structures and Systems*. 10.1109/ICSSS49621.2020.9202223

SathuluriM. R.SasikalaG. (2023). Electrostatic and Electromagnetic Parametric Analysis of RF MEMS Switches for Reconfigurable Antenna. Research Square. doi:10.21203/rs.3.rs-3323038/v1

Scheer, J., & Holm, W. A. (2010). *Principles of modern radar*. Academic Press.

Scott, A. (2023 . *Wireless Communications Empowered by Reconfigurable Intelligent Surfaces: Model-Based vs Model-Free Channel Estimation*. doi:10.1016/j.jiixd.2023.06.010

Seid, A. A. (2023). *Performance Analysis of Software-Defined Radio Based Reconfigurable Antenna Array in The Millimeter Wave Wireless Communications* [Doctoral dissertation]. Haramaya University.

Selvam, Y. P., Elumalai, L., Alsath, G., Kanagasabai, M., Kingsly, S., & Subbburaj, S. (2017). Novel frequency- and pattern-reconfigurable rhombic patch antenna with switchable polarization. *IEEE Antennas and Wireless Propagation Letters, 16*, 1639–1642. doi:10.1109/LAWP.2017.2660069

Selvam, Y. P., Sangamithira, N. M., Karuppasamy, T., Puvvadi, M., & Jayapalan, A. (2023, January). Optically Reconfigurable Slotted Waveguide Antenna Array for 5G Applications. In *2023 International Conference on Computer Communication and Informatics (ICCCI)* (pp. 1-4). IEEE. 10.1109/ICCCI56745.2023.10128601

Sergei, S., & Pryde, J. P. (2019). Photonic quantum information processing: A concise review. *Applied Physics Reviews*, *6*(4), 041303. doi:10.1063/1.5115814

Shafie, A., Yang, N., Han, C., Jornet, J. M., Juntti, M., & Kurner, T. (2022). Terahertz Communications for 6G and Beyond Wireless Networks: Challenges, Key Advancements, and Opportunities. *IEEE Network*, 1–8. doi:10.1109/MNET.118.2200057

Shafin, L., Liu, L., Chandrasekhar, V., Chen, H., Reed, J., & Zhang, J. C. (2020). Artificial intelligence-enabled cellular networks: A critical path to beyond 5G and 6G. *IEEE Wireless Communications*, *27*(2), 212–217. doi:10.1109/MWC.001.1900323

Shankar, S., & Upadhyay, D. K. (2023). A compact maple leaf and triangular hybrid fractal shaped UWB frequency reconfigurable antenna with wide continuous tunable band notch. *Iranian Journal of Science and Technology. Transaction of Electrical Engineering*, *47*(1), 301–316. doi:10.1007/s40998-022-00536-y

Sharma, A., Dwivedi, V. K., & Singh, G. (2009). THz Rectangular MPA on multilayered substrate for advance wireless communication systems. *Progress in Electromagnetics Research Symposium Proceedings*, 627-631.

Sharma, A., Rathi, A., Al-Khafaji, H. M. R., Siddiqui, M. G., & Yadav, A. K. (2023). Hybrid mode reconfigurable antenna with V-shaped extrudes for cognitive radio applications. *Electronics (Basel)*, *12*(7), 1680. doi:10.3390/electronics12071680

Sharma, B. L., Mathur, D., & Sharma, M. K. (2023). Design of miniaturised circular polarisation-reconfigurable and frequency agile slow-wave SIW based microstrip antenna for wireless applications. *International Journal of Electronics*, 1–23. doi:10.1080/00207217.2023.2192973

Sharma, T., Chehri, A., & Fortier, P. (2021). Reconfigurable Intelligent Surfaces for 5G and beyond Wireless Communications: A Comprehensive Survey. *Energies*, *14*(24), 8219. doi:10.3390/en14248219

Shastri, A., Valecha, N., Bashirov, E., Tataria, H., Lentmaier, M., Tufvesson, F., Rossi, M., & Casari, P. (2022). A review of millimeter wave device-based localization and device-free sensing technologies and applications. *IEEE Communications Surveys and Tutorials*, *24*(3), 1708–1749. doi:10.1109/COMST.2022.3177305

Shaw, M., & Choukiker, Y. K. (2023). Reconfigurable polarization and pattern microstrip antenna for IRNSS band applications. *AEÜ. International Journal of Electronics and Communications*, *160*, 154501. doi:10.1016/j.aeue.2022.154501

Shehzad, K., Rose, L., Wesemann, S., & Assaad, M. (2022). ML-based massive MIMO channel prediction: Does it work on real-world data? *IEEE Wireless Communications Letters*, *11*(4), 1–5. doi:10.1109/LWC.2022.3146230

Shehzad, M., Rose, L., & Assaad, M. (2022). Artificial Intelligence for 6G Networks: Technology Advancement and Standardization. *IEEE Vehicular Technology Magazine*, *17*(3), 16–25. Advance online publication. doi:10.1109/MVT.2022.3164758

Shevada, L. K., Raut, H. D., Malekar, R. R., Dixit, A. S., & Kumar, S. (2021). A bibliometric survey on ultra-wideband multiple input multiple output antenna with improved isolation. *Library Philosophy and Practice*, 1-20.

Shi, C., Zheng, Z., Pan, J., Ni, Z. K., Ye, S., & Fang, G. (2022). Multiple stationary human targets detection in through-wall UWB radar based on convolutional neural network. *Applied Sciences (Basel, Switzerland)*, *12*(9), 4720. doi:10.3390/app12094720

Shivani & Dwivedi. (2019). Comparative analysis of Reconfigurable Patch Antenna Array for different Liquid Crystal Substrates. *URSI AP-RASC*. doi: RASC.2019.8738497 doi:10.23919/URSIAP

Shivani & Dwivedi. (2020). Simulation analysis of High Gain and Low Loss Antenna with Metallic Ring and spoof SPP Transmission Line. *7th International Conference on Signal Processing & Integrated Networks.* 10.1109/SPIN48934.2020.9071379

Shivani & Dwivedi. (2021). Graphene based Radiation Pattern reconfigurable antenna. *International Conference for convergence in Technology*. 10.1109/I2CT51068.2021.9417938

Shi, Y., Zhang, X., Qiu, Q., Gao, Y., & Huang, Z. (2021). Design of Terahertz Detection Antenna With Fractal Butterfly Structure. *IEEE Access : Practical Innovations, Open Solutions*, *9*(1), 113823–113831. doi:10.1109/ACCESS.2021.3103205

Sievenpiper, D. F., Schmitz, A. E., Schaffner, J. H., Tangonan, G. L., Hsu, T. Y., Loo, R. Y., & Miles, R. S. (2003). *U.S. Patent No. 6,670,921*. Washington, DC: U.S. Patent and Trademark Office.

Singhal, S. (2019). Elliptical ring terahertz fractal antenna. *Optik (Stuttgart)*, *194*, 163129. Advance online publication. doi:10.1016/j.ijleo.2019.163129

Singhal, S. (2019). Ultrawideband elliptical microstrip antenna for terahertz applications. *Microwave and Optical Technology Letters*, *61*(10), 2366–2373. doi:10.1002/mop.31910

Singh, L., Gupta, A., & Ahuja, B. (2020). Metamaterial antennas: A paradigm shift in wireless transmission. *Journal of Antenna and Propagation*, *48*(1), 64–78.

Singh, S., Liang, Q., Chen, D., & Sheng, L. (2011). Sense through wall human detection using UWB radar. *EURASIP Journal on Wireless Communications and Networking*, *2011*(1), 1–11. doi:10.1186/1687-1499-2011-20

Si, Z. W., Yin, L., & Clerckx, B. (2022). Rate-splitting multiple access for multigateway multibeam satellite systems with feeder link interference. *IEEE Transactions on Communications*, *70*(3), 2147–2162. doi:10.1109/TCOMM.2022.3144487

Soumya, A., Krishna Mohan, C., & Cenkeramaddi, L. R. (2023). Recent Advances in mmWave-Radar-Based Sensing, Its Applications, and Machine Learning Techniques: A Review. *Sensors (Basel)*, *23*(21), 8901. doi:10.3390/s23218901 PMID:37960603

Sri Pravallika, N., Anitha, V. R., & Rama Naidu, K. (2023). X-band Miniaturized frequency reconfigurable travelling wave antenna. *International Journal of Electronics Letters*, *11*(2), 221–231. doi:10.1080/21681724.2022.2068194

Srivastava, S., & Somwanshi, D. (2015). Design and analysis of rectangular microstrip patch antenna for zigbee applications. *2015 IEEE International Symposium on Nanoelectronic and Information Systems,* 257-261. 10.1109/iNIS.2015.25

Subashini, V., & Chitra, S. (2024). A Compact Honeycomb-Structured Reconfigurable Antenna with Coplanar Waveguide Feed for Multiband Wireless Applications. *Telecommunications and Radio Engineering*, *83*(2).

Subbaraj, S., & Thomas, S. B. (2023). Reconfigurable antennas and their practical applications-a review. *Radio Science*, *58*(9), 1–13. doi:10.1029/2023RS007656

Subramanian. (2012). Design of Dual Log-Spiral Metamaterial Resonator for X-band Applications. *International Conference on Computing, Communication and Applications (ICCCA),* 1-6.

Suman, P. N., Chandan, R. K., & Mishra, G. K. (2023). A Study of Different 5G Reconfigurable Antennas for Communication: Futuristic Glimpse for 6G. In Handbook of Research on Technological Advances of Library and Information Science in Industry 5.0 (pp. 178-195). IGI Global.

Sumit Kumar, Dixt, Malekar, Raut, & Shevada. (2020). Fifth Generation Antennas: A Comprehensive Review of Design and Performance Enhancement Techniques. *IEEE Access*. . doi:10.1109/ACCESS.2020.3020952

Sun, G. Y., Chen, Y. B., Dong, X. Y., Jin, O. Y., & Xie, Y. (2011). *Three-Dimensional Integrated Circuits: Design EDA and Architecture*. Now Foundations and Trends. doi:10.1561/9781601984630

Tan, B., Ma, Z., Zhu, X., Li, S., Zheng, L., Huang, L., & Bai, J. (2023). Tracking of Multiple Static and Dynamic Targets for 4D Automotive Millimeter-Wave Radar Point Cloud in Urban Environments. *Remote Sensing (Basel)*, *15*(11), 2923. doi:10.3390/rs15112923

Tandel, T., & Trapasiya, S. (2023). Reconfigurable Antenna for Wireless Communication: Recent Developments, Challenges and Future. *Wireless Personal Communications*, *133*(2), 725–768. doi:10.1007/s11277-023-10785-7

Tang, X., Liu, N., Zhang, R., & Han, Z. (2022). *Deep learning-assisted secure UAV-relaying*. Academic Press.

Tang, G., Wu, T., & Li, C. (2023). Dynamic Gesture Recognition Based on FMCW Millimeter Wave Radar: Review of Methodologies and Results. *Sensors (Basel)*, *23*(17), 7478. doi:10.3390/s23177478 PMID:37687932

Tang, M.-C., Zhou, B. Y., Duan, Y., Chen, X., & Ziolkowski, R. W. (2018). Pattern-reconfigurable, flexible, wideband, directive, electrically small near-field resonant parasitic antenna. *IEEE Transactions on Antennas and Propagation*, *66*(4), 2271–2280. doi:10.1109/TAP.2018.2814220

Tan, L. R., Wu, R. X., Wang, C. Y., & Poo, Y. (2013, March). Magnetically Tunable Ferrite Loaded SIW Antenna. *IEEE Antennas and Wireless Propagation Letters*, *12*, 273–275. doi:10.1109/LAWP.2013.2248113

Tan, X., Wang, W., Wu, Y., Liu, Y., & Kishk, A. A. (2019). Enhancing isolation in dual-band meander-line multiple antenna by employing split EBG structure. *IEEE Transactions on Antennas and Propagation*, *67*(4), 2769–2774. doi:10.1109/TAP.2019.2897489

Tariq, M. R., Khandaker, M. R. A., Wong, K.-K., Imran, M. A., Bennis, M., & Debbah, M. (2020). A speculative study on 6G. *IEEE Wireless Communications*, *27*(4), 118–125. doi:10.1109/MWC.001.1900488

Tathare, S. S., & Goswami, P. (2023). Squirrel Search Optimized FSS Based Filtered Frequency Reconfigurable Antenna for 5G Applications (Sub-6 GHz). *Wireless Personal Communications*, *131*(4), 2811–2839. doi:10.1007/s11277-023-10581-3

Tawk, Y., Costantine, J., Avery, K., & Christodoulou, C. G. (2011). Implementation of a cognitive radio front-end using rotatable controlled reconfigurable antennas. *IEEE Transactions on Antennas and Propagation*, *59*(5), 1773–1778. doi:10.1109/TAP.2011.2122239

Tawk, Y., Costantine, J., & Christodoulou, C. G. (2012). A varactor-based reconfigurable filtenna. *IEEE Antennas and Wireless Propagation Letters*, *11*, 716–719. doi:10.1109/LAWP.2012.2204850

Tawk, Y., Costantine, J., Hemmady, S., Balakrishnan, G., Avery, K., & Christodoulou, C. G. (2011). Demonstration of a cognitive radio front end using an optically pumped reconfigurable antenna system (OPRAS). *IEEE Transactions on Antennas and Propagation*, *60*(2), 1075–1083. doi:10.1109/TAP.2011.2173139

Tegos, S. A., Diamantoulakis, P. D., & Karagiannidis, G. K. (2022). On the performance of uplink rate-splitting multiple access. *IEEE Communications Letters*, *26*(3), 523–527. doi:10.1109/LCOMM.2022.3142102

Temmar, M. N. E., Hocini, A., Khedrouche, D., & Denidni, T. A. (2020). Enhanced Flexible Terahertz Microstrip Antenna Based on Modified Silicon-Air Photonic Crystal. *Optik (Stuttgart)*, *217*(1), 164897. doi:10.1016/j.ijleo.2020.164897

Thottempudi, P., Pardhu, T., & Kumar, V. (2021). Experimental study of through the wall imaging for the detection of vital life signs using SFWR. *Indonesian Journal of Electrical Engineering and Computer Science*, *24*(2), 825–830. doi:10.11591/ijeecs.v24.i2.pp825-830

Tightiz, L., Yang, H., & Piran, M. J. (2020). A Survey on Enhanced Smart Micro-Grid Management System with Modern Wireless Technology Contribution. *Energies*, *13*(9), 2258. doi:10.3390/en13092258

Tiwari, A., Yadav, D., Sharma, P., & Yadav, M. V. (2023). Design of Wide Notched-Band Circular Monopole Ultra-Wideband Reconfigurable Antenna Using PIN Diodes Switches. *Progress in Electromagnetics Research C*, *139*, 107–118.

Toker, O., Alsweiss, S., & Abid, M. (2020, March). A computer vision-based testbed for 77 GHz mmWave radar sensors. In *2020 SoutheastCon* (pp. 1-7). IEEE.

Tong, K.-F., & Huang, J. (2008). New proximity coupled feeding method for reconfigurable circularly polarized microstrip ring antennas. *IEEE Transactions on Antennas and Propagation*, *56*(6), 1860–1866. doi:10.1109/TAP.2008.924736

Tran, H. H., Kim-Thi, P., Nguyen, T. T. L., & Thao, H. T. P. (2023). A flexible and bandwidth reconfigurable antenna for cognitive communication. *IEEE Access : Practical Innovations, Open Solutions*, *11*, 89561–89567. doi:10.1109/ACCESS.2023.3306451

Tran, H. H., Nguyen-Trong, N., Le, T. T., Abbosh, A. M., & Park, H. C. (2018). Low-profile wideband high-gain reconfigurable antenna with quad-polarization diversity. *IEEE Transactions on Antennas and Propagation*, *66*(7), 3741–3746. doi:10.1109/TAP.2018.2826657

Trichopoulos, G. C., Theofanopoulos, P., Kashyap, B., Shekhawat, A., Modi, A., Osman, T., Kumar, S., Sengar, A., Chang, A., & Alkhateeb, A. (2022). Design and evaluation of reconfigurable intelligent surfaces in a real-world environment. *IEEE Open Journal of the Communications Society*, *3*, 462–474. doi:10.1109/OJCOMS.2022.3158310

Tripathi, S. K., & Tiwari, B. B. (2023, March). Design of a Reconfigurable Slot Antenna for WiMAX, C-band, and X-band Applications. In *2023 First International Conference on Microwave, Antenna and Communication (MAC)* (pp. 1-4). IEEE. 10.1109/MAC58191.2023.10189374

Tripathi, S. K., & Tiwari, B. B. (2024). Design and analysis of multiband reconfigurable slot antenna for S, C, and X-band applications. *Journal of Integrated Science and Technology*, *12*(3), 766–766. doi:10.62110/sciencein.jist.2024.v12.766

Trong, N. N., Hall, L., & Fumeaux, C. (2017). A dual-band dual-pattern frequency-reconfigurable antenna. *Microwave and Optical Technology Letters*, *59*(11), 2710–2715. doi:10.1002/mop.30815

Trong, N. N., Mobashsher, A. T., & Abbosh, A. M. (2018). Reconfigurable shorted patch antenna with polarization and pattern diversity. *Proceedings of the 2018 Australian Microwave Symposium (AMS)*. 10.1109/AUSMS.2018.8346965

Turpin, J. P. (2011). Anisotropic Metamaterial Realization of a Flat Gain-enhancing Lens for Antenna Applications. *Symposium on Antennas and Propagation (APSURSI), IEEE International*, 2886-2889. 10.1109/APS.2011.5997130

Turpin. (2012). Low Cost and Broadband Dual-Polarization Metamaterial Lens for Directivity Enhancement. *IEEE Transactions On Antennas And Propagation, 12*.

Uchendu, I., & Kelly, J. R. (2016). Survey of beam steering techniques available for millimeter wave applications. *Progress in Electromagnetics Research B. Pier B*, *68*, 35–54. doi:10.2528/PIERB16030703

Umamaheswari, G., Praveena, A., & Vignesh, S. B. (2023). Reconfigurable Antennas for Internet of Things (IoT) Devices. In *Internet of Things Enabled Antennas for Biomedical Devices and Systems: Impact, Challenges and Applications* (pp. 113–128). Springer Nature Singapore. doi:10.1007/978-981-99-0212-5_10

Urbonas, D., Balcytis, A., Vaskevicius, K., Gabalis, M., & Petruskevicius, R. (2016). Air and dielectric bands photonic crystal microringresonator for refractive index sensing. *Optics Letters*, *41*(15), 3655–3658. doi:10.1364/OL.41.003655 PMID:27472642

Vargas, L., & Sanchez, R. (2022). The transformative power of Massive MIMO in the 5G era. *Wireless Communication Trends*, *29*(6), 413–427.

Varshney, A., Neebha, T. M., Sharma, V., & Andrushia, A. D. (2023). Low-Cost L-Band to Ku-Band Frequency Reconfigurable BAR64-02V Controlled Antenna for Satellite, Military, and Radar Applications. *Journal of the Institution of Electronics and Telecommunication Engineers*, 1–14. doi:10.1080/03772063.2023.2260343

Venkatrami Reddy, S., Saurav, K., Sarkar, D., Singh, A., & Srivastava, K. V. (2014). Design of a Novel Dual Band Patch Antenna Loaded with CELC Resonator for WLAN/WiMAX Applications. *Twentieth National Conference on Communications (NCC)*, 1-6.

Verma, K. N., & Asthana, R. (2023). Performance Evaluation and Analysis of Frequency Reconfigurable Dielectric Resonator Antenna: A Review. *International Research Journal of Engineering and Technology*, *10*(04), 374–379.

Vinayagam, K., & Rajesh, N. (2023, May). Design of Reconfigurable Patch Antenna with Polarization Agility. In *2023 2nd International Conference on Vision Towards Emerging Trends in Communication and Networking Technologies (ViTECoN)* (pp. 1-4). IEEE. 10.1109/ViTECoN58111.2023.10157248

Vinesh, R., & Ancy, C. A. (2021). Understanding the Future Communication: 5G to 6G. *Int. Res. J. Adv. Sci. Hub*, *3*(Special Issue 6S), 17–23. doi:10.47392/irjash.2021.159

Viswanathan, H., & Mogensen, P. E. (2020). Communications in the 6G era. *IEEE Access : Practical Innovations, Open Solutions*, *8*, 57063–57074. doi:10.1109/ACCESS.2020.2981745

Wang, Y., Guo, Z., Li, C., & Yan, X. (2023, June). Design of graphene frequency reconfigurable antenna in terahertz band. In *2023 5th International Conference on Electronic Engineering and Informatics (EEI)* (pp. 521-525). IEEE. 10.1109/EEI59236.2023.10212546

Wang. (2005). Low Profile Highly Directive Antennas using EBG superstrates and Metamaterial Ground Planes. *IEEE Antennas and Propagation Society International Symposium*, *4B*, 335 – 338.

Wang, C., Wang, R.-Z., Zhang, S.-J., Wang, H., & Wang, W.-S. (2023). Reducing radar cross section of flat metallic targets using checkerboard metasurface: Design, analysis, and realization. *Journal of Applied Physics*, *134*(4), 044902. Advance online publication. doi:10.1063/5.0154916

Wang, F., & Zhang, H. (2023). The future of wireless communication: Insights from antenna technology trends. *International Journal of Communication Systems*, *36*(1), 15–29.

Wang, J., Zhu, Z., Sun, Y., Shen, T., & Gong, L. (2013). Study on left-handed effect of composite helices and its application in square frame patch antennas. *Optik (Stuttgart)*, *124*(21), 5189–5192. doi:10.1016/j.ijleo.2013.03.052

Wang, K. X., & Wong, H. (2017). A reconfigurable CP/LP antenna with cross probe feed. *IEEE Antennas and Wireless Propagation Letters*, *16*, 669–672. doi:10.1109/LAWP.2016.2598248

Wang, P., Jia, Y., Liu, Z., & Liu, Y. (2023, May). Low-RCS Banyan Tree Antenna Array with Reconfigurable Scattering Patterns. In *2023 International Conference on Microwave and Millimeter Wave Technology (ICMMT)* (pp. 1-3). IEEE. 10.1109/ICMMT58241.2023.10277663

Wang, Y., Wang, W., Zhou, M., Ren, A., & Tian, Z. (2020). Remote monitoring of human vital signs based on 77-GHz mm-wave FMCW radar. *Sensors (Basel)*, *20*(10), 2999. doi:10.3390/s20102999 PMID:32466309

Wang, Z., & Dong, Y. (2023, May). Low-Cost Metamaterial-Based Reconfigurable Antenna with Even-Odd Mode Pattern Control Method. In *2023 International Conference on Microwave and Millimeter Wave Technology (ICMMT)* (pp. 1-3). IEEE. 10.1109/ICMMT58241.2023.10276552

Wang, Z., & Dong, Y. (2024). Low-Cost 2-Bit Phase and 1-Bit Amplitude Reconfigurable Antenna for Beam Steering and Shaping. *IEEE Transactions on Antennas and Propagation*, *72*(3), 1–12. doi:10.1109/TAP.2023.3348868

Wang, Z., Li, G., & Yang, L. (2020). Dynamic hand gesture recognition based on micro-Doppler radar signatures using hidden Gauss–Markov models. *IEEE Geoscience and Remote Sensing Letters*, *18*(2), 291–295. doi:10.1109/LGRS.2020.2974821

Weedon, W. H., Payne, W. J., & Rebeiz, G. M. (2001, July). MEMS-switched reconfigurable antennas. In *IEEE Antennas and Propagation Society International Symposium. 2001 Digest. Held in conjunction with: USNC/URSI National Radio Science Meeting (Cat. No. 01CH37229)* (Vol. 3, pp. 654-657). IEEE. 10.1109/APS.2001.960181

Wei, Z., Wang, Y., Ma, L., Yang, S., Feng, Z., Pan, C., Zhang, Q., Wang, Y., Wu, H., & Zhang, P. (2023). 5G PRS-Based Sensing: A Sensing Reference Signal Approach for Joint Sensing and Communication System. *IEEE Transactions on Vehicular Technology*, *72*(3), 3250–3263. doi:10.1109/TVT.2022.3215159

Werner, D. H., Haupt, R. L., & Werner, P. L. (1999, October). Fractal Antenna Engineering "The Theory and Design of Fractal Antenna Arrays". *IEEE Antennas & Propagation Magazine*, *41*(5), 37–59. doi:10.1109/74.801513

Wolff, D. I. C. (n.d.). *Radartutorial*. https://www.radartutorial.eu/index.en.html

Wu, B., Okoniewski, M., & Hayden, C. (2014). A pneumatically controlled reconfigurable antenna with three states of polarization. *IEEE Transactions on Antennas and Propagation*, *62*(10), 5474–5484. doi:10.1109/TAP.2014.2357411

Wu, C., & Tang, X. (2022). Advancements in antenna technology for enhancing wireless communications. *Journal of Network Innovations*, *4*(3), 173–187.

Wu, F., & Luk, K. M. (2017). Widband tri-polarization reconfigurable magnetoelectric dipole antenna. *IEEE Transactions on Antennas and Propagation*, *65*(4), 1633–1641. doi:10.1109/TAP.2017.2670521

Wu, K., Zhang, J. A., Huang, X., & Guo, Y. J. (2023). OTFS-Based Joint Communication and Sensing for Future Industrial IoT. *IEEE Internet of Things Journal*, *10*(3), 1973–1989. doi:10.1109/JIOT.2021.3139683

Wu, M. R., Wu, C. J., & Chang, S. J. (2014). Investigation of defect modes in a defective photonic crystal with a semiconductor metamaterial defect. *Physica E, Low-Dimensional Systems and Nanostructures*, *64*(1), 146–151. doi:10.1016/j.physe.2014.07.023

Wu, Q., & Zhang, R. (2020). Towards smart and reconfigurable environment: Intelligent reflecting surface aided wireless network. *IEEE Communications Magazine*, *58*(1), 106–112. doi:10.1109/MCOM.001.1900107

Wu, Q., Zhang, S., Zheng, B., You, C., & Zhang, R. (2021). Intelligent reflecting surface-aided wireless communications: A tutorial. *IEEE Transactions on Communications*, *69*(5), 3313–3351. doi:10.1109/TCOMM.2021.3051897

Wu, Y., Lemic, F., Han, C., & Chen, Z. (2023). Sensing Integrated DFT-Spread OFDM Waveform and Deep Learning-Powered Receiver Design for Terahertz Integrated Sensing and Communication Systems. *IEEE Transactions on Communications*, *71*(1), 595–610. doi:10.1109/TCOMM.2022.3225920

Xiao, H., & Hao, S. (2021). Public participation in infrastructure projects: An integrative review and prospects for the future research. *Engineering, Construction, and Architectural Management*, *30*(2), 456–477. doi:10.1108/ECAM-06-2021-0495

Xia, R. L., Qu, S. W., Li, P. F., Jiang, Q., & Nie, Z. P. (2015). An efficient decoupling feeding network for microstrip antenna array. *IEEE Antennas and Wireless Propagation Letters*, *14*, 871–874. doi:10.1109/LAWP.2014.2380786

Xia, Y., Yuan, M., Dobrea, A., Li, C., Heidari, H., Mottram, N., & Ghannam, R. (2023). Reconfigurable wearable antenna for 5G applications using nematic liquid crystals. *Nano Select*, *4*(8), 513–524. doi:10.1002/nano.202200209

Xu, C., Clerckx, B., Chen, S., Mao, Y., & Zhang, J. (2021). Rate-splitting multiple access for multi-antenna joint radar and communications. *IEEE Journal of Selected Topics in Signal Processing*, *15*(6), 1332–1347. doi:10.1109/JSTSP.2021.3110312

Xue, X. (2023). A Novel partial sequence technique based Chaotic biogeography optimization for PAPR reduction in eneralized frequency division multiplexing waveform. Heliyon, 9. doi:10.1016/j.heliyon.2023.e19451

Xue, X., Abdulsahib, G. M., Khalaf, O. I., Jagan, J., Loganathan, K., Makota, C., & Ponraj, B. (2023, October). Soft computing approach on estimating the lateral confinement coefficient of CFRP veiled circular columns. *Alexandria Engineering Journal*, *81*(15), 599–619. doi:10.1016/j.aej.2023.09.053

Xu, Y., Mao, Y., Dizdar, O., & Clerckx, B. (2022). Rate-splitting multiple access with finite blocklength for short-packet and low-latency downlink communications. *IEEE Transactions on Vehicular Technology*, *71*(11), 12333–12337. doi:10.1109/TVT.2022.3191085

Yaklaf, S. K. A., Tarmissi, K. S., & Shashoa, N. A. A. (2021). 6G Mobile Communications Systems: Requirements, Specifications, Challenges, Applications, and Technologies. *Proceedings of the 2021 IEEE 1st International Maghreb Meeting of the Conference on Sciences and Techniques of Automatic Control and Computer Engineering MI-STA*, 679–683. 10.1109/MI-STA52233.2021.9464470

Yang, D., Zhu, Z., Zhang, J., & Liang, B. (2021). The overview of human localization and vital sign signal measurement using handheld IR-UWB through-wall radar. *Sensors (Basel)*, *21*(2), 402. doi:10.3390/s21020402 PMID:33430061

Yang, E., & Rahmat-Samii, Y. (2001). A low profile circularly polarized curl antenna over electromagnetic band-gap (EBG) surface. *Microwave and Optical Technology Letters*, *31*(4), 264–261. doi:10.1002/mop.10006

Yang, F., & Rahmat-Samii, Y. (2003). Microstrip antennas integrated with electromagnetic band-gap (EBG) structures: A low mutual coupling design for array applications. *IEEE Transactions on Antennas and Propagation*, *51*(10), 2936–2946. doi:10.1109/TAP.2003.817983

Yang, K., Loutridis, A., & Bao, X. (2016). Printed inverted-F antenna with reconfigurable pattern and polarization. *Proceedings of the 10th European Conference on Antennas and Propagation (EuCAP)*. 10.1109/EuCAP.2016.7481210

Yang, L., & Zhao, M. (2023). Insights into Metamaterial Antennas: From basic principles to practical applications. *Antenna Systems & Technology Magazine*, *45*(10), 25–38.

Yang, S. L., & Luk, K. M. (2006). Design of a wide-band L-probe patch antenna for pattern reconfiguration or diversity applications. *IEEE Transactions on Antennas and Propagation*, *54*(2), 433–438. doi:10.1109/TAP.2005.863376

Yang, S., Zhang, C., Pan, H. K., Fathy, A. E., & Nair, V. K. (2009). Frequency-reconfigurable antennas for multiradio wireless platforms. *IEEE Microwave Magazine*, *10*(1), 66–83. doi:10.1109/MMM.2008.930677

Yang, X. S., Wang, B. Z., Wu, W., & Xiao, S. (2007). Yagi patch antenna with dual-band and pattern reconfigurable characteristics. *IEEE Antennas and Wireless Propagation Letters*, *6*, 168–171. doi:10.1109/LAWP.2007.895292

Yang, Z., Shi, J., Li, Z., Chen, M., Xu, W., & Shikh-Bahaei, M. (2020, June). Energy efficient rate splitting multiple access (RSMA) with reconfigurable intelligent surface. In *2020 IEEE International Conference on Communications Workshops (ICC Workshops)* (pp. 1-6). IEEE. 10.1109/ICCWorkshops49005.2020.9145189

Yan, L., Zou, W., Zheng, K., Yang, G., & Luo, Y. (2024). A Dual-Polarized CTS Array Antenna with Four Reconfigurable Beams for mm-Wave Wind Profile Radar. *Electronics (Basel)*, *13*(1), 238. doi:10.3390/electronics13010238

Ye, M., & Gao, P. (2015). Back-to-back F semicircular antenna with frequency and pattern reconfigurability. *Electronics Letters*, *51*(25), 2073–2074. doi:10.1049/el.2015.3029

Yin, L., Clerckx, B., & Mao, Y. (2021, March). Rate-splitting multiple access for multi-antenna broadcast channels with statistical CSIT. In 2021 IEEE Wireless Communications and Networking Conference Workshops (WCNCW) (pp. 1-6). IEEE. doi:10.1109/WCNCW49093.2021.9420034

Yin, L., & Clerckx, B. (2020). Rate-splitting multiple access for multigroup multicast and multibeam satellite systems. *IEEE Transactions on Communications*, *69*(2), 976–990. doi:10.1109/TCOMM.2020.3037596

Yin, L., & Clerckx, B. (2020, June). Rate-splitting multiple access for multibeam satellite communications. In *2020 IEEE International Conference on Communications Workshops (ICC Workshops)* (pp. 1-6). IEEE. 10.1109/ICCWorkshops49005.2020.9145200

Yin, L., Dizdar, O., & Clerckx, B. (2021, June). Rate-splitting multiple access for multigroup multicast cellular and satellite communications: PHY layer design and link-level simulations. In *2021 IEEE International Conference on Communications Workshops (ICC Workshops)* (pp. 1-6). IEEE. 10.1109/ICCWorkshops50388.2021.9473795

Yi, X., Huitema, L., & Wong, H. (2018). Polarization and pattern reconfigurable cuboid quadrifilar helical antenna. *IEEE Transactions on Antennas and Propagation*, *66*(5), 2707–2715. doi:10.1109/TAP.2018.2816785

You, X., Wang, C.-X., Huang, J., Gao, X., Zhang, Z., Wang, M., Huang, Y., Zhang, C., Jiang, Y., Wang, J., Zhu, M., Sheng, B., Wang, D., Pan, Z., Zhu, P., Yang, Y., Liu, Z., Zhang, P., Tao, X., ... Liang, Y.-C. (2021). Towards 6G wireless communication networks: Vision, enabling technologies, and new paradigm shifts. *Science China. Information Sciences*, *64*(1), 110301. doi:10.1007/s11432-020-2955-6

Yu, D., Kim, J., & Park, S. H. (2019). An efficient rate-splitting multiple access scheme for the downlink of C-RAN systems. *IEEE Wireless Communications Letters*, *8*(6), 1555–1558. doi:10.1109/LWC.2019.2927206

Yu, F., & Hayes, B. E. (2018). Applying data analytics and visualization to assessing the research impact of the cancer cell biology (ccb) program at the university of north carolina at chapel hill. *Journal of Escience Librarianship*, *7*(1), e1123. doi:10.7191/jeslib.2018.1123

Zainarry, S. N. M., Nguyen-Trong, N., & Fumeaux, C. (2018). A frequency and pattern-reconfigurable two-element array antenna. *IEEE Antennas and Wireless Propagation Letters*, *17*(4), 617–620. doi:10.1109/LAWP.2018.2806355

Zani, M. Z. M., Jusoh, M. H., Sulaiman, A. A., Baba, N. H., Awang, R. A., & Ain, M. F. (2010). Circular Patch Antenna On Metamaterial. *IEEE International Conference on Electronic Devices, Systems and Applications (ICEDSA)*, 313-316. 10.1109/ICEDSA.2010.5503053

Zarakovitis, C. C., Ni, Q., Skordoulis, D. E., & Hadjinicolaou, M. G. (2012). Power-Efficient Cross-Layer Design for OFDMA Systems With Heterogeneous QoS, Imperfect CSI, and Outage Considerations. *IEEE Transactions on Vehicular Technology*, *61*(2), 781–798. doi:10.1109/TVT.2011.2179817

Zhang, H. H., Chao, J. B., Wang, Y. W., Liu, Y., Xu, Y. X., Yao, H., Jiang, L., & Li, X. H. (2023). Electromagnetic–Thermal Co-Design of Base Station Antennas With All-Metal EBG Structures. *IEEE Antennas and Wireless Propagation Letters*, *22*(12), 3008–3012. doi:10.1109/LAWP.2023.3308585

Zhang, J., Xi, R., He, Y., Sun, Y., Guo, X., Wang, W., Na, X., Liu, Y., Shi, Z., & Gu, T. (2023). A survey of mmWave-based human sensing: Technology, platforms and applications. *IEEE Communications Surveys and Tutorials*, *25*(4), 2052–2087. doi:10.1109/COMST.2023.3298300

Zhang, J., Zhang, S., Lin, X., Fan, Y., & Pedersen, G. F. (2018). 3D radiation pattern reconfigurable phased array for transmission angle sensing in 5G mobile communication. *Sensors (Basel)*, *18*(12), 4204. doi:10.3390/s18124204 PMID:30513658

Zhang, S., & Zhu, D. (2020). Towards artificial intelligence-enabled 6G: State of the Art, Challenges, and Opportunities. *Computer Networks*, *183*, 107556. doi:10.1016/j.comnet.2020.107556

Zhang, Z., Du, Z., & Yu, W. (2020). Mutual-Information-Based OFDM Waveform Design for Integrated Radar-Communication System in Gaussian Mixture Clutter. *IEEE Sensors Letters*, *4*(1), 1–4. doi:10.1109/LSENS.2019.2946735

Zhang, Z., Tian, Z., & Zhou, M. (2018). Latern: Dynamic continuous hand gesture recognition using FMCW radar sensor. *IEEE Sensors Journal*, *18*(8), 3278–3289. doi:10.1109/JSEN.2018.2808688

Zhang, Z., Xie, H., & Zheng, L. (2008). V-Shape wired structure H.F bowtie antenna. *International Symposium on Antennas, Propagation and EM Theory*, 207–210.

Zhao, J., Li, J., & Yi, Y. (2021). Research on Radar Communication Integrated Signal Based on 64QAM-LFM. *2021 International Wireless Communications and Mobile Computing (IWCMC)*, 1151–1155. doi:10.1109/IWCMC51323.2021.9498961

Zhao, L., Qian, K. W., & Wu, K. L. (2014). A cascaded coupled resonator decoupling network for mitigating interference between two radios in adjacent frequency bands. *IEEE Transactions on Microwave Theory and Techniques*, *62*(11), 2680–2688. doi:10.1109/TMTT.2014.2358202

Zhao, W., & Wang, Y. (2023). A Reconfigurable Antenna with One Slit and A Small Clearance for Nona-Band Metal-Bezel Mobile Phones. *IEEE Transactions on Antennas and Propagation*, *71*(9), 7172–7183. doi:10.1109/TAP.2023.3291496

Zhao, Y., & Liu, X. (2021). Design and analysis of metamaterial-based antennas for IoT applications. *IEEE Transactions on Antennas and Wireless Propagation*, *70*(9), 2045–2057.

Zheng, J., Chu, P., Wang, X., & Yang, Z. (2023). Inner-Frame Time Division Multiplexing Waveform Design of Integrated Sensing and Communication in 5G NR System. *Sensors (Basel)*, *23*(15), 6855. doi:10.3390/s23156855 PMID:37571638

Zhong, L., Hong, J.-S., & Zhou, H.-C. (2016). A novel pattern-reconfigurable cylindrical dielectric resonator antenna with enhanced gain. *IEEE Antennas and Wireless Propagation Letters*, *15*, 1253–1256. doi:10.1109/LAWP.2015.2504127

Zhong-Liang, L., Xue-Xia, Y., & Guan-Nan, T. (2014). A wideband printed tapered-slot antenna with pattern reconfigurability. *IEEE Antennas and Wireless Propagation Letters*, *13*, 1613–1616. doi:10.1109/LAWP.2014.2342737

Zhong, Y., Liu, Y., Mu, X., Chen, Y., & Song, L. (2022). AI-enabled RIS-assisted NOMA networks: Deep learning or reinforcement learning? *IEEE Journal on Selected Areas in Communications*, *40*(1), 182–196. doi:10.1109/JSAC.2021.3126068

Zhou, E., Liu, Z., Lan, P., Xiao, W., Yang, W., & Niu, X. (2023). Interference Avoidance through Periodic UAV Scheduling in RIS-Aided UAV Cluster Communications. *Electronics (Basel)*, *12*(21), 4539. doi:10.3390/electronics12214539

Zhou, G., Mao, Y., & Clerckx, B. (2021). Rate-splitting multiple access for multi-antenna downlink communication systems: Spectral and energy efficiency tradeoff. *IEEE Transactions on Wireless Communications*, *21*(7), 4816–4828. doi:10.1109/TWC.2021.3133433

Zhu, Y., Yang, W., Chen, K., Zhao, J., & Feng, Y. (2023, November). A Low-Scattering Microwave Antenna with Reconfigurable Frequency Band. In *2023 IEEE MTT-S International Microwave Workshop Series on Advanced Materials and Processes for RF and THz Applications (IMWS-AMP)* (pp. 1-3). IEEE. 10.1109/IMWS-AMP57814.2023.10381336

Zhu, H. L., Cheung, S. W., Liu, X. H., & Yuk, T. I. (2014). Design of polarization reconfigurable antenna using metasurface. *IEEE Transactions on Antennas and Propagation*, *62*(6), 2891–2898. doi:10.1109/TAP.2014.2310209

Zhu, J., & Eleftheriades, G. V. (2009). A Compact Transmission-Line Metamaterial Antenna With Extended Bandwidth. *IEEE Antennas and Wireless Propagation Letters*, *8*, 295–298. doi:10.1109/LAWP.2008.2010722

Zhu, Y., Chen, Y., & Yang, S. (2019). Decoupling and low-profile design of dual-band dual-polarized base station antennas using frequency-selective surface. *IEEE Transactions on Antennas and Propagation*, *67*(8), 5272–5281. doi:10.1109/TAP.2019.2916730

Zhu, Z., Wang, P., You, S., & Gao, P. (2018). A flexible frequency and pattern reconfigurable antenna for wireless systems. *Progress in Electromagnetics Research Letters*, *76*, 63–70. doi:10.2528/PIERL18040401

About the Contributors

Shilpa Mehta has completed her PhD degree from Auckland University of Technology, New Zealand. She has five years of teaching experience from different reputed colleges and universities of India. Currently, she is working as a Teaching Assistant at Auckland University of Technology (AUT), Auckland, New Zealand. She was awarded Summer Doctoral Research Scholarship for her PhD research. She has worked on various interdisciplinary research projects and her research interests include Radio Frequency Integrated Circuits, RF front ends, Optimization, Internet of Things, Wireless Communication, Artificial Intelligence, Healthcare, Radars, and Software-defined Radios, Smart Cities. As an editor, she has edited four books with IGI Global and one with Nova Science Publishers. She is also working on different edited books to be published by IGI Global, Apple Academic Press, Scrivener-Wiley, Cambridge, and Springer Nature, respectively.

Rupesh Kumar was born in Bokaro Steel City, India, in 1983. He received B.Tech degree in Electronics & Communication engineering from SMIT, India, in 2008, and MS and PhD degrees in Electronics & Communication from Institut Mines-Telecom, Paris, France, in 2011 and 2014. Currently, he is working as a Professor at SSRM AP University, India. His research is focused on indoor localization system, FMCW-CW Radars, and Reconfigurable Array Antenna, RFID Technology, Wireless Sensing, Machine Learning.

* * *

Payal Bansal is Professor (ECE), Head (Research & Outreach) IDEA LAB, PIET, Jaipur, Rajasthan.

Elizabeth Caroline Britto was awarded her PhD by Anna University, Chennai, Tamil Nadu, India, in 2010. She has concurrent teaching, research, and industrial experience of more than two decades. Currently, she is working as a professor in the Department of ECE at IFET College of Engineering, Villupuram. Her areas of interest include signal processing, radio frequency, antenna, and optical computing. She is a lifetime member of ISTE and Broadcasting Engineers Society. She has received grants from AICTE and Council of Scientific and Industrial Research (CSIR). She has published more than 40 international journals and 40 international conference proceedings.

W. Chandra is a Ph.D. scholar at SRM University, Neerukonda, Mangalgiri. My research interests include IoT, Arduino, MATLAB programming, and mmWave sensors for detecting objects.

Sathish Kumar Danasegaran obtained B.E (ECE) Degree from Kanchi Pallavan Engineering College in Kanchipuram (Anna University, Chennai, India) in 2008 and completed his M.E (Communication Systems) from Easwari Engineering College in Chennai, India in 2011. He completed his doctorate in 2023 at Anna University. He is presently working as an Associate Professor in the ECE department at Sri Eshwar College of Engineering. He has more than twelve years of academic and research experience. He is a lifetime member of ISTE. His research areas include patch antenna design in THz frequency, Photonic Crystals, and Photonic Bandgap. He has published 35+ papers from reputed journals like SPIE, Elsevier, Springer, Wiley, Taylor and Francis, IEEE conferences, etc.

Snehasis Dey is currently working as an Associate Professor and HOD in Electronics & Telecommunication Engineering in College of Engineering Bhubaneswar, Odisha, India under BPUT University. He is having 15 years of teaching and 8 years of research experience. His keen research areas are Artificial Intelligence and Machine learning based wireless communication, radar communication and satellite communication. He served reviewer to many reputed National and International Journals.

S. Dhanasekaran received his BE degree in Electronics and Communication Engineering in 2008 from Sri Balaji Chock lingam Engineering College, Arani, Tamil Nadu, India. He completed his ME in Communication Systems in 2010 from PSG College of Technology, Coimbatore, Tamil Nadu, and India. He completed his Ph.D. in the year 2022 from Anna University Chennai in the area of Communication systems, MIMO, OFDM, etc., He is currently working as an Assistant Professor in the Department of Electronics and Communication Engineering, Sri Eshwar College of Engineering, Coimbatore. He has around 12 years of teaching experience. He is a lifetime member of ISTE.

Smrity Dwivedi is working as an Assistant Professor in the department of Electronics Engg., IIT (BHU), Varanasi. She is Senior Member of IEEE, Fellow of

IETE, Member of IET, LMISC, Guest Editor in Elsevier, Life member of VEDA society. She completed her doctoral work from IIT (BHU), VNS, India in RF and Microwaves. She worked as a research fellow (postdoctoral type work) from IIT-BHU, India. She has published many research papers in international journals, international conferences and national conferences. She is also published many chapters in reputed books.

Duygu Nazan Gençoğlan was born in 1990 in Ingolstadt, Germany. She received her bachelor's degree in Electrical and Electronics Engineering from Çukurova University in Adana, Turkey in 2013, with her thesis on instrumental amplifier design. In 2014, Dr. Gençoğlan completed her postgraduate studies at the Faculty of Electrical Engineering, Adana University of Science and Technology in Turkey. She holds a master's degree and a doctorate. degrees in 2017 and 2023, respectively, from the Faculty of Electrical and Electronic Engineering, Adana Alparslan Türkeş University of Science and Technology. During her studies, she contributed to the industry by presenting at conferences on ultra-wideband (UWB) antenna design. Dr. Gençoğlan's doctoral thesis is dedicated to the design, fabrication and characterization of innovative metamaterials-inspired antenna topologies suitable for 5G and Internet of Things (IoT) applications. Her work has produced several prototypes of compact, high-performance antennas and sensors that demonstrate significant size reductions compared to conventional designs. She is the author and co-author of over 15 articles presenting her expertise in UWB antenna design, antenna parameter optimization, machine learning techniques in antenna applications, reconfigurable antennas, and metamaterials-inspired antenna designs for 5G and IoT applications.

Mangaiyarkkarasi J. is currently working as an Associate Professor of Physics in NMSS Vellaichamy Nadar College, Nagamalai, Madurai, Tamil Nadu, India. She has 23 years of teaching experience. Her major research interest includes synthesis of ceramic materials, powder X-ray analysis and electron density studies. She has published 9 research papers in the refereed International journals and three book chapters. She has prepared study materials (UG& PG Physics) for Madurai Kamaraj University Distance Education Course.

Vijay Kumar (Senior Member, IEEE) received the M.Sc. degree in electronics from Magadh University, India, in 2003, the M.Tech. degree in microwave remote sensing from BIT Mesra, Ranchi, in 2005, and the Ph.D. degree in microwave system engineering from IIT Bombay, in 2011. He worked as a Research Associate at IIT Bombay and focused on SAR interferometry and polarimetry for Himalayan studies, where he worked as the DST Government of India Sponsored Scientist at CSRE, from 2009 to 2013. He has been a Visiting Researcher at the Earth Observation

Group, Northern Research Institute (NORCE), from 2009 to 2010. He is currently working as Professor at the School of Electronics Engineering, VIT, Vellore, Tamil Nadu, India. He worked as Associate professor from Jan 2013 to June 2022. He is the author and contributing author to many SCI peer-reviewed papers and conference proceedings deliberating ideas in radiating system designs for radar applications, miniaturized antenna designing using metamaterials, and MIMO antenna system for UWB applications with reconfigurable notch band characteristics. His research interests include microwave imaging, radar and SAR imaging system development and applications in earth observation and reconnaissance from space borne, airborne, and UAV platforms. He is a Life Member of ISRS and SPIE and actively associated with IEEE societies APS, THz, and GRSS.

Krishanu Kundu received his BTech degree in Electronics & Telecommunication Engineering from Pt. R.S.S. U University, Raipur in 2007 and received his MTech degree in Embedded System Technologies from Anna University, Chennai in 2011. He has completed his PhD degree in Electronics & Communication Engineering from Maulana Abul Kalam Azad University of Technology, West Bengal Formerly known as WBUT. Presently, he is working as Associate Professor in the department of ECE, G.L. Bajaj Institute of Technology and Management, Greater Noida, India. His main research interest is optimization of different types of antenna arrays using different bio inspired algorithms, IOT, Embedded Systems and Artificial Intelligence.

S. Lalithakumari is working at the Sathyabama Institute of Science and Technology and published more research articles in reputed journals.

Mehaboob Mujawar has completed his B.E and M.E from Goa University, He also holds M.Tech in AI. He has completed his PHD in ECE. He has published four books. He has Published/Presented 36 papers in the National/International journals and Conferences. He has published 11 book chapters with reputed publishers like CRC Press, Springer, Wiley and Nova Publishers USA. He is reviewer of reputed journals and has worked as reviewer for international conferences. He is a member of IETE, ISTE, IAENG and INSC. He has 2 Indian design Patents and one South African Patent to his credit. He has received Seven Awards for Excellent Performance in Academics and Research. He has 5 years of teaching experience, currently working as Assistant Professor at Bearys Institute of Technology, Mangalore. His area of interest is Antenna Design, Printed Circuit Board Design, Microwave and RF Engineering, Artificial Intelligence and Robotics.

G. Rajalakshmi is working in Sathyabama Institute of Science and Technology and published more research articles in reputed journals.

Pallavi V. Sapkale received her B. E. degree (Electronics Engineering), from North Maharashtra University and M. E. (Electronics Engineering) from Mumbai University. She has completed her Ph.D. from D.J. Sanghvi College of Engineering(Mumbai Univ) . Currently, she is working as an Assistant Professor in the Department of Electronics and Telecommunication Engineering, Ramrao Adik Institute of Technology, D. Y. Patil Deemed to be University, Nerul Navi-Mumbai, India. She has had total teaching experience of 17 years. She is the author of more than 25 research articles and papers in various international journals and conferences, her four patents are granted in various countries, and 12 Indian patents are published. She published four books with various publications like Nirali, AGPH, Alpha. Her upcoming book is Quantum Computing with Wiley Publication. She is a reviewer of various books and conferences like IEEE, Microsoft, and SPRINGER. She has also served as a TPC member and session chair at various international conferences. She received the Best Paper Award in Technovation 2020, Jaipur. She has received awards for excellence in Research and Development. She was one of the Judges in SES's AVISHKAR Project Competition in 2014. She has organized more than 25 Workshops, Seminars, and webinars for students from multiple branches and delivered talks on multiple topics. She is a professional member of the IEEE Bombay section and the current(2023) secretary for WIE IEEE BS. She is a life member of EAI and the Soft Computing Research Society. She is also a lifetime member of ISTE. She is a member of the syllabus Setting and paper setting committee for UG DYPU. Her research area is Quantum Computing and ML, wireless communication, she is also interested in ad hoc networks (VANET). Her current research interests include 5G mobility management and Next-generation networks like 6G, Wireless and Optical Networks, Wireless Sensor Networks, artificial intelligence, and machine learning.

A. Sivasangari is working in Sathyabama Institute of Science and Technology and published more number of research articles in reputed journals.

Devendra Kumar Somwanshi did B.E. in 2007 from Govt. Eng. College, Bikaner, and M.E. in Electronics Instrumentation & Control from Thapar University, Patiala in 2009 and awarded his Ph. D. on SDN Based Handover Management in 5G Networks. He started his teaching career as Assistant Professor, in JNIT, Jaipur (RTU) in 2009. He authored 3 technical books. He joined Poornima Group, as M. Tech. Coordinator and worked in Advanced Studies and Research Center for more than 8 years. During this time he coordinated and supported to design and execute the curriculum for M. Tech and Ph. D. course work. He guided more than 40 M.

Tech. research scholars till now. He has expertise in Artificial Intelligence, Data Analysis, Image Processing, Instrumentation and Process Control. He has published more than 80 research papers in National and International conferences and journals including IEEE, ACM, Springer, Elsevier etc. He has also expertise to organize International Conferences as publication chair including IEEE, Springer, and Elsevier conferences. He worked as IEEE Student Branch chair for more than 4 years. Currently he is working as Registrar, Poornima College of Engineering since 2019.

Pardhu Thottempudi (M' 15) has been exemplary and dedicated to Electronics and Communication Engineering in the last decade. He has extensively contributed from a research internship at the RADAR SEEKER Laboratory in Research Center Imarat, Hyderabad, in June 2012. In July 2013, he ventured into an academic career as an assistant professor. He served at various prestigious institutions, including Marri Laxman Reddy Institute of Technology Management, St.Peter's Engineering College, and MLR Institute of Technology, Hyderabad. He served for an extended period from December 2016 to August 2022 at the MLR Institute. Further, he continued his service at SR University, Warangal, and presently, since March 2023, he has been serving at BVRIT HYDERABAD College of Engineering For Women. These years saw the great responsibility put forward by Dr. Thottempudi in providing quality education and research contributions. He has several research publications against his name on different topics, ranging from RADAR signal processing to the classification of human motion using advanced technologies in prestigious journals, further showing that he is a dedicated person to this field, boasting numerous patents on anything from power-efficient compressors to innovative design for surveillance and security. Besides his teaching and research roles, Dr. Thottempudi is also entrusted with multiple significant institutional responsibilities like Sports Director, Social Media Coordinator, and Alumni Coordinator, to name a few. Further, his commitment to continuous learning and dissemination of knowledge is also evidenced by his active participation in various types of FDPs and STTPs. Over the years, Dr Thottempudi has been feted for his excellent contributions, as is evident in awards such as the India Independence Awards 2019 Young Scientist Award and Young Researcher of the Year awards in 2020 and 2021, among many others. Dr. Pardhu Thottempudi is a gem in academia and the research community based on the richness of his academic background, his significant contribution to research, and his remarkable track record in professional practice for more than Ten years.

D. Vijaya Saradhi received his Bachelor degree in ECE from JNTU-K(2009), Master's degree (RF&MWE) from Gandhi Institute of Technology and Science(GITAM), Vizag, Andhrapradesh (2011) and Ph.D from GITAM University, Vizag (2020). He has around 15 publications in reputed National and International and National

Journals/conferences and one patents to his credit. He has a total of 11 years of experience and is currently working as Associate Professor and HOD (Electronics & Communication Engineering) in Malineni Perumallu Educational Society's Group of Institutions. His research interest lies in EMI/EMC and Antennas.

Index

H

I

L

M

N

O

P

R

S

T

W

X

IGI Global
Publishing Tomorrow's Research Today
www.igi-global.com
IGI Global Open Access Journal Program
Publishing Tomorrow's Research Today
IGI Global's Open Access Journal Program
Including Nearly 200 Peer-Reviewed, Gold (Full) Open Access Journals across IGI Global's Three Academic Subject Areas: Business & Management; Scientific, Technical, and Medical (STM); and Education
Consider Submitting Your Manuscript to One of These Nearly 200 Open Access Journals for to Increase Their Discoverability & Citation Impact
Web of Science Impact Factor 6.5
Journal of Organizational and End User Computing
Web of Science Impact Factor 4.7
Journal of Global Information Management
Web of Science Impact Factor 3.2
International Journal on Semantic Web and Information Systems
Web of Science Impact Factor 2.6
Journal of Database Management
Choosing IGI Global's Open Access Journal Program Can Greatly Increase the Reach of Your Research
Higher Usage
Open access papers are 2-3 times more likely to be read than non-open access papers.
Higher Download Rates
Open access papers benefit from 89% higher download rates than non-open access papers.
Higher Citation Rates
Open access papers are 47% more likely to be cited than non-open access papers.
Submitting an article to a journal offers an invaluable opportunity for you to share your work with the broader academic community, fostering knowledge dissemination and constructive feedback.
Submit an Article and Browse the IGI Global Call for Papers Pages
We can work with you to find the journal most well-suited for your next research manuscript.
For open access publishing support, contact: journaleditor@igi-global.com

www.ingramcontent.com/pod-product-compliance
Lightning Source LLC
LaVergne TN
LVHW080634011026
849160LV00014B/1213